TOURING CHINA

A volume in the series

Histories and Cultures of Tourism
Edited by Eric G. E. Zuelow

A list of titles in this series is available at cornellpress.cornell.edu.

TOURING CHINA

A History of Travel Culture, 1912–1949

Yajun Mo

Foreword by
Eric G. E. Zuelow

CORNELL UNIVERSITY PRESS
Ithaca and London

First published 2021 by Cornell University Press

Library of Congress Cataloging-in-Publication Data

Names: Mo, Yajun, 1980– author.
Title: Touring China : a history of travel culture, 1912–1949 / Yajun Mo ; foreword by Eric G.E. Zuelow.
Description: Ithaca [New York] : Cornell University Press, 2021. | Series: Histories and cultures of tourism | Includes bibliographical references and index.
Identifiers: LCCN 2021007640 (print) | LCCN 2021007641 (ebook) | ISBN 9781501760624 (hardcover) | ISBN 9781501761041 (paperback) | ISBN 9781501760631 (pdf) | ISBN 9781501760648 (epub)
Subjects: LCSH: Tourism—China—History—20th century.
Classification: LCC G155.C55 M6 2021 (print) | LCC G155.C55 (ebook) | DDC 915.104/4—dc23
LC record available at https://lccn.loc.gov/2021007640
LC ebook record available at https://lccn.loc.gov/2021007641

For my parents, who made all this possible

Contents

Foreword

In 2012, China moved past the United States as well as every other country on earth to become the leader in outbound tourism. It was a significant moment. Since World War II, the United States had easily led this category, so China's arrival as the top tourist-producing nation represented a remarkable shift. China has only widened the gap since, sending $277.35 billion abroad in 2018; the United States was a distant second with $144.47 billion.[1]

If one were to rely on the current historiography of modern tourism, the appearance of Chinese as avid tourists evidently came from nowhere, as they only suddenly discovered a passion for leisure travel. In tourism history circles, aside from representing non-Western locales as potential destinations, the narrative has largely been dominated by Europe and the United States. The majority of historians tell some version of the same story. Tourism originated with the grand tour in the 1700s. In the century after, fueled by what they'd learned from travel writing, maps, paintings, and a new breed of guidebook, more and more people from across the social spectrum grew interested in seeing what they were told ought to be seen. Railways and steamships made scheduling easier and shortened travel times, making the pastime affordable. Working-class groups, especially in Britain, saw tourism as a way to provide leisure for their members. At the same time, entrepreneurs such as Thomas Cook, by far the most famous early travel agent, arranged package holidays and inexpensive day trips, both for profit and in pursuit (at least in the case of Cook himself) of high-minded ideals about the potential of tourism to make people better.[2]

Although most might not immediately think of it as such, tourism represents an important exercise of power that manifests in a variety of ways. It highlights particular historical narratives, rife with selective forgetting and remembering. Developers, politicians, and others point to specific locations said to embody an identity: vital destinations for community members as pilgrimage sites, but also for outsiders who are keen to know the "real" place that they are visiting. It can be used to improve underdeveloped regions, an economic and political tool. Tourism can even be used to highlight difference, instructing the tourist in how to behave while using the host community as a counterexample. Of course, the dichotomy between hosts and guests showcases haves and have-nots. Workers and locals are frequently poor, tourists wealthy, and the results troubling.

On the basis of most existing scholarship, one might be forgiven for thinking that tourism never amounted to more than a European imposition in colonized or partially colonized countries, places where imperialists exploited local communities that were left powerless in the face of an oppressive colonial gaze. Much of the literature adopts Edward Said's notion of orientalism. The colonized often seem to be voiceless, their histories, cultures, and agency defined for them by Europeans intent on utilizing any and every form of power they could muster in order to exercise control and to make a profit.[3]

Such one-sided Eurocentrism matters quite a lot. Over the past two or three decades, historians uncovered a striking fact about tourism: it is anything but a trivial pastime and is closely intertwined with virtually everything that makes us modern. Tourism helped to shape our ideas about health, served as a powerful political tool, emerged as a massive economic engine, and played a part in shaping how we see the natural world while at the same time exerting an often negative environmental impact. Leisure travel was frequently at the vanguard of new technological innovations such as steam transportation, bicycles, automobiles, and aircraft. The story of tourism is the story of modern history, both a product of the changes listed above and a contributing factor to them.[4] How one recounts this narrative matters because it has dramatic implications for how we understand the last three hundred years. When countries other than those in Europe or the United States are ignored, or are presented in a one-dimensional way, their significant role is minimized or erased, corrupting our understanding of the past.

At the same time, tourism is best thought of in a broad, transnational way. Not only do tourists move across borders, but tourism developers seem always to be in contact with their counterparts abroad. Myriad groups inform one another. They borrow ideas, make changes, and inspire further developments in still other places. If the narrative we tell eliminates groups or individuals involved in such transnational engagements, then we may miss the presence of such relationships altogether.

Yajun Mo's engaging and deeply thought-provoking new study of tourism, nation, and identity in China during the 1920s and 1930s represents an important corrective. It is the story of how Chinese consciously used tourism to create the China they wanted. According to Mo, China as it existed between 1842 and 1943 is best described as a "semicolonial setting" in which "multiple imperialist powers fragmented China's political sovereignty and influenced economic, social, and cultural development." Some areas were governed outright by foreign powers—Hong Kong and Taiwan—while others were "forced open and maintained extraterritoriality in a series of treaty ports, leased territories, and other classes of foreign occupancies. These places, numbering more than two hundred, were partially under foreign jurisdiction and accommodated foreign settlements and enterprises." Chinese enjoyed greater control over their country than was true in a colony such as India, but they still faced significant limitations and challenges. Sovereignty was certainly not unfettered.

Tourism arrived as a consequence of European and US imperialism. Colonizers, apparently never anxious to accept and enjoy a place as it is, quickly carved out areas for themselves. As they did in many other occupied countries, they built hill stations, sanatoriums, and resorts which made it possible to remain separate from the locals. For Euro-Americans, escaping the "repulsive sights and sounds and nauseating smells" of Chinese cities was deemed essential.[5] Likewise, facilities such as tennis courts and pools were segregated and strictly off-limits to Chinese. Such sites not only allowed physical separation but also utilized familiar architectural styles and provided leisure activities in environments deemed healthful for sensitive Euro-American bodies and immune systems.

Transportation developed in an uneven way, disproportionately centered in the east, Manchuria, and along the Yangzi and Pearl Rivers, leaving inland provinces and frontier regions underserved and more inaccessible. As Mo writes, it was "jurisdictionally fragmented between Chinese sovereign territories and transportation infrastructure on the one hand and semicolonial ports and colonial and semicolonial transportation zones on the other." The network certainly facilitated the development of tourism, but it could not help but shape how that process occurred. Foreign tourists could come ashore, utilizing a growing network of Thomas Cook and Son offices as well as the spreading transportation network to tour beauty spots and tourist sites celebrated in English-language guidebooks, the first of which was published in 1894. The aforementioned European resorts followed soon thereafter.

Affluent Chinese made use of transportation offerings as well, visiting pilgrimage sites such as the birthplace of Confucius or the Great Wall, an emerging national symbol when it first became accessible in 1909. Yet transportation was segregated into "Chinese" and "foreign" classes. Europeans and Americans enjoyed much more luxury.

From this unequal footing, Chinese adopted the European and American import of tourism, turning it to their own ends and presenting a China, as well as a nationalist agenda, that was of their own making. They created their own images, maps, and travel writing to showcase landscapes, historic sites, and diverse cultures which served to begin repairing the fracture generated by outsiders. Tourism helped to create "the nation as a whole." According to Mo, "How *lüxing*, or travel, became a means for Chinese to reify their national space was inseparable from the development of tourism industry and the rise of a nationalist travel culture in the first half of the twentieth century." Tourism was a way for Chinese to showcase their modernity, to minimize political divisions, and to assert a claim to their cultural patrimony.

A significant part of this story centers on the efforts of Chen Guangfu, a Shanghai banker educated in the United States, who endeavored to create a business that combined "financial products with travel services." The result was the China Travel Service, a travel agency that organized trips, published a magazine and guidebooks, and inspired others to pursue exploring China. On one level, this was very much a business venture. Chen realized that he could not compete with massive banks in large cities, but if he established a "concrete identity for his bank," focusing on "second-tier cities," he could distinguish his business, carving out a successful operation in underserved areas. He had another concern as well, a "racial consciousness" born of discrimination while traveling in the United States and while working in semi-colonial organizations dominated by the West. He found the size of foreign investment in China troubling and felt a sense of urgency about promoting travel in order to counteract foreign dominance. The two goals fitted neatly together, one serving the other to create "a geographical circuit different from those designed by giant financial institutions."

Although realizing his dream was by no means easy, partly because much of China was ruled by competing warlords, Chen soon began opening offices in Northern China and beyond. He encouraged, publicized, and utilized a growing rail network to connect North and South China, and capitalized on the growth of self-organized travel clubs to expand his efforts. Within only ten years, China Travel Service had forty-three branches in coastal and inland China, along with offices in Hong Kong, Singapore, and the United States (which focused on helping Chinese students traveling to and from China).

Part of Chen's success stemmed from a close relationship to the Nationalist government, which was keen to expand into warlord-controlled and peripheral areas in order to grow its influence, while also celebrating the growth of a transportation network that made expanded control feasible. The China Travel Service offered a means by which ordinary Chinese could survey the results. Likewise, China Travel Service luxury hotels stood as symbols, "projecting a positive image to domestic citizens and foreign travelers," while at the same time offering a striking contrast to segregated Euro-American facilities. When Japan launched its military advance into China, tourism infrastructure made it possible not only to move the government to a safer wartime location but also to promote patriotism and resistance. Tourism was an important form of nationalist politics, offering Chinese "a shared expression of their national space."

Tourism publications, including the China Travel Service's official magazine, *China Traveler* (edited by Zhao Junhao, another central character in the story of Republican China's tourism efforts), maps (such as the ones that both begin and end Yajun Mo's narrative), photographs, and news articles all promoted destinations in China. Scientific adventures to frontier regions, especially the Northwest, further captured people's imaginations and provided travel writers with an engaging story to tell. When intrepid archaeologists, linguists, folklorists, and anthropologists arrived at important, if then little-known, historic sites only to find that foreign expeditions had been and gone, taking priceless relics with them, it increased urgency: the "hinterland" should be visited and China's patrimony saved. The notion that it was vital to see China and to rescue it from foreign interference made for great copy and a strong nationalist message.

Mo's narrative about China sits easily in the context of what we know took place in other countries at the same time. During the interwar years, regimes of all ideological stripes made use of tourism. The Nazis' Kraft durch Freude program might be the best known, with its organized trips, propaganda publications, and growing infrastructure development efforts,[6] but it was far from alone. French socialists encouraged rational leisure, which would make people happier and healthier, by building facilities and encouraging tourism.[7] The Soviets strove to create a type of leisure that was self-consciously separate from bourgeois examples, going so far as to prescribe holidays based on individualized health requirements.[8] Americans encouraged travel because it would promote patriotism while devoting government funds to putting people back to work building campgrounds and parks.[9] The Irish encouraged people to "See Ireland First," and they worked at creating a vision of the Emerald Isle that was in line with a new postcolonial reality.[10]

This book fits neatly into the story of nations and nationalism as well, even as it helps to create a more useful and comprehensive narrative. As with tourism, the story of modern nations tends toward Eurocentrism. Modern nations are often depicted as products of modernization, an essential identity designed to unify a new breed of worker in an increasingly urbanized, industrialized, and homogenized world. Once one country nationalized, the story goes, others soon followed because it was the most effective way to organize a society in an age of industrial capitalism.[11]

Not every scholar is keen to accept such a view. Partha Chatterjee, writing about India, argues that the nation was more than a foreign import. Anticolonial nationalism "creates its own domain of sovereignty, an area for self-expression. It was a means by which the colonized could assert themselves, utilizing nationalism to declare "the domain of the spiritual its own territory" and refusing "to allow the colonial power to intervene in that domain." The colonized exhibited agency, preventing the colonizer from entering the "'inner' domain of national culture."[12] A dearth of scholarship on tourism in colonial India obscures how leisure travel might have functioned within Chatterjee's framework, yet it certainly played a part. The first Thomas Cook tours were not populated by Europeans, after all, but by Indians who soon started to publish their own guidebooks in order to highlight unique traditions and important sites for Indian travelers.[13]

Yajun Mo provides us with arguably the most comprehensive account of the linkages between tourism and nationalism in a colonial (or semicolonial) setting yet available, a model for historians working on the histories of tourism, colonialism, and nationalism. She shows how Chinese took something that might easily be interpreted as a tool of oppression and turned it to their own ends, reimagining leisure travel as a way to assert their own power and to regain control of their territory. It proved a vital tool for imagining a Chinese nation.

Following World War II, China plunged first into civil war—with leisure travel largely limited to organized trips to Taiwan—and then into an age of Communism. Under the new regime, and in keeping with other similar countries, the focus of tourism shifted from the bourgeoisie toward workers' rest, relaxation, and improvement. The China Travel Service did not long survive this transition, but the "legacies of the Republican-era tourism and travel culture" remained important. Familiar sites remained pilgrimage destinations. The travel network, still somewhat disjointed, continued to expand; the "radius of tourism" was further extended.

In the end, Yajun Mo makes a compelling case: "Simply put, modern travel made modern China." The process by which this happened was "multilayered and multidirectional." It is very much a Chinese story, and yet it is one that takes place against a larger backdrop and that has implications that promise to inform our understanding not only of modern China, but of nationalism, tourism, and even anticolonial struggle more broadly. In so doing, Mo significantly deepens our understanding of the twentieth-century world: a world partly made by tourism.

Eric G. E. Zuelow

Acknowledgments

This book is about journeys, both physical and intellectual. Writing it has taken me to different places, and along the way, I have accumulated many debts. At the University of California Santa Cruz, my mentor Gail Hershatter has been a personal and intellectual beacon of light for me. Through countless hours of one-on-one classes and conversations about research, writing, and life, Gail not only taught me to be an inquisitive scholar and disciplined writer but also showed me how to be a mindful person. To this day, Gail remains a trusted critic of my work and a reliable source of support in my life. As I was preparing this book for publication, she read over and commented on the entire manuscript in great detail. And when the pressure of publishing morphed into self-doubt, her vote of confidence kept me going. No acknowledgement can fully express the endless gratitude I have for the care and labor she has poured into my growth.

Emily Honig joined forces with Gail as my advisor after she became a faculty member in the history department at UC Santa Cruz. Her accounts of doing archival research and interviews in Shanghai in 1979–81 fascinated me and made me see that fieldwork is creative work. Her careful reading of my work also showed me that clear writing is clear thinking. Alan Christy urged me to think beyond national borders and ask questions that might otherwise go unasked. Together with Noriko Aso, Alan encouraged me to explore the transnational history of travel between China and Japan. Their encouragement and recommendations led me to spend a year at the Inter-University Center for Japanese Language Studies in Yokohama. In my class with Minghui Hu, I learned that classics in the field of Chinese studies were worth reading, but we should read them from our own vantage point. Edmund (Terry) Burke III showed me what world history could offer in pedagogical and research methodologies. Among other history professors I had the honor of learning from at UC Santa Cruz, Dana Frank and Alice Yang were my earliest guides to graduate school and academic careers, for which I am indebted to them.

Beyond the history department, Neferti Tadiar, who was teaching in the history of consciousness department at UC Santa Cruz at the time, introduced me to postcolonial and Marxist theories and made me realize that they were neither intimidating nor impenetrable. I took another "HisCon" course on cultural studies with James Clifford, who offered invaluable insights on my research. Based on his own experiences of researching and writing about travel, Jim's advice remained useful throughout the time I worked on this project. Although I never took any classes with her, Lisa Rofel at UC Santa Cruz was and still is an important mentor to me. At the Chinese University of Hong Kong (CUHK), Leung Yuen-sang and Yip Hon-ming shepherded me through the master's program there and encouraged me to pursue a PhD in history. I thank them for their confidence in me.

The friendship and camaraderie with my fellow graduate students also sustained me during my studies at UC Santa Cruz. I have benefitted from my fellow China historians in training: Ana Maria Candela, Shelly Chan, Angelina Chin, Alexander Day, Xiaofei Gao, Fang Yu Hu, Wenqing Kang, Amanda Shuman, Xiaoping Sun, and Jeremy Tai. Although most of us overlapped with each other for only one to two years (if not less), friendships were quick to form in the trenches of graduate school. Over the years, our friendships have deepened as we have kept in touch about each other's lives and work, offering help in preparing for the job market and interviews, and catching up intermittently at conferences. As I am writing this piece on Thanksgiving Day in the United States, I am reminded that many of my earliest lessons about American culture—many of which were food-related—were proffered by my classmates. Among them, I thank Nellie Chu, Urmi Engineer, Michael Jin, Peter Leykam, Eliza Martin, and Dustin Wright for their friendship.

After leaving the West Coast, I have found an intellectual home on the East Coast at Boston College and its history department. Sarah Ross has been the most steadfast senior mentor any junior faculty member could dream of. From responding to my email inquiries religiously to offering commiserations whenever they were needed, Sarah advised me on how to navigate an academic career with discipline, wit, and pragmatism. Other senior colleagues—particularly Julian Bourg, Robin Fleming, Prasannan Parthasarathi, Virginia Reinburg, Dana Sajdi, Conevery Valencius, Franziska Seraphim, and Ling Zhang—offered to read my book proposal and different chapters and imparted their insights on writing and publishing. I also thank Kevin Kenny, now at New York University, for supporting this project when he was the chair of the history department at Boston College. Penelope Ismay and Priya Lal made sure that I felt welcomed from the time of my arrival in the department. Beyond the history department, I appreciate the opportunity to work with my colleagues in the Asian Studies Program. Led by Franziska Seraphim, the program's multidisciplinary colleagues, including Aurelia Campbell, Lydia Chiang, Julia Chuang, Andrew Grant, Ingu Hwang, Fr. Joseph You Guo Jiang, David Johnson, Tina Klein, Fang Lu, David Mozina, Prasannan Parthasarathi, and Ling Zhang have created a tight-knit and supportive community for Asia-related researchers on campus. Their support was indispensable to the completion of this book.

Beyond my home institution, the Fairbank Center for Chinese Studies at Harvard University has been another hospitable environment in which I could improve this project. Like many other China scholars in the Boston area, I relish the opportunity to attend lectures, conferences, and public events organized by the Fairbank Center, where stimulating conversations often lead to new ideas. I thank the director, Michael Szonyi, and the staff, especially Julia Cai, Nick Drake, and James Evans, for operating a vibrant and welcoming academic hub for China scholars. This book benefited from several lectures in the Modern China History Series at the Fairbank Center, organized by Arunabh Ghosh. I thank Arunabh and the speakers, particularly Denise Ho, Richard So, Shellen Wu, and, Peter Zarrow for their openness in discussing varied research topics with me after their formal talks. Under the auspices of the Fairbank Center, my colleague Ling Zhang helped organize an informal workshop on my book manuscript. Sei-Jeong Chin, Arunabh Ghosh, Peter Perdue, Philip Thai, and Ling Zhang attended the workshop, and their careful reading and constructive comments helped me hone the central arguments in this book.

Colleagues and friends from other institutions also provided advice, encouragement, and other assistance for this book. Alexander Day offered to read and comment on the entire book manuscript, for which I am deeply grateful. I thank Emma Teng at MIT for her encouragement when she learned that her book on travel writing about Taiwan inspired me to write a chapter on Taiwan. When I was living in New York, Robert Barnett, Kevin Landdeck, Eugenia Lean, William Charles (Chuck) Wooldridge, and other organizers and participants in the Modern China Seminar at Columbia University welcomed me to their monthly gatherings. In California, I am indebted to the conversations with professors at other UC campuses and Stanford University. Among them, I thank Beverly Bossler, Andrew Jones, Susan Mann, Tom Mullaney, and William Schaefer for giving me opportunities to seek their advice on my project. I thank Tom Mullaney for arranging for a review of my work on his website. I also thank Jenny Chio for her thoughtful review of my manuscript.

Many pieces of work that have gone into this book were also presented at conferences, and my book has benefited from comments from my fellow panelists, discussants, and other conference participants. Among them, I particularly thank Jessamyn Abel, Jennifer Altehenger, Janet Chen, Zhihong Chen, Sei-Jeong Chin, Yongtao Du, Arunabh Ghosh, Miriam Gross, Rebecca Karl, Han Sang Kim, Elisabeth Köll, Margaret Kuo, Jeff Kyong-McClain, Yu-ting Lee, Christopher Leighton, Amy Marshall, Kate McDonald, Ke Ren, Jeremy Thai, Weili Ye, Zhang Ke, and Qian Zhu. The "Writing Studio on Zoom," organized by Chris Jones in Arizona and Conevery Valencius in Boston, is a godsend. The virtual writing space allowed me to stick to a daily routine and helped alleviate the feeling of isolation in 2020, making finishing this book during a pandemic possible. The warm support from Emily Baum, Buyun Chen, Xin Fan, Miao Feng, Zachary Howlett, Colleen Laird, Andrew Liu, Lex Jing Lu, Amy Marshall, Eugenio Menegon, Rebecca Nedostup, Weilin Pan, Steven Pieragastini, Ying Qian, Caroline Reeves, Ke Ren, Victor Seow, Kyle Shernuk, Lihong Shi, Susan Smulyan, Brigid Vance, Leslie Wang, Y. Yvon Wang, Xin Wen, Yulian Wu, Ying Xiao, Shirley Ye, Wen Yu, Aslı Zengin, Qian Zhu, Ya Zuo, and many other friends and colleagues reminds me that I belong to a wonderful scholarly community in and beyond New England. For my brief time at Long Island University's Post campus, I thank Jeanie Attie, Jay Diehl, Sara Gronim, Willie Hiatt, and Molly Tambor. I also thank Geoff Cock, Wesley Dick, Chris Hagerman, Deborah Kanter, and Marcy Sacks, who made my transition from graduate school to professional life at Albion College an easy one.

Back when we could still meet in person, gathering for food and drink, or just to talk, Julia Chuang, Arunabh Ghosh, and Philip Thai unfailingly lifted my spirits. They also helped make this book better in various ways, from reading drafts to helping with images. I am looking forward to our postpandemic reunions. David Mozina is my comrade in arms in book writing. Together with Ling Zhang, his formidable partner, David inspired me to find my own voice while being unafraid of the painful passage I had to travel through to find it. When the pandemic delayed the last round of reviews and revisions of this book, David delivered freshly baked bread to boost my morale. I have cherished his unwavering support even more than his delectable sourdough bread.

This book was also made possible with the support of many institutions and foundations. A yearlong leave in 2017–18, funded by Morrissey College of Arts and Sciences at Boston College and International Center for Studies of Chinese Civilization (ICSCC) at Fudan University, allowed me to focus on writing. At Boston College, I thank Dean Gregory Kalscheur for granting the extra leave time and continuing to support this project. A subvention granted by the Dean's Office of the Morrissey College of Arts and Sciences at Boston College also provided financial support to defray some of this book's production costs. At Fudan, my thanks go to Zhang Qing, Zhang Ke, and Huang Chen and other staff at ICSCC for facilitating my productive residence at the center. Several research trips to mainland China and Hong Kong in 2013–16 were funded by a Research Expense Grant at Boston College, a Faculty Research Grant of the College of Liberal Arts and Sciences at Long Island University, and a Faculty Travel Grant at Albion College. The Ito Foundation USA Grant supported my advanced Japanese-language studies at the Inter-University Center in Yokohama. A Mok Hing Cheong Postgraduate Scholarship supported my study at CUHK.

I am also indebted to the librarians and archivists who assisted my research over the years. At Boston College's O'Neill Library, Julia Hughes, Anne Kenny, and Brittany "Bee" Lehman offered critical support in securing access to sources, databases, and images for this book. In Shanghai and Nanjing, where I collected much of the primary sources that form the basis of this book, the staff at the Shanghai Archives, the Second Historical Archives of China in Nanjing, Shanghai Library, and Fudan University Library deserve my heartfelt thanks. In Hong Kong, the librarians at CUHK's University Library and the University of Hong Kong's Fung Ping Shan Library pointed me in the right direction when I was lost. I could not have carried out my work in the United States without the help from the staff at the Yenching Library and Widener Library at Harvard University, the C. V. Starr East Asian Library at Columbia University, the Asia Library at the University of Michigan, the Hoover Institution Library and Archives at Stanford University, the C. V. Starr East Asian Library at UC Berkeley, and the McHenry Library at UC Santa Cruz. The staff at the New York Public Library's research division helped track down a few sources for this project. Special thanks go to the staff at Boston Public Library's Central Library in Copley Square, where I spent the winter months of 2018–19 revising this book.

At Cornell University Press, I thank Emily Andrew for soliciting this book for the press's new Histories and Cultures of Tourism book series. With the expert assistance of Allegra Martschenko and Alexis Siemon, Emily arranged for review by three attentive anonymous readers. I am grateful for their diligent, engaged, and timely reports. I am also thankful to the series editor, Eric Zuelow, for showing interest in this project early on. His continuing engagement and goodwill lifted me up at various stages of this project. I also thank him for writing the foreword. I thank Karen Hwa, senior production editor, who masterfully guided this book through production. Thanks as well to

Florence Grant, whose copyediting prowess vastly improved the manuscript. I acknowledge the help of Lorri Hagman at the University of Washington Press and Lucy Rhymer at Cambridge University Press. I am indebted to Eleanor Goodman—poet, wordsmith, and a dear friend—for polishing my writing. Her meticulous work made the final product more readable and elegant. I am grateful to Gregory Epp, an exceptional developmental editor, for offering valuable advice when I consulted him on the revision of the manuscript. Mike Bechthold created the map in this book. Rachel Lyon produced its index.

Since we met, Irene Gates has known about "the book." Living through every unexpected setback, she has remained unflinching in her support for me. Without her labor of love, this book would not exist. I also thank Charles Gates for reading and commenting on an early version of the introduction. His comments helped improve the introduction immensely. Caroline Gates kindly answered my questions about the history of travel in the context of France, for which I am grateful. On the home front, I thank all the kids and adults at 11 BVP in Cambridge, who provided joy, friendship, and care, particularly during the difficult months of 2020.

My mother, Miao Jiling, and my father, Mo Xiheng, are my rock. I thank them for instilling in me a love of reading and writing, for forgiving me when I spent little time at home during my research trips in China, and for never failing to show me their unconditional love. Although they will not be able to read it in English, I dedicate this book to my parents. After all, they made all this possible.

TOURING CHINA

Introduction

Tourism, Travel Culture, and the Making of the Chinese National Space

In October 1934, the photojournalism magazine *Dazhong huabao* published a two-page spread titled "Beautiful China" to commemorate the twenty-third anniversary of the founding of the Chinese republic.[1] Serving also as the final installment of a yearlong travel column, Travelogues of China (*Zhongguo youji*), the spread looks at first glance like a typical map of China framed by its idealized begonia-leaf shape (figure 1). However, the map does not contain any of the standard cartographic symbols or place names. With only the provincial boundaries drawn, the magazine instead filled each space—corresponding with a given province—with a black-and-white photograph. All taken by the column's author and renowned photojournalist Wang Xiaoting (known as "Newsreel" Wong in the West), these images featured some of the most recognizable travel destinations in China. Referencing a glossary at the bottom right-hand side, readers could identify these iconic attractions individually, along with their "matching" provinces. The Temple of Heaven in Beijing symbolized Hebei Province, for example, while Yunnan Province in China's southwestern corner was represented by the geological wonder of its Stone Forests. Collectively, as the accompanying English caption summarized, these "famous impressions that represent our thirty provinces . . . form the leaf-shaped map of this land of ageless splendor."[2]

The ingenious visual rendition of the "Beautiful China" spread was half reality and half myth. By 1934, modern transportation and the tourism business were indeed developed enough in China proper—a region encompassing the eighteen Chinese provinces of the former Qing empire—for many scenic attractions to be accessible to tourists. In contrast, the provinces in what is known as China's Inner Asian periphery—a vast area that included Tibet, Xinjiang, and Mongolia—were neither under the direct control of the Chinese government nor within the reach of leisure travelers. This juxtaposition of reality and myth reveals much about a Republican-era vision that viewed postimperial China as a coherent and homogeneous nation-state, like many other modern nations in the world. As the photo spread and travel column demonstrate, traveling to and witnessing the diverse natural landscapes, historical sites, and local peoples throughout China not only made different parts of the country legible but also rendered "the nation as a whole" (*quanguo*) imaginable.

How *lüxing*, or travel, became a means for Chinese to reify their national space was inseparable from the development of the tourism industry and the rise of a nationalist travel culture in the first half of the twentieth century.

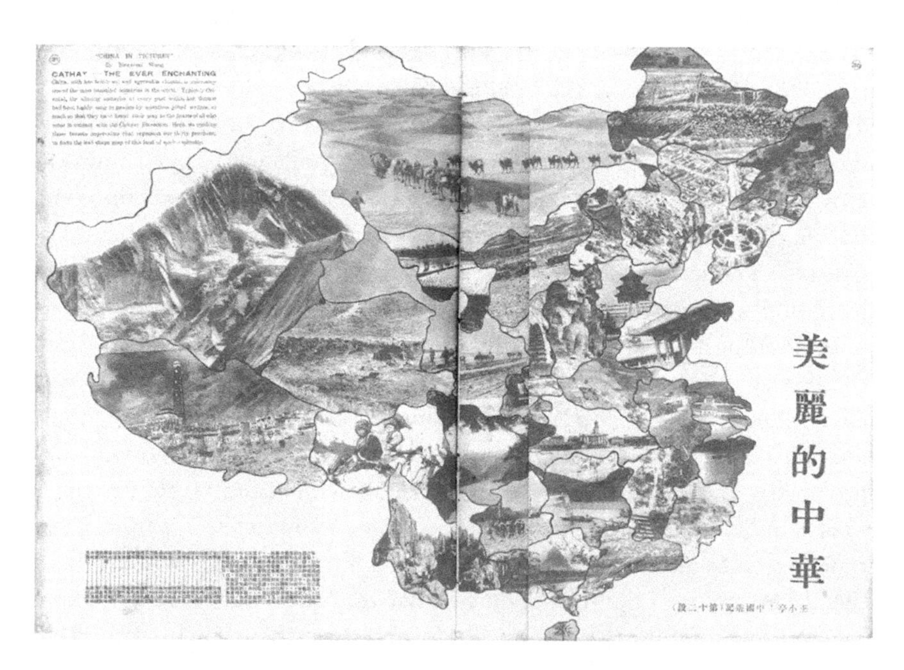

FIGURE 1. "Beautiful China." *Source:* Wang Xiaoting, "Meili de Zhonghua," *Dazhong huabao* [The cosmopolitan], no. 12 (October 1934): 30–31.

This was a tumultuous period, during which China transitioned from empire to nation-state. In the last decades of the Qing dynasty (1644–1911), foreign powers forced a series of unequal treaties on China after a string of diplomatic and military conflicts. Opening up treaty ports like Shanghai to foreign trade and settlement, these treaties also granted political, commercial, and missionary privileges to Western powers and later Japan, undermining China's sovereignty. The treaty system and domestic upheavals significantly weakened the Qing empire, which crumbled in 1911. The subsequent founding of the republic did not end the chaos. Power struggles among militants during the warlord period (1916–27) and then between the Nationalists and the Communists in the Nationalist era (1927–49) exacerbated the political and economic fragmentation of the nation. Continued frontier crises in the Qing peripheries (notably Tibet, Xinjiang, Mongolia, and Manchuria) and war with Japan also threatened China's territorial integrity. While different aspects of Western modernity continued to spread in China via treaty ports, the privileges afforded to foreigners sparked anti-imperialist and nationalist movements among Chinese citizens. This paradox of Republican China was the condition from which modern Chinese tourism and travel culture emerged.[3]

Despite the political turbulence and economic hardship, domestic travel in China began to grow in volume and geographic scope.[4] From treaty ports to the interior, the expansion of steam-powered transportation improved travel infrastructure and stimulated the nascent tourism industry. With the increasing appeal of commodified travel, middle- and upper-class urbanites started to leave the city for the clean air of the countryside, to hike in the mountains and visit the seaside, or to tour historic landmarks in different parts of the country. Meanwhile, the more adventurous travelers and learned professionals took part in scientific and journalistic explorations to far-flung regions. The Japanese invasion and occupation of China's eastern seaboard from 1937 to 1945 also made travel to the Southwest—the "home front" of Free China—more common.

Rather than the idealized visual representation of travel in "Beautiful China" presented by "Newsreel" Wong, travel culture—by which I mean the practices and beliefs surrounding the production and consumption of travel—in modern China was composed of a mosaic of tourist activities, frontier explorations, and wartime travel. Together, they allowed Chinese citizens to redefine postimperial China in a period of dramatic political change. These diverse forms of modern travel also contributed to the imagination of China as a congruent national space. The fascinating accounts of these journeys constitute the central case studies in this book.

How Modern Travel Made Modern China

Sightseeing was a commonplace practice in imperial China. Touring Hangzhou, for instance, can be dated back to the Northern Song dynasty (960–1127).[5] While travel for pleasure was not new, the rise of modern infrastructures during the first half of the twentieth century provided Chinese with new means, motives, and opportunities for travel. Mass transportation transformed individual travelers into group tourists. New ideas of wellness and leisure propelled people up mountains and onto beaches. As modern transportation made long-distance vacationing feasible, commercial guidebooks and travel magazines made it desirable. These modern infrastructures—from the physical and tangible to the social and invisible—shaped modern tourism in China.

In the Republican period (1912–49), Chinese, especially upper- and middle-class urbanites, pursued fun and adventure through travel. Yet they also understood that travel and tourism were deeply connected with state and nation building in China. As an industry, tourism contributed to the development of a national economy and helped project an attractive image of China to the world. Whereas urban consumers laid claim to a modern identity and asserted their national pride through tourism, adventurous travelers and wartime refugees viewed their travel in the periphery as a way of asserting sovereignty at a time when China's territorial unification was uncertain. In tourism and travel culture, Chinese found a shared expression of their national space.

The Arrival of Modern Travel in China

Industrial transport in the form of the steamship and the railroad arrived in China on the heels of Western imperialism.[6] By the end of the nineteenth century, not only had a broad spectrum of Chinese—from top officials and elite businessmen to urban students and mobile laborers—used steamships and trains in their travel for study, work, and business, but foreign-owned travel agencies, a relatively novel business institution even in the West, had also appeared in Beijing, Hong Kong, and Shanghai, catering to international tourists.[7] But how exactly did these new technologies and institutions come to change travel in China?

To begin with, modern transportation had a dramatic impact on leisure travel. A 1918 guidebook by the Shanghai-Nanjing and Shanghai-Hangzhou-Ningbo Railways boasted that unlike travelers in the past, who endured long treks to enjoy a mere day or two of sightseeing, passengers riding their trains could "watch the clouds at Mount Zhong in the morning, drink water from the Huangpu River at noon, and listen to the bell at Nanping in the evening."[8] The speed of the new mode of transport allowed tourists to fit as many sites as possible into a short period. Economic fares on mass transportation systems also made group travel a new trend. In peak travel seasons, steamship and railway companies routinely provided extra capacity to meet the demand for group outings to scenic spots outside the city.

With this growing demand, China's modern travel industry as such emerged during the tumultuous warlord period. One key player was the China Travel Service (CTS). CTS's predecessor, the Travel Department of Shanghai Commercial and Savings Bank, was founded in 1923 by the US-educated Shanghai banker Chen Guangfu (1880–1976). It was born out of Chen's vision to combine financial products with travel services, akin to the business model of American Express. However, unlike American Express, whose business expansion was aided by the United States' rise as a world power, Chen's small Travel Department had to compete with foreign travel agencies in China during the chaotic warlord period. Without much help from the weak central regime, the Travel Department gained its footing by prioritizing the needs of Chinese travelers and promoting domestic tourism. The success of this strategy laid the foundation for CTS, which was separated from the bank in 1926.

Unlike the warlord regime, the Nationalist government worked closely with CTS. During the Nanjing Decade (1927–37), the central government and provincial authorities relied on CTS for crucial travel services along the state transportation infrastructure. This close collaboration allowed CTS to offer more comprehensive travel services and build up tourism facilities throughout the newly integrated national transportation grid on China's eastern seaboard. Accordingly, the agency effectively expanded the reach of the Chinese tourism network, or what I call modern China's "tourism radius," from coastal areas to inland provinces.

Perhaps the most telltale sign of the popularity of tourism was the maturation of travel clubs in urban centers. During the Nanjing Decade, the Unison Travel Club, one of the most successful travel clubs in modern China, attracted thousands of members in Shanghai and surrounding areas and coordinated hundreds of group tours to popular destinations in the lower Yangzi region. As China's tourism radius grew to encompass much of China proper, the well-off members of the Unison Travel Club traveled on week-long itineraries to North and South China in the 1930s. For affluent urban consumers, tourism became a symbol of their cosmopolitan identity and a ritual marking their citizenship.

Modern Travel in China's Peripheries

The prevalence of tourism was complemented by other travel trends, including frontier travels. Although China's tourism radius remained largely within China proper, throughout the Republican period, the far-flung frontier regions—the Northwest, the Southwest, and Manchuria and Taiwan—also attracted diverse travelers from the east coastal region precisely because they lay beyond the tourism radius.

As modern travel became an indispensable part of learned professions, scientists and journalists began to travel extensively for fieldwork and exploration. Journeying beyond the well-trodden routes followed by tourists, they ventured into remote areas and brought new knowledge and information back to their professional circles and the general public. Joining an expedition to Dunhuang led by the American archaeologist Langdon Warner in 1925, the Beijing University scholar Chen Wanli became the first Chinese academic to visit the Mogao grottoes, a world-famous Buddhist site on the Silk Road in northwestern Gansu Province. Understanding that today's exploration might yield tomorrow's tourism, transport managers and tourism promoters made note of the itineraries of scientific explorers and journalists. In the Northwest, they designed itineraries, offered discounts, and developed tourism products for adventurer-tourists as soon as railways extended westward in the mid-1930s.

In addition to a spirit of adventure and scientific curiosity, the quest to forge a connection with territories geographically and politically distant from the nation's center proved to be an important motivator for travelers as well. Wartime tourism in the Southwest and postcolonial tourism in Taiwan testified to the symbolic power of travel as a way to exercise sovereignty over China as a whole. This desire to assert control over the borderlands through travel was also shared by armchair travelers. From Chen Wanli's travelogue, readers in Beijing and Shanghai could learn about the expedition's arduous treks and dramatic encounters with villagers in the Northwest as the scholars tried to safeguard valuable historical artifacts from suspicious locals. Ordinary readers hankered after the texts and images recording such exciting travels, and this was reflected in popular print media at the time.

Modern Travel and Modern Print Media

Like modern transport, modern print media—newspapers, magazines, and mass-produced books—played a crucial role in the proliferation of modern travel and tourism. Foreshadowing what we call "travel journalism" today, guidebooks, travel magazines, and travel columns in Republican China introduced new ways of sharing reliable information with tourists and business travelers. Leading publishing companies considered travelers' handbooks key to their competitiveness among urban consumers with increasing mobility. CTS devoted considerable effort to publishing tourist guidebooks and a monthly travel magazine, *Lüxing zazhi* (*China Traveler*). Through these print media, CTS publicized the company's travel services and facilities and promoted tourist sites through travelogues penned by real-life travelers.

The popularity of travel-related print media, especially guidebooks and travel columns for China as a whole, engendered a panoramic view of the country. Like Wang Xiaoting's travel column that opens this book, by featuring tourist destinations throughout the country, travel print media demonstrated the link between the commercial growth of tourism and the spatial claim of the nation-state. Compared with maps and geography textbooks, scenic guides and travelogues made China's territorial image more concrete.[9] Even though travel routes and must-see landscapes remained two-dimensional in print, they carried the promise that one could follow them to experience China's landscapes for oneself.

Travel writings about the peripheries of China were also popular among urban readers. In journalist periodicals, travel columns and special reportages penned by explorers and journalists on the road became common features. Photographs of remote destinations proved to be a mainstay in pictorial magazines. One of the most circulated magazines in the Republican period, *Liangyou* pictorial, organized a "Nationwide Photographic Tour," during which their journalists and photographers traveled throughout China and sent back their writings and photographs to be serialized in the magazine.[10] Through these travel reports and images, the reading public in the core regions of China actively engaged with the outlying territories before they gained the opportunity to tour them personally. These vicarious journeys in turn transformed these outlying regions into powerful locations in the imaginary geography of the fledgling nation-state.

Nation Formation through Tourism and Travel Culture

At the center of all of this burgeoning activity were three main groups of actors: transport and tourism promoters, travel and tourism consumers, and travel culture producers (i.e., publicists, writers, and publishers who produced and marketed expertise in travel and tourism). These groups were often overlapping, and they played a key role in the nation formation of modern China at a time when the infrastructure of the nation-state—including a strong and stable national government, a nationwide transportation and communication network, and a national market—was barely in place. Although China went through a series of territorial and political disintegrations after the fall of the Qing empire, the understanding that the republic had "inherited" the entire Qing realm as its national space was widely shared among the urban bourgeoisie. This shared belief could not have conveyed or propagated itself without a medium. And what exactly was that medium? As will become clear in the following chapters, tourism and travel culture became the linchpin in piecing together the seeming incongruence between the political reality of fragmentation and the idealized phantasm of the wholeness of modern China.

Tourism and China's Semicolonial Condition

This book belongs to a relatively short but dynamic lineage of tourism history, though it is perhaps an uneasy fit. With a few exceptions, tourism history as a field remains a largely Euro-American-centric affair.[11] China appears only as a destination for colonial travelers and international tourists in the late nineteenth and early twentieth centuries; Chinese participation in tourism in China does not begin until the post-Mao reform era (1978–present).[12] In addition, although scholars in various disciplines have examined travel culture in imperial China from different angles, the evolution of tourism and travel culture in the Republican period, especially their place in the nationalist culture of modern China, has only received preliminary interest.[13] This book will challenge this limited view of Chinese tourism history by broadening our understanding of modern Chinese travel culture.

As part of the budding global phenomenon of tourism, China's tourism development bore critical similarities to the historical trajectories of tourism in Western industrial nations. It was, for example, also a product of the industrialization of transportation and a rising consumerism in the urban sector.[14] The tourism industry in modern China, too, was deeply involved in the state projects of modernization and nation building, which aimed to formulate a territorialized national identity.[15] However, as the treaty system had shaped China into a semicolonial country (a concept to be discussed in chapter 1), unlike many industrial nations in the West, China bore the brunt of the violent push for economic, political, and cultural dominance by various Euro-American and Japanese powers. At the same time, unlike Korea, Vietnam, or India, China never lost its sovereignty entirely, so its status vis-à-vis the imperial powers was not straightforwardly compatible with the classic colonial model. Although China's compromised sovereignty limited the development of native industries including tourism, without a single colonial overlord, the scope of colonial tourism in China was constrained. As this book will demonstrate, this distinctive semicolonial condition had broad implications in tourism and travel culture in modern China.

The chapters that follow trace how the economic, political, and cultural forces stemming from China's semicolonial condition manifested themselves in many seemingly opposing directions. For example, for many "travel-minded" Chinese, no matter whether they were visiting nearby summer resorts at Moganshan or joining the CTS-organized European tour to see the Berlin Olympic Games, tourism was associated with Western modernity, and being a tourist was to claim a cosmopolitan identity.[16] In this context, the emergence of modern tourism in China resonates with tourisms that evolved in typical colonial settings, in which leisure travel and vacationing were introduced to the locals through the establishment of hill resorts and seaside sanatoriums.[17] Although such facilities were often created by and for the foreign settlers, ironically, their exclusivity attracted the locals, who were eager to learn what the fuss was all about.[18]

Yet tourism development and the emergence of a nationalist travel culture were also linked to the anxieties engendered by the same colonial condition. More than just an imitation of certain elements of Western colonial modernity, Chinese business elites and nationalist consumers viewed their participation in the native tourism industry as a way of resisting foreign control over China's industrial transportation sector, preventing further territorial encroachments from the Western and Japanese powers, and fighting against colonial racism. For instance, when provincial governments in the lower Yangzi region were busy integrating the transportation trunk lines in 1934–35, they also prioritized the modernization of travel infrastructure in the Yellow Mountains and other celebrated travel destinations in order for them to compete with tourist resorts developed by foreigners.[19]

Unlike the case studies of colonial tourism in China, a history of Chinese domestic tourism reveals unexpected effects of colonialism on travel and tourism. China's semicolonial condition, for example, is reflected in the spatial characteristics of tourism mobility. As will be addressed in detail in chapter 4, for instance, for a traveler from Shanghai to visit Kunming in the southwestern province of Yunnan in the 1920s or 1930s, the most common route was not a direct overland domestic route. Instead, a triangulated course of steamship and railway journeys was preferred, from semicolonial Shanghai to British colonial Hong Kong, and on to Haiphong in French colonial Vietnam, where one could take a French colonial railway to Kunming. This journey from one domestic destination to another required travelers to traverse Chinese borders twice, thereby encountering a range of border control practices determined by the colonial powers. Such experiences prescribed the way modern Chinese practiced and envisioned their physical mobility. Nevertheless, these constraints and complications did not stifle the enthusiasm for travel among Chinese business elites and the urban bourgeois class. On the contrary, their access to modern tourism services, which helped them maneuver around these limitations, distinguished their modern identity from their less mobile fellow countrymen. At the same time, the lack of an integrated national polity and market also prompted them to produce and imagine a national space through travel.

Although this introduction refers to domestic tourism, domestic journeys in modern China were often transnational in nature, as seen in the aforementioned journey from Shanghai to Kunming. In the first half of the twentieth century, coastal Chinese relied on a hybrid network of modern transportation apparatuses—combining colonial, state, and private lines—to journey through China. Linking semicolonies in China to European or Japanese colonies in Asia and to China's sovereign territories, this network made domestic tourism in China possible and gave rise to a tourism radius stemming from the urban centers in the east toward the western edges of China proper. It also made domestic tourism transnational, either geographically or jurisdictionally. This transnational aspect of domestic mobility would eventually allow many coastal Chinese to retreat to inner and western China via Hong Kong and Vietnam when coastal China was occupied during the Second Sino-Japanese War, fundamentally shifting the national travel network in a time of crisis.

During the Republican period, the territorial boundaries of China were hardly settled, and therefore what did and did not count as domestic remained blurry. New boundaries could appear, while established borders might disappear. As will be discussed in chapter 5, when Manchuria was occupied by the Japanese in the early 1930s, travel arrangements between other parts of China and Manchuria became complicated because domestic travel turned transnational with the appearance of the new border. Conversely, when Taiwan, a Japanese colony since 1895, was returned to China in 1945, it became a new domestic tourist destination. Situated beyond the tourism radius of modern China, territories that were inherited from the Qing empire held an indispensable place in touristic and travel imaginaries. Mimicking the colonial logic of appropriating colonies for the metropole, Chinese travelers also viewed tourism and travel as a crucial way to domesticate these frontiers into

China's national space. In this sense, travel and tourism functioned as a tool for nation building in semicolonial China.

Tourism and China's National Space

The parallel rise of tourism and nationalism is not a uniquely Chinese story. Scholars of nationalism and tourism in other countries have traced how capitalist, socialist, and fascist states promoted domestic tourism for nationalist ends.[20] In Republican China, however, the central state—either warlord or Nationalist—was too weak to invest in tourism. The absence of state leverage meant that tourism development in modern China did not engender a hegemonic nationalist message. Instead, tourism and modern travel allowed Chinese people to express their nationalist sentiments in many forms. Guidebook publishers and travelogue authors, for example, used nation-centric travel narratives to combat China's postimperial fragmentation. At the same time, tourism and travel permitted a common expression of China's spatial claim at a time when Chinese borders remained elastic. As domestic tourism was dependent upon the boundaries of the nation, the expansion of the tourism radius during the Nanjing Decade helped the Nationalist state bolster its territorial control in China proper. The popularity of frontier travel narratives demonstrates that preserving China's national space, based on the Qing contours, garnered wide support.

This construction of China's national space manifested in two distinct ways. Modern travel helped inscribe local and regional destinations with national importance, connecting them to the national space. Tourists making pilgrimages to landmarks like the Yellow Mountains or the Great Wall, for instance, viewed these locations as powerful symbols of the national landscape and sanctified sites of national history. Frontier explorations to Inner Mongolia and Xinjiang reinforced the economic, scientific, and political value of these borderlands to the nation. Wartime travel narratives about the Southwest were redolent with the centrality of the area to the national survival and revival. Travels to Manchuria and Taiwan, parts of the former Qing territories seized by the Japanese empire, reminded Chinese travelers of the traumatic national humiliation of imperialist aggression. Through these itineraries, the abstract national space became physically tangible and emotionally accessible.

In addition to imbuing the national space with substance, modern travel also gave shape to that space. With the expansion of people's spatial horizons, a panoramic view of the country was adopted by transportation companies, the tourism industry, major publishing houses, and other corporations with national ambitions. For example, travel agencies and publishers were eager to define their products, such as travel services, tourist resorts, and guidebook and travelogue publications, in explicitly national terms. CTS eagerly extended its network nationwide during the Nanjing Decade. Featuring the "five largest summer resorts in China" or serializing photo albums of "China's famous mountains" in newspapers and magazines, publishers recast regional destinations within a national scope.[21] This panoramic view was encapsulated in the prevalence of the term *quanguo*—the nation as a whole. Publications of various kinds began to use *quanguo* to signify their broad coverage of China's travel destinations. As travel agencies and tourist clubs expanded, adventurous individuals and groups began to attempt to travel around the entire country. As will be discussed in chapter 2, *Liangyou* pictorial organized a "Nationwide Photographic Tour" in 1932–33. Besides travel columns serialized in the magazine, Liangyou Printing and Publishing Company also issued a photo album titled *Zhonghua jingxiang* (the original English title was *China as She Is: A Comprehensive Album*) in 1934.[22] With dramatic close-up and wide-angle photos of tourist attractions and iconic landscapes all around China, the album offered the most comprehensive view of *quanguo* to date and became an instant commercial success.

Although domestic travel in this era could and often did reveal China's compromised sovereignty and fragmentated national territory, modern travel was also envisioned as a way to ameliorate, if not rectify, these exposed fault lines. In his "Inaugural Preface" to *China Traveler*—the official magazine of CTS—Chen Guangfu proclaimed that "[t]hose who have set foot on the entire nation (*quanguo*) will naturally abandon their provincialism."[23] While pointing out the lack of modern transportation in inland provinces during their nationwide tour, *Liangyou* journalists and photographers used their journeys to establish tangible links between the core and the periphery of the national space.

The use of the term "national space" is inspired by the idea of "core-periphery." Just as the flow of international tourists is often from the global core to the periphery, in the case of domestic tourism in modern China, tourists tended to move from urban centers to peripheral areas. The extensions of railways from coastal treaty ports to inland cities also accentuated the core-periphery structure at a national level. For example, while the Nationalist government promoted collective travel to Manchuria shortly after it had nominally unified China in 1927–28, the Chinese government railways such as the Ping-Sui and Longhai Railways advocated exploratory tourism in the Northwest to adventurous students and urban youth. By constantly searching for what Louis Turner and John Ash have termed the "pleasure periphery," they extended the tourism radius from the area immediately surrounding modern China's industrial core to the underindustrialized peripheries.[24]

This scrutiny of the construction of China's national space through tourism and travel culture is also stimulated by the attention to space in studies of nationalism.[25] The highly influential concept of geobody denotes an idealized image of the territory of a nation, such as those depicted on national maps.[26] "Geobody" has been a useful term in elucidating the spatial practices involved in the making of modern China's nationhood, from cartography to geography textbooks.[27] However, unlike the solidified notion of a geobody, the notion of national space as used here will stress the more fluid expression of place and space in tourism and travel culture. Just as the magazine spread at the beginning of this introduction demonstrated, travelers and travel promoters identified key components of the national space by giving it specific place-based content. Whereas modern textbooks and maps instilled in Chinese citizens an idea of China's geobody, tourism practices gave a concrete reality to that abstract concept. The function of tourism and travel in consolidating the national space was so apparent that central and local governments and state-owned transportation institutions all actively supported the expansion of the CTS network throughout the Nationalist era.

Just as travelers and tourists are not simply mindless receivers of the information, values, signs, and symbols packaged by the tourism industry, national citizens are not passive recipients of official narratives of national history, territory, or identity.[28] In Republican China, tourism development allowed business elites to join hands with politicians and intellectuals in the mission of molding China's new selfhood after the fall of the Qing empire, and it provided a forum for urban populations to contemplate what shape a postimperial China should take. Rather than trying to serve as an encyclopedic history of tourism and travel in modern China, this book focuses on a selection of itineraries of physical and intellectual journeys spanning China proper and the periphery. These tourist journeys, frontier explorations, and wartime travels reveal the expansion of the spatial horizons of Chinese citizens and the formation of a nationalist travel culture in modern China. Both developments contributed to the construction of China's national space, and collective identity, in the first half of the twentieth century.

Organization of the Book

This book is divided into five chapters. The first two chapters center on the emergence and widening of the tourism radius of modern China. Examining travel as a business, chapter 1 traces how CTS and the Unison Travel Club spearheaded new forms of production and consumption via tourism. Without any state-sponsored national tourist institutions, private companies and civic organizations took up the responsibility for developing and promoting domestic tourism while projecting their desire for a unified national market onto a nationalist tourism industry.

Parallel to the proliferation of tourism entrepreneurship and tourism consumption was the efflorescence of a tourism-centered travel culture in China's coastal urban societies. Chapter 2 discusses this phenomenon by scrutinizing the production of guidebooks, travel magazines, and travel columns by print capitalism and tourism companies in modern China. The diverse travel narratives in commercial print media, it argues, gave shape to the notion of *quanguo*, the nation as a whole.

Moving beyond the tourism radius of modern China, the last three chapters examine touristic and nontouristic mobilities in the geographical and spatial peripheries: the Northwest (chapter 3), the Southwest (chapter 4), and Manchuria and Taiwan (chapter 5). With limited opportunities for organized tourism, these regions were primarily associated with modes of travel other than for leisure, such as scientific explorations, business and investigative journeys, and wartime escape. While the causes of their marginalization in domestic tourism varied from region to region, there were persistent efforts to extend the tourism radius to these regions, and thereby to domesticate them.

By tracking the margins of consumerist tourism, the last three chapters also elucidate how certain forms of mobility challenged, negotiated, and reinforced the narrative of the oneness of China's coastal regions, inland provinces, and frontier regions. Crossing the 1949 divide, the book ends with a conclusion that sketches the afterlife of the Republican-era tourism industry and travel culture in the early 1950s and emphasizes their persisting influences in modern China.

This book progresses from China's core to its peripheries and delineates how travel and tourism facilitated the construction of a Chinese national space by its citizens. This construction became a central force in China's transition from empire to nation-state. Far from a fait accompli, China's transition to nation-state remains unfinished, despite the seeming longevity and stability of China's geobody in recent centuries. The pressing question of what shape a Chinese nation-state should take was, and still is, subject to ongoing debate.

CHAPTER 1

Travel as a Business

The Making of Modern Chinese Tourism

On August 12, 1923, in the local news section of the *Shenbao* newspaper, a brief announcement appeared with the headline "Shanghai Commercial and Savings Bank adds a Travel Department."[1] This new department planned to sell tickets for different Chinese government railways and to offer pick-up services at train stations and steamship docks. Initiating these services in Shanghai and Hangzhou, the Shanghai Commercial and Savings Bank (hereafter Shanghai Bank) intended to gradually expand their Travel Department to the bank's other locations. Although easily missed amid other miscellaneous local news, this short paragraph recorded the beginning of China's modern tourism business. From this Travel Department emerged the CTS, which by the end of the Nanjing Decade had developed into the largest travel agency operated by and for Chinese nationals.

Shanghai Bank's Travel Department did not emerge out of nowhere but rather was located within the new economic, cultural, and social life in early Republican Shanghai. Not only did travel feature prominently in the work and life of the city's business elites, but it also became a preferred leisure activity for working urbanites, to whom the Unison Travel Club offered both travel knowledge and travel opportunities.[2] Beginning in the early 1920s, the club organized excursions and weekend getaways, providing an escape from the urban industrial lifestyle. It was so popular that the club attracted more than five thousand members before the Japanese invasion of China in 1937.[3]

Travel as a business emerged in China in the second half of the nineteenth century. By the early 1900s, foreign travel agencies appeared in treaty port cities. Meanwhile, Western expatriates replicated tourism habits at home by establishing seaside and hillside resorts for their leisure needs. While travel services and opportunities became more available for foreigners in China, there was a growing unmet demand among Chinese nationals. Embracing travel as part of their own modern identity, the rising Chinese bourgeois class had to face China's fragmented sovereignty and virulent racism as they pursued fun and adventure through modern travel. During the tumultuous warlord period, the modern transportation systems were unreliable, and Chinese were unwelcomed in foreign resort spaces. Accordingly, corporate institutions and voluntary associations from the private sphere stepped in to fill the vacuum. Shanghai Bank's Travel Department, the first travel agency operated by and for Chinese nationals, emerged from the private financial sector, while clubs like the Unison Travel Club were first initiated by ordinary people with a common interest.

These inchoate travel institutions found a more supportive environment under the Nationalist government, which accelerated the centralization of China's transport infrastructure during the Nanjing Decade. Taking an active role in the Nationalists' nation-state building efforts, CTS expanded its network and services. In collaboration with the government railways, CTS sold all-inclusive package tours to a variety of domestic destinations, while the Unison Travel Club organized hundreds of group tours for its members. These tours gradually increased in distance and range, and transformed Chinese travel culture. No longer associated with the gentry literati of the imperial period, travel culture in the Republican era was defined by the urban bourgeoisie, whose travel experiences came to be imbued with a new national vision.

The Emergence of Travel as a Business in Semicolonial China

The industrialization of travel in China was entangled with semicolonialism. By semicolonialism, I refer to the condition of China between 1842 and 1943, during which the presence of multiple imperialist powers fragmented China's political sovereignty and influenced its economic, social, and cultural development. Besides the colonies—such as Hong Kong and Taiwan—which were Chinese territories ceded completely to foreign powers, some European countries, the United States, and Japan also forced open and maintained extraterritoriality in a series of treaty ports, leased territories, and other classes of foreign occupancies. These places, numbering more than two hundred, were partially under foreign jurisdiction and accommodated foreign settlements and enterprises.[4]

Selected for their strategic locations—on the coast, along major rivers, and near international boundaries—treaty ports served as the key nexus that "undergirded a modern transport infrastructure in China."[5] From China's coastline to the inland waters, as well as to colonial railways and their surrounding transportation zones, foreign control over the modern transport sector alarmed Chinese elites. In the last decades of the nineteenth century, political and business leaders began developing China's own modern shipping and railway industries. As a result, the modern transport sector in China was operated through a Sino-foreign joint effort conditioned by semicolonialism.[6]

How did this new industrial transport infrastructure reshape travel in China? First, it expanded people's physical mobility. From the 1860s to the 1890s, both Chinese and Western travelers had increasingly relied on steamships for long-distance travel.[7] The speed and convenience of railway travel made it a preferred mode in the densely populated Yangzi and Pearl River deltas by the early Republican period.[8] Second, the semicolonial nature of the new transportation infrastructure also transformed the geography of travel. In North China, coastal steamship navigation and railways replaced the declining Grand Canal as the main transportation lines for modern travelers. While river navigation continued to be an important part of travel in southeastern China, the main linkage points moved from river junctions to places where railways intersected with water routes.[9] Moreover, because the geographical scope and the key termini of this modern transport network hinged on semicolonial treaty ports, the access to modern travel was geographically uneven throughout China. Modern travel was more accessible in eastern coastal regions, Manchuria, and along the Yangzi and Pearl Rivers than in inland provinces and on the western frontiers. It was also jurisdictionally fragmented between Chinese sovereign territories and transportation infrastructure on the one hand, and semicolonial treaty ports and colonial and semicolonial transportation zones on the other. The new geographical and jurisdictional aspects of this modern transport infrastructure reshaped travel in modern China.

This budding industrial transportation system stimulated tourism to China. Foreign tourists began to come ashore in the late nineteenth century. Thomas Cook—considered the father of modern tourism—visited Hong Kong and Shanghai in 1873 on the first around-the-world tour organized by Thomas Cook and Son.[10] Even though the time Cook spent in China was minimal, his company's package tour nonetheless brought China into the global network of tourism.[11] By 1900, Thomas Cook not only opened branches in Shanghai, Hong Kong, and Beijing but also offered a range of travel services in China, Korea, and Japan. Around the same time, the Travel Department of the American Express Company also set up offices in Shanghai and Hong Kong. Similarly, the International Sleeping Car Company and the Japan Tourist Bureau entered the Chinese market through Manchuria, where the railway interests were divided between Russian and Japan.[12]

The late Qing and early Republican periods also witnessed increasing tourism among foreign residents in China. Following the footsteps of Chinese, foreigners frequented local scenic spots and regional tourist sites. An 1894 English guidebook published for Western settlers in Shanghai featured tourist destinations as far as the gorges of the Upper Yangzi River.[13] Before long, well-known Chinese scenic spots could no longer satiate their appetite, and Western-style hillside and seaside resorts began to pop up. Known as "Peking-sur-Mer" or "Tientsin-by-the-Sea," the beach resort at Beidaihe attracted influential Western sojourners in Tianjin and Beijing. Meanwhile, Western missionaries established mountain resorts at Moganshan and Guling (at Mount Lu) in Eastern China. Mainly seasonal sites meant for leisure, these newly "discovered" vacationlands became more accessible when railway and steamship lines were extended and local transportation and communication systems were improved. By the 1910s, these resorts attracted thousands of expatriate tourists every summer.[14]

Chinese leisure travel practices began to change as well. The widening of steam-powered transport networks significantly shortened the travel time to popular attractions. For travelers from Shanghai, going to the long-celebrated West Lake in Hangzhou required a five-day journey by junk in the 1850s. Steamships shortened the journey to twenty-four hours in the late nineteenth century, and a new railway further condensed the trip to under six hours in 1910.[15] Traveling faster also meant traveling farther. With the completion of the Shanghai-Nanjing and Tianjin-Pukou Railways—the two major trunk lines connecting the lower Yangzi region and North China—tourists from Shanghai could reach Confucius's hometown of Qufu and Mount Tai in Shandong in twenty-four hours. By the 1910s, what were once hard-to-reach pilgrimage sites became feasible options for urbanites in the lower Yangzi region.[16] Modern transportation also transformed previously unreachable locations into iconic attractions. The Great Wall, now considered the national symbol of China, only became accessible to tourists in 1909. The completion of the Beijing-Kalgan Railway tunneled through the rugged mountains north of Beijing and brought tourists to the Juyong Pass, where they could climb a stretch of wall built in the Ming dynasty.[17]

Although the late Ming boom in private travel has been considered the harbinger of tourism in China, recreational sightseeing in the seventeenth century only involved localized business innovations catering to the literati and other admirers of their culture.[18] To meet the needs of larger numbers of travelers, the modern tourism business that emerged around 1900 adopted the institutional framework of a multisectional industry. As we will see in this chapter, the tourism industry in twentieth-century China encompassed nearly all the practicalities of travel. Rather than localized businesses sprouting up around popular sites, the modern travel industry in China developed a larger geographic reach over the nation-state. Like the emergence of the tourism industry in Europe, this development was the result of industrial capitalism.[19]

However, China's semicolonial condition meant that its tourism development differed from that in the West. Describing the steamship passenger trade in the late Qing and Republican periods, Anne Reinhardt has demonstrated that steamship passengers were segregated between "Chinese" and "foreign" classes of travel, whether the company be Chinese, European, or Japanese. With foreigners having access to better travel services and more luxurious amenities, Chinese and foreign passengers not only occupied separate physical spaces but also inhabited hierarchical social spaces based on racial categories.[20] In the railway sector, as the lending countries of railway loans assumed control over the construction and management of their invested projects, various Chinese government railways were managed separately by foreigners without a centralized system before the Nanjing Decade.[21] The unequal dynamic between foreign railway staff and Chinese passengers was so well-known that it appeared in plots in late Qing exposé fiction. For example, in *Chiren shuomeng ji* (A fool's tale of his dream), authored by Lüsheng (the pen-name means "traveler"), the reform-minded protagonists are forced to take a train from Tanggu to Tianjin after the steamship comprador refused to dock in Tianjin due to the low tide. On the train, the foreign conductor extorts an extra three yuan per suitcase from them, even though he left some other luggage alone simply because there was a note in French attached to it.[22] Foreign managerial personnel also tended to pay less attention to the needs of Chinese travelers. In steamship cabins and railway cars, Chinese travelers faced the everyday effects of foreign colonial encroachments.

Similarly, foreign tourism spaces and businesses were rife with racial discrimination. Besides avoiding the summer heat, Euro-Americans considered visiting summer resorts as a way to escape the "repulsive sights and sounds and nauseating smells" of cities congested by the Chinese population.[23] Tourism, therefore, was also a means for Western expatriates to establish racial boundaries. With their own self-governing bodies collecting "taxes" from their summering communities, these Western tourism spaces in China were able to coordinate transportation and communication, maintain hygiene at recreational grounds, and oversee road construction and other public services. By carving out isolated milieus to replicate the natural and social environments of the upper-middle class in their homelands, foreign sojourners asserted racial boundaries that differentiated Western-style tourism spaces from the rest of China. As a Chinese observer wrote in 1916, among hundreds of residences at Moganshan resort, only two belonged to Chinese. And Chinese were barred from using the tennis courts or the swimming pool, even though Moganshan was not a foreign concession and its day-to-day operations were carried out by Chinese porters, servants, and shopkeepers.[24] Staying at the Guling resort during her 1917 visit to Mount Lu, Lü Bicheng, an acclaimed woman poet and writer, depicted an essentially foreign space where the only Chinese presence was sedan chair carriers and the local woodcutters.[25]

Nevertheless, the aura of these foreign tourism spaces imprinted on the consciousness of Chinese urbanites in two distinct ways. On the one hand, with their association with sports, wellness, and other elements connected to the Western notion of leisure, these tourism spaces symbolized the idea of being modern.[26] In treaty port cities, Western industrial capitalism not only reshaped people's work regimens but also transformed their leisure practices. For the Chinese middle class, tourism, like other foreign forms of recreation (such as dancing and going to movies), became desirable because partaking in it represented one's access to Western "modern" culture.[27] Xu Ke, who penned a series of Chinese guidebooks to Moganshan, Beidaihe, and Jigongshan (a Western-style mountain resort in Henan Province) for the Commercial Press during the warlord period, highlighted that without adventurous Westerners, these beautiful sites would not have been "discovered." He admonished Chinese to rid themselves of their "passivity" and follow Westerners in embracing the spirit of tourism.[28] On the other hand, many Chinese were alarmed that these resorts were de facto foreign enclaves. In *Eastern Miscellany*, an influential intellectual magazine, articles calling for the "recovery" of these summer resorts appeared as early as 1908. One author recommended that the Chinese government and native business owners should replace foreigners as the main service providers for summering communities.[29] Viewed as Western microcosms demonstrative of foreign extraterritorial rights in China, these Western-style summer resorts became a contentious issue and a rallying cry for the recovery of treaty rights.

Substantial changes came to Moganshan in 1920, when the Shanghai-Hangzhou Railway took a series of actions to challenge the foreign exclusivity at the mountain resort. First, the railway issued "through tickets" to tourists traveling from Shanghai that included railway, steamship, and sedan chair services.[30] Such services incentivized Shanghai tourists to combine a visit to West Lake in Hangzhou with a trip to Moganshan.[31] Second, the railway purchased the Moganshan Hotel from its German owner. Renamed the Railway Hotel, this modern lodge began to welcome Chinese tourists alongside Western visitors.[32] Lastly, the state railway authorities demanded that the Moganshan Summer Resort Association—a Western self-governing organization—transfer the ownership of public infrastructure and the administrative duties of local transport services to them. These moves to enhance services for Chinese travelers to Moganshan suggest that in the early 1920s, Chinese transport managers and entrepreneurs began to realize the significance of travel services in breaking down the racial boundaries in modern travel and tourism. And so the Shanghai Bank Travel Department was born.

In a description of what prompted him to found the Travel Department in 1923, the Shanghai Bank's owner, Chen Guangfu, recalled an unpleasant encounter he had at Thomas Cook and Son in Hong Kong. Chen went to book tickets for a business trip but ended up leaving without making a purchase. The Western clerk there was too preoccupied chatting with a Caucasian woman to acknowledge him. Humiliated, Chen speculated that it was because he, the customer, was "not of their race."[33]

A Jiangsu native born in 1881, Chen Guangfu belonged to the generation of elites who came of age in a more outward-looking China. Leaving his hometown in his teens, Chen apprenticed at a customs broker firm in the treaty port city of Hankou and learned some English. This allowed him to clerk at the Hankou Customs Office, a semicolonial institution dominated by Western agents, and work as an English translator at the Hanyang Arsenal, a late Qing institution founded during the Self-Strengthening Movement. In 1904, he was selected to join the Qing delegation to the World's Fair in Saint Louis. Instead of returning home after the visit, Chen stayed behind to pursue an education in the United States. He eventually graduated from the Wharton School of Business at the University of Pennsylvania and returned to China in 1909. Straddling the realms of semicolonial clerkship, reformist statesmen, and Westernized new elite, Chen Guangfu's early experiences eventually steered him to a career in business in the Republican period. In 1915, Chen opened the Shanghai Commercial and Savings Bank, one of the earliest privately owned modern Chinese banks.[34]

Chen Guangfu's experience in the United States also formed his racial consciousness. The World's Fair at Saint Louis was a formative experience that left many members of the Chinese delegation with a deep dismay over

China's inferior place on the global stage.[35] As a Chinese studying in the United States under the US Exclusion Act, Chen also grappled with Western ideas of freedom and democracy while struggling with racial discrimination.[36] His racial consciousness was manifested in his reaction to the clerk at Thomas Cook and Son. Viewing this experience as discrimination, Chen also voiced his concerns over China's economic sovereignty. "It is already shocking to see how enormous foreign investments in China are," Chen elaborated later in a speech, "and then [foreigners] have even gotten involved in China's domestic travel business, while our fellow countrymen wallow in backwardness."[37] Establishing a Travel Department in his bank was propelled by this sense of urgency.

On various occasions, Chen Guangfu repeated the same genesis story—the encounter at Thomas Cook and Son—in slightly different versions.[38] His unpleasant experience pointed to everyday discrimination on a personal level and a pervasive power differential between Chinese businessmen and foreign companies. His emphasis on the tensions between native business interests and foreign economic encroachment also served another purpose; namely, it allowed Chen to appeal to the popular "national product movement" (*guohuo yundong*). As Karl Gerth has explained, the national product movement of the 1920s exhorted Republican Chinese citizens to buy *guohuo* (national products) in order to strengthen China's nascent native industries, as well as to resist encroaching imperialist powers.[39] In terms of this movement, manufactured consumer goods like matches, cigarettes, and garments occupied the most contested domains. However, by playing up an anti-imperialist origin story and branding his financial and travel businesses as services provided by and for Chinese nationals, Chen Guangfu signaled that Shanghai Bank and its Travel Department were selling their services as quintessential "national products" just like those physical commodities.

This strategy also allowed Chen Guangfu to create a concrete identity for his bank among many domestic and foreign competitors. Known as the "little Shanghai Bank," Chen's private bank was opened with very little initial capital. Surviving the brutal Shanghai financial world by encouraging lower-income urbanites to open savings accounts with as little as a one-yuan deposit, Shanghai Bank was continually searching for new ways to expand. As Chen later commented, during the early years, Shanghai Bank could not compete with foreign banks in the treaty ports, while in Beijing and other administrative centers, it could not keep up with the big national banks.[40] Therefore, seeking a potential market in second-tier cities became a viable strategy, and the tourism business was a crucial step in exploring new territory. Geographical expansion was not new to financial institutions; Shanghai

Bank's efforts to expand beyond the major cities mirrored the development of both the Bank of China and the Bank of Communications.[41] By conjoining tourism and financial businesses, however, Chen Guangfu envisioned a geographical circuit different from those designed by giant financial institutions. Instead of regional systems based on the treaty port system or Chinese administrative units, Shanghai Bank prioritized the flow of travel among the increasingly mobile urban professionals and middle-class businessmen, turning its travel service into an effective venue to expand its client base.

Opening a travel agency in semicolonial China was by no means easy. Even though the nationalization and centralization of the railways in China had been underway since the establishment of the republic, foreign influence in railway management continued in the 1920s.[42] When Chen proposed his plan to establish a travel agency at a national railway meeting in 1923, the foreign railway managers objected to it on the grounds that tourism services were available via foreign firms. Chinese managers, on the other hand, supported Chen's proposal and outvoted their foreign colleagues to grant Shanghai Bank the right to sell tickets on government railways—a task that had been entrusted only to foreign travel agencies.[43]

Even as the state railway companies opened the door for Chen's Travel Department, railway imperialism in Manchuria turned out to be an insurmountable obstacle. Transformed from the isolated Qing homeland into an international railroad thoroughfare, Manchuria in the early twentieth century was split into two spheres of influence. In the south, there was the Japanese-controlled South Manchuria Railway, which also managed the colonial railway in Korea and oversaw a through-rail connection from Changchun to Pusan. In the north, the Soviet-controlled Chinese Eastern Railway—a strategic line Russia had constructed to create a shortcut that connected to the Trans-Siberian Railroad—linked Manchuria to the "Russian Far East" and to the region bordering Xinjiang, China's northwestern frontier. Chen Guangfu recognized the pivotal role Manchuria played in East Asian and Eurasian transportation networks. He planned to open the Travel Department in three locations in Manchuria, but his ambitions were hampered when Shanghai Bank failed to gain commissions from either the South Manchuria Railway or the Chinese Eastern Railway. An effective agent of Japanese imperialist power, the South Manchuria Railway allowed only the Japan Tourist Bureau to provide services along its routes. At the same time, the European-run International Sleeping Car Company, entrusted by the Chinese Eastern Railway, handled ticket bookings from Changchun to Harbin; travelers could also book passenger tickets on the Chinese Eastern Railway at Thomas Cook and Son and the American Express Company.[44] It was only through the warlord Zhang Zuolin, who funded some rival rail lines in Manchuria, that the Shanghai Bank Travel Department was able to open an office in Shenyang (also known as Mukden) in 1925.[45] Like the railway sector, the steamship passenger business had also been dominated by foreign agents in the warlord period. Foreign managers and captains of steamship lines delegated the Chinese cargo and passenger businesses to compradors, who often took their cuts from the proceeds, adding considerable surcharges to passenger tickets.[46] The Travel Department, however, decided to undercut this semicolonial convention by selling steamship tickets directly from steamship companies to their travelers.[47]

Besides ticket sales, Shanghai Bank's Travel Department offered travel-related financial services, including traveler's checks and foreign currency exchange. Aiming to attract more banking customers through travel services and tourism ventures, the fledgling Travel Department's simple business model soon carved out a niche market among Chinese travelers. In its first year of operation, the Travel Department already attracted considerable business. According to a 1924 report, one of the bestsellers at the Travel Department was the "Through Travel Ticket" between Shanghai, Beijing, and Hankou. Combining railway and steamship fares, this through service not only offered significant discounts but also permitted travelers, in two months' time, to board trains between Shanghai, Beijing, and Hankou and take the steamers between Hankou and Shanghai. During the peak travel season in the spring, the high demand for tickets to popular tourist destinations also drove customers to the Travel Department, whose advance ticket sales allowed busy urbanites to avoid crowded ticket booths at train stations and steamship companies.[48]

The attention to travelers' needs was reflected in the selection of Travel Department locations. Within a year of its opening, three Travel Departments appeared in Nanjing, Zhenjiang, and Bengbu in northern Jiangsu.[49] Clustered together in the vicinity of Nanjing, these locations might seem redundant, but they made sense to rail travelers in the early 1920s. Both minor treaty ports on the Yangzi River, Zhenjiang and Bengbu were the penultimate stops on the Shanghai-Nanjing and Pukou-Tianjin railways, which intersected in the Nanjing area. For people traveling on these two interconnected transportation arteries connecting North and South China, transferring from one line to another was difficult because Pukou was located on the northern side of the Yangzi River opposite Nanjing, and so required crossing the river. The Travel Department staff from the Zhenjiang branch, for example, often tended to the needs of travelers by handling luggage transfers in Nanjing and Pukou, and providing seat information on connecting trains.[50]

The expansion of the Travel Department also extended Shanghai Bank's influence into a broader geographical region. In recounting the first five years of the travel agency's evolution, Zhu Chengzhang, the manager of the Travel Department, proudly listed the existing and pending branch offices along seven major railway lines. In some cases, the travel agency preceded the bank. "It's not wrong to call the travel agency the vanguard of the bank," Zhu commented.[51] By catering to a more mobile clientele, it publicized the bank's financial services to a wider business demographic. "Everyone who knows the Travel Department will know Shanghai Bank," Chen Guangfu proclaimed.[52]

During the 1920s, however, the Travel Department did not yield any profits, even as it sold more than a million tickets within its first five years.[53] When questioned about its worth, Chen Guangfu highlighted the intangible assets it had created. The Travel Department generated "goodwill," as Chen put it, which was "more important than money."[54] Later recollections of the travel agency's early days were peppered with stories about creating "goodwill" by fulfilling customers' needs regardless of the cost. When two clients came in to purchase through tickets from Shanghai to New York, instead of sending them to Thomas Cook and Son, which issued such through tickets directly, Zhu Chengzhang drove there and bought the tickets himself in order not to disappoint the customers. The Travel Department also paid attention to cultivating their brand. From giving out ticket covers with their logo to sending staff in uniform to railway stations and ship docks, the department projected an image of professionalism and high visibility.[55]

However, warlord warfare across China hampered the effective operation of the Travel Department. The corrupt militarist regimes were unwilling and unable to maintain basic travel services along government railway lines.[56] During a conflict, warlords often interrupted regular rail services to transport soldiers. As an agency affiliated with a bank, the Travel Department also became a target for extortion from the militants, who frequently harassed financial institutions for funds. When the Nationalist Party's military campaign against northern warlords escalated in 1926–27, the Travel Department had to close down most of its branches outside Shanghai.[57] The political fragmentation discouraged the development of the travel industry even as the demand for tourism services expanded among urban Chinese.

The Industrial Life and Travel Clubs

Around the time of the establishment of the Travel Department, Shanghai also witnessed the emergence of self-organized travel clubs. Although traveling in groups was a common practice in literati circles and among religious pilgrims in the imperial era, committing long periods of time to travel was no longer an option for the urban bourgeoisie in the early twentieth century.

After the establishment of the republic, the adoption of the solar calendar and especially the use of the week transformed the cycle of work and leisure in the cities.[58] This new temporal rhythm of on-time and off-time created a novel periodicity of everyday life, rendering the weekend a sacrosanct period for leisure among white-collar urbanites. The introduction of public holidays offered another impetus for travel in groups. In semicolonial treaty ports, public holidays included those from both Western and Chinese traditions. Depending on what kind of institutions they served, white-collar workers in Shanghai, for example, could have days off on Christmas and Easter, or on the "Old Calendar New Year" and the Qingming Tomb Sweeping Day in April. New Year's Day was a public holiday for all. Combined with weekends, these holidays created longer breaks, allowing the urban middle class to participate in more elaborate trips.

Meanwhile, public institutions and civil organizations supplanted the traditional gentry and religious groups as the main facilitator of group travel. Modern schools organized school trips, explicitly linking travel and a well-rounded education.[59] Workplaces and other social groups considered group tours a way to enhance internal cohesion and lift morale among coworkers and members. The Shanghai YMCA, for example, routinely organized excursions. Whereas sports clubs like the Pure Martial Athletic Association (*Jingwu tiyu hui*) connected touring to physical fitness, savings associations such as the Thrift and Savings Society (*Jiande chuxu hui*) identified travel in groups as a healthy lifestyle choice and an economical leisure activity.[60]

Among all the civil groups, however, the Unison Travel Club (*Yousheng lüxing tuan*) stood out. It was one of the most successful organizations in Republican China that singled out tourism as its primary mission.[61] The Unison Travel Club began to formalize its organization in the early 1920s, and its first members' meeting was held in January 1922, when the club's declaration, motto, club flag, and the designs of its membership card and badge were announced.[62] In 1925, the club's core members launched its first "membership recruitment conference," during which new members were solicited through existing members' social networks.[63] It was a resounding success, adding more than 370 new members to the club. Although the rise in membership diminished the intimate atmosphere, the club's increasing popularity signals that domestic tourism was evolving into a common civic activity in Shanghai and its surrounding areas.

Although the majority of the Unison Travel Club members are untraceable, the club's newsletter, *Unison*, offers some clues. For example, the "news from members" section reported on club members' positions or business ventures alongside their touring activities. In 1931, to complement the launch of the Unison Travel Club's first long-distance tour to North China, a special issue was published with information about sixty tour participants.[64] *Unison* also included similar information about its committee members.[65] From this sample of roughly one hundred members, we can piece together the basic composition of the club members. With only a few mentions of women members, this sample indicates that the club consisted mainly of male urbanites from their mid-twenties to late forties. They tended to be doctors, stockbrokers, managers in banks and insurance companies, clerks in public administrative offices, white-collar workers at foreign-owned commercial firms, and business owners and supervisors at Chinese trading and industrial enterprises. Probably skewing toward its more affluent members, the available data also suggests that workplace networks were one of the main sources for recruiting new members. While white-collar company men were the mainstay in the sample, considering the range of its membership dues and the general income level in 1930s Shanghai, people from the lower-middle to middle class, such as school teachers, could also afford to join.[66]

Beginning in the early Republican period, railway companies looked to tourism as a way to expand their passenger business.[67] One important innovation they adopted was the "special tourism trains" (*youlan zhuanche*), which were operated exclusively for day trippers and vacationers during the busy tourism seasons. A pioneer in this venture, the Shanghai-Hangzhou-Ningbo Railway started a special tourism train in 1916, serving sightseers to Haining. Every year around the autumnal equinox, thousands of people would line up on the shores of the Qiantang River in Haining to witness the unusual natural phenomenon of a tidal bore. On the three days with the most impressive waves, the Shanghai-Hangzhou-Ningbo line replaced some of its freight trains with passenger trains to bring in visitors from Shanghai. The railway company secured boat transfers and an enclosure along the Qiantang River for its passengers. The excursion fare—priced at 8.5 yuan in 1927—included round-trip transportation, three meals, and tea or coffee.[68] Families, schools, and workplace organizations took advantage of the special tourism trains.

Given the large size of their touring groups, the Unison Travel Club began to negotiate with transportation companies directly for group discounts and chartering services. They also collaborated with workplace- or school-based groups to bring the social experience of collective travel to a broad spectrum of people. In 1929, for instance, the Unison Travel Club and the Pure Martial Athletic Association partnered with the General Association of Customs Workers and the alumni organization of the Finance and Tax College to organize a group trip to Chongming, which brought hundreds of excursionists to the island near Shanghai on a chartered ship.[69]

Promoting weekend and holiday group tours, the Unison Travel Club also incorporated a variety of leisure practices, such as horseback riding and hiking, as well as new technologies like bicycles, airplanes, and photography, into the club's tourist activities.[70] By fashioning group tours into a popular urban bourgeois practice, the Unison Travel Club not only helped their members embrace the industrial work-leisure regimes but also molded a nationalist tourist culture alongside the expanding native travel industry.

Building a Chinese Travel Business

If the weakness of the warlord regime in resisting foreign control over China's transportation sector stimulated Chen Guangfu to brand the Travel Department as a Chinese travel business, his aspirations became actions during the Nanjing Decade. With an anti-imperialist spirit, the establishment of the Nationalist government in 1927 ushered in a new era for Chinese capitalists. Emboldened by this watershed moment, Chen Guangfu separated the Travel Department from Shanghai Bank to form an independent travel agency. He named it *Zhongguo lüxingshe*, or "China Travel Service" in English, anticipating the emergence of a national travel network and tourism market. His forward-looking vision paid off. Within ten years, CTS opened forty-three new branches in coastal and inland China, along with other offices in Hong Kong, Singapore, and the United States.[71] With the expansion of CTS, the neologism *lüxingshe* became the standard word for travel agencies in the Chinese-speaking world.

Several factors contributed to the fast expansion of CTS. First was Chen's personal ties to the Nationalist government. Chen Guangfu was not a stranger to the party's patriarchs. He met Sun Yat-sen, the founding father of the party, during the Saint Louis World's Fair in 1904 and befriended Kong Xiangxi, a fellow Chinese student studying in America, who later became a key figure in the Nationalist government. Kong Xiangxi and the matriarch of the powerful Song family—the mother of the Song sisters, who married Kong Xiangxi, Sun Yat-sen, and Chiang Kai-shek respectively—were among the early investors in the Shanghai Bank.[72] As soon as he entered Shanghai in March 1927, the Nationalist Party leader Chiang Kai-shek approached Shanghai financial elites seeking their support. Chen's connections with the party elders made him a trustworthy ally. As the founding member of the Shanghai

Bankers Association, Chen convinced several banks to lend Chiang 2 million yuan in silver dollars. After the inauguration of the Nanjing government, Chen was appointed the director of the Jiangsu-Shanghai Finance Committee, which issued 136 million yuan worth of treasury bonds to bankroll the new government. These efforts secured Chen a seat on the boards of major national banks and made him a director of the Nationalists' central bank.[73] Although he declined to officially join the government, Chen Guangfu represented Nanjing in several international conferences and bilateral loan negotiations in the 1930s and 1940s, becoming an unofficial spokesperson for China's financial interests in the West.[74] As Chen cultivated personal and professional ties to the Nationalist state, Shanghai Bank joined China's financial heavyweights, and CTS came to dominate the travel and tourism industry during the Nationalist era.

The Nationalists, after their ascent to the national stage, also consciously exerted governmental influence on the tourism industry and the protection of travel resources. Shortly after the inauguration of the new government, the Ministry of Transportation and Communication issued a provisional regulation regarding the tourism industry (*lüxingye*). It stipulated that any financial or other enterprises providing tourism services, such as ticket booking and luggage pick-up, must register with the ministry for a tourism industry license before signing contracts with government railways for ticket sales. The company must also have a minimum of 50,000 yuan in capital investment.[75] In fact, CTS was the first travel agency registering for a business license under the category of "tourism industry."[76] In 1928, the Ministry of the Interior promulgated new regulations to protect "scenic spots, historical sites, and historical relics" (*mingsheng guji guwu*) and encouraged provincial and municipal governments to conduct surveys of scenic and historical sites in their jurisdictions.[77]

But the most important catalyst of CTS expansion and the growth of the tourism industry in general was the integration of the railway network and the construction of motorways in China proper. During the Nanjing Decade, the Nationalists gradually interlinked the north-south and east-west rail arteries by rehabilitating and extending existing lines and constructing connecting lines between major routes. The completion of the Canton-Hankou line in the 1936, for example, created a second major north-south thoroughfare. In the east-west direction, China possessed a second through route after connecting the existing rails and extending the Longhai line further west to Xi'an. With encouragement from the central government, provincial governments built new lines to provide shortcuts between busy arteries.[78] Almost nonexistent before 1927, the motorway system in China also underwent significant expansion. Three major trans-provincial highways increased the highway mileage from 18,000 miles in 1927 to 69,000 miles by the end of 1936.[79] Extending the existing system in the southeastern provinces—the stronghold of the Nationalists—westward into Yunnan, the Nanjing-Kunming highway was completed in 1937, linking the national capital to the southwestern frontier. A geographical expression of the Nationalist state-building ambitions, these new developments turned China's modern transportation grid into an adequately connected interprovincial network.

Unlike the warlord government, the Nationalists carried out these projects without incurring major foreign debts. With their desire for a stronger government that would protect their interests against foreign privilege, Chinese bankers felt obliged to participate in financing these projects. Yet the interdependence between the state and the native capitalists went beyond funds. The development of CTS exemplified the significant role played by private businesses in the Nationalist state-building efforts. With few resources to provide travel services, the railway and highway authorities welcomed CTS to open branches along their routes. In return, CTS gained government support in its nationwide expansion. In fact, one can track the progress of China's transportation development in the 1930s simply by tracing the openings of CTS branches. When the Longhai Railway extended westward in 1932–35, the CTS opened offices in Shanzhou, Tongguan, and Xi'an in four years, shadowing the progression of the rail extension. The opening of a CTS branch in Xugou in 1933 corresponded to the opening of the Lianyungang seaport.[80] When the Canton-Hankou line was still under construction, CTS provided motorcar and sedan chair services to bridge the gap between the existing rail terminals.[81] Understanding how the development of CTS was tied to these construction projects, Chen Guangfu sometimes suggested new locations for the travel agency even before the infrastructure was fully completed.[82]

By the early 1930s, CTS had finally begun to turn a profit. In 1936, the year before the Japanese invasion of China, the surplus reached an unprecedented 156,670 yuan, and its number of employees climbed to 394 in early 1937.[83] By the early 1930s, CTS clients could purchase tickets from nearly all the major steamship lines, including the British Blue Funnel Line, the US Pacific Mail Steamship Company, the Imperial German Mail Line, and the Italian-owned Lloyd Triestino.[84] Collaborations between the CTS and foreign travel agencies were also normalized. In 1928, an agreement with Thomas Cook and Son allowed CTS customers to obtain assistance from Thomas Cook branches around the world.[85] As the interwar period witnessed a boom in international travel, CTS was invited to the "Europe-Asia Through Traffic

Conference" and the "East Asia Tourism Conference" in the 1930s, during which CTS, Thomas Cook and Son, the Japan Tourist Bureau, and Intourist Co. (the state tourist company of the Soviet Union) signed deals allowing them to refer clients to each other.[86] From an inconspicuous local travel service shunned by semicolonial transportation institutions and international tourism giants, CTS matured into a force to be reckoned with.

Among its many new ventures, the largest investment of CTS was in hotels. The extension of transportation lines exposed the inadequacy in modern lodging facilities in many places. During an inspection tour to Xuzhou—a city where two major railway lines intersected—Chen Guangfu noticed that many third-class passengers slept at the train station even in cold weather.[87] Chen suggested to the CTS manager in Xuzhou that they open a hotel to serve such customers. The Xuzhou Guesthouse, which opened in the winter of 1931, became the first guesthouse CTS built.[88] Subsequently, CTS built a chain of guesthouses at inland railway transits or remote scenic areas. Just as CTS's name helped popularize the term *lüxingshe*, calling its guesthouses *zhaodaisuo* (lit. "a service place") created a neologism that is still commonly used in mainland China and Taiwan today. Envisioned as a new type of lodging distinguishable from the Western-style luxury hotels in treaty ports, the CTS *zhaodaisuo* cultivated a reputation for being hygienic and comfortable, yet reasonably priced. Installed in "out-of-way places," as one reporter put it in an English-language newspaper, where living conditions were "too much for city tender-feet," these modern lodgings provided "reasonable comfort, but absolutely no disorder of any kind is permitted."[89] Here, the "disorder" referred to the vices of opium smoking, gambling, and prostitution which were common at the old-fashioned Chinese inns. Such activities were prohibited at CTS guesthouses to ensure a peaceful environment.

Besides these budget accommodations, CTS also built three luxury hotels catering to the highest echelons of the Chinese social hierarchy as well as to distinguished international visitors. Among them, its flagship hotel in Nanjing, the Nationalist capital, was the grandest. Investing 500,000 yuan in the project, CTS commissioned the Shanghai-based Allied Architects Firm to design the building. A three-story, L-shaped modern structure with an elegant modernist look, the Metropolitan Hotel in Nanjing was equipped with luxury amenities ranging from tennis courts to rooftop gardens and a state-of-the-art service staff, making it the most exclusive hotel in the nation's capital.[90] Hosting high-level officials and foreign dignitaries, the Metropolitan Hotel functioned as the de facto state guesthouse for the Nanjing government (figure 2). Similarly, in Nanchang, which in 1933–34 became the headquarters for the Nationalists' anti-Communist campaigns, CTS built the

FIGURE 2. "The Metropolitan Hotel in Nanjing." *Source:* "Nanjing shoudu fandian," *Lüxing zazhi* [China traveler] 9, no. 9 (September 1935): 3.

Hongdu Hotel to accommodate the Nationalist politicians and military leaders who frequented the city alongside Chiang Kai-shek. CTS also opened the forty-six-room Western Capital Hotel in November 1935, when Xi'an was designated as an "alternate capital" (*peidu*). With much of its equipment imported from Shanghai, the facility became the first modern hotel in Xi'an, as the ancient capital became the epicenter of the campaign to develop the Northwest, a topic to be discussed in more detail in chapter 3.

As deluxe venues in key cities for the Nationalist government, these CTS hotels, albeit privately owned, served as official facilities of the Nationalist state, projecting a favorable image to domestic citizens and foreign visitors. Some details of the infamous Xi'an Incident and its aftermath further illustrate the ties between CTS facilities and the Nanjing government. When Chiang Kai-shek flew to Xi'an in December 1936 in an attempt to convince the "Young Marshal," Zhang Xueliang, to fight the Communists, Zhang, frustrated by Nanjing's inaction against the Japanese, kidnapped Chiang, triggering a shocking series of events later known as the Xi'an Incident. The kidnapping did not in fact happen in Xi'an, as Chiang was captured at his lodging—the CTS guesthouse at Huaqingchi—on the outskirts of the city. A historic royal villa best known for its mineral hot springs and tales of the Tang dynasty (618–907) consort Yang Yuhuan, Huaqingchi had been mismanaged and was on the verge of total dilapidation when the Shaanxi provincial government enlisted CTS to run the property. After careful repairs and upgrades, the Huaqingchi Guesthouse reopened as several separate compounds for distinguished visitors. At the same time that Zhang Xueliang detained Chiang Kai-shek at Huaqingchi, his subordinates also lay siege to the CTS-owned Western Capital Guesthouse in Xi'an, where Chiang's entourage was staying, trapping even unrelated travelers inside the building for nearly a week.[91] After the incident came to an end with Chiang's release, Zhang Xueliang was ultimately put under house arrest for more than half a century. The initial two "prison" locations where Zhang was kept were also CTS guesthouses.[92] The fact that several CTS facilities were involved in this political turmoil reveals the blurry boundaries between the official and commercial purposes of these spaces.

The expansion of the CTS network during the Nationalist era was indicative of the close ties between the travel industry and nation-state-building projects. As the travel agency extended its facilities wherever Nanjing extended its governmental reach, the agency served almost exclusively urban travelers from the core regions of the state, whose access to inland regions depended upon modern facilities and services. If the Travel Department helped Shanghai Bank develop an alternative geographical circuit from those of its more powerful competitors in the financial market, CTS became an indispensable tool for the incipient Nationalist state to fashion a new national space that was not fragmented by semicolonialism.

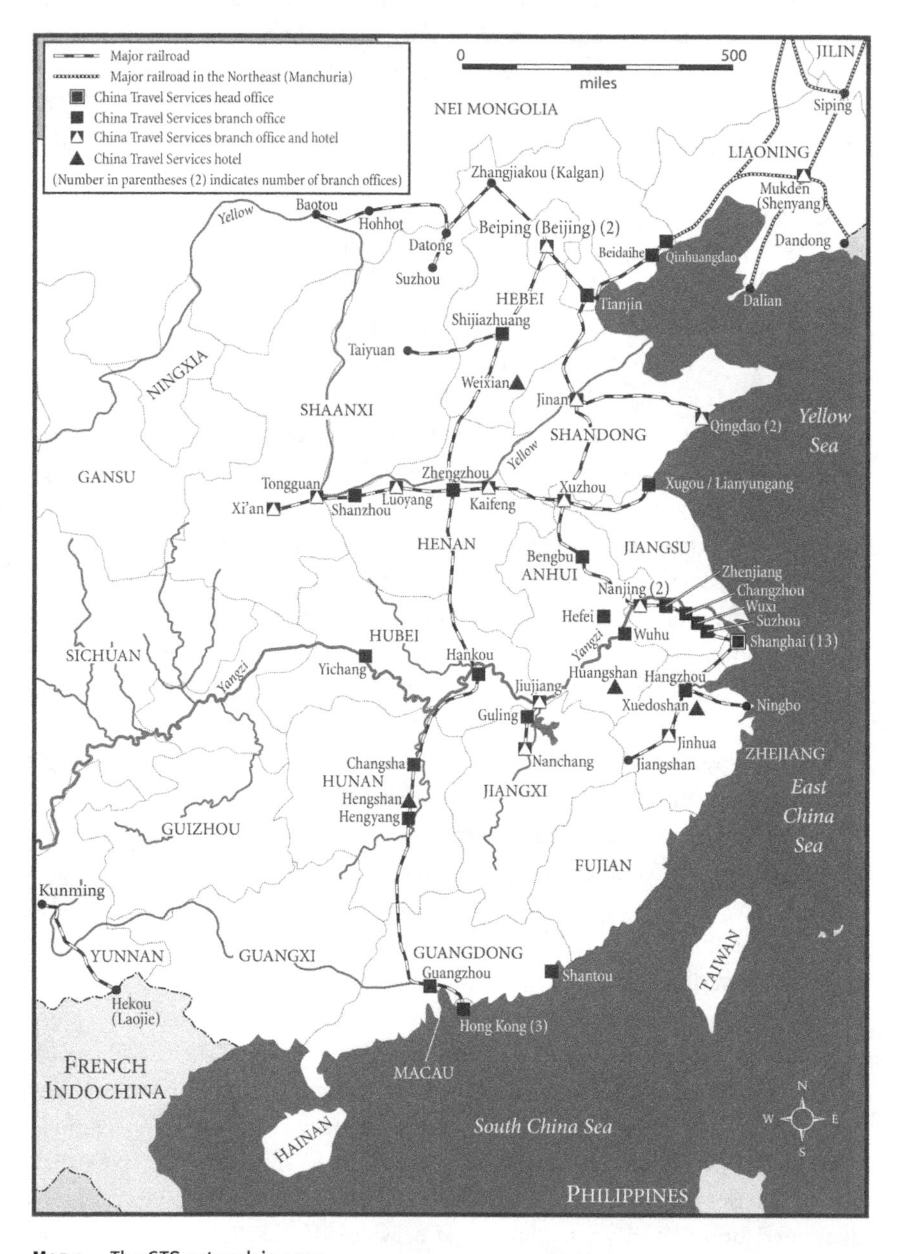

MAP 1. The CTS network in 1937

Serving Chinese Travelers

With an expanding outreach, CTS became even more committed to its mission to serve the travel needs of Chinese during the Nanjing Decade. According to an internal CTS publication, the agency's goal was "to build a travel business from a Chinese standpoint." The author argued that many Chinese had viewed travel as a hardship not simply because it was a common expression in traditional Chinese travel culture. Another equally important factor was the vacuum in travel services catering to Chinese nationals. The purpose of a Chinese travel agency was therefore "to encourage their interest in tourism and eliminate the difficulties in their travels."[93]

For more internationally mobile Chinese travelers, CTS's connection with international circuits of travel allowed them to access convenient travel services even beyond Chinese borders. In the early 1930s, CTS ventured overseas, setting up liaison offices in San Francisco and Seattle as well as a branch in Singapore.[94] These overseas locations were chosen to serve two distinct groups of ethnic Chinese journeying in opposite directions. The two offices on the US West Coast mainly served Chinese students studying abroad. CTS offered a single package that included assistance in obtaining travel documents, purchasing foreign currencies, and booking transportation tickets to and within the United States. In the 1920s and 1930s, CTS even chartered an entire steamer every year for students crossing the Pacific in summer, while clerks from its liaison offices in the United States assisted them upon their arrival, ensuring that they reached the colleges and universities they planned to attend.[95]

The Singapore branch targeted the well-off *huaqiao* (Chinese diaspora) communities who sought travel agency assistance in making plans to visit Nationalist China. Many of the overseas Chinese had never before been to China and did not dare to go back after "hearing that there are many kidnappers and bandits in China."[96] The CTS staff in Singapore addressed these misgivings by introducing famous scenic spots in China and providing packaged tours. Besides the traditional sites like gardens and temples, the CTS tours for overseas Chinese groups highlighted the new constructions of the Nationalist state or places symbolizing Chinese modernity and nationalism. Receiving the first overseas Chinese tourist group in 1934, for example, CTS representatives escorted the group to Shanghai's new civic center, the destroyed Wusong forts on the outskirts of Shanghai, and the headquarters of Shanghai Bank. Located away from the foreign concessions, Shanghai's new civic center was built under the Nationalist state's Greater Shanghai Plan, symbolizing a new vision of Shanghai without foreign domination.

At the Wusong forts, which were bombed during the Japanese attack on Shanghai in 1932, *huaqiao* tourists were told stories about "the heroic defense efforts of Chinese soldiers" and reminded of the threat of Japanese imperialism. At Shanghai Bank, they were shown "how a modern Chinese bank operates behind the scenes."[97] Neither the glitter of the *yangchang* (foreign zones) nor the traditional vibe of the "native city," these sites presented Shanghai as a city both modern and Chinese, while highlighting the ever-present menace of imperialism.

Overseas Chinese could also entrust their children to CTS as they sent them back to China for a Chinese education. With new "supervision departments" in its Shanghai, Beijing, Hong Kong, and Guangzhou branches, CTS could be engaged to find accommodations, help them register for school, manage education funds, and send report cards to their parents to document progress.[98] Although some of these services were not conventional travel services, ensuring that these youngsters' sojourns in China would be well managed and their educational activities carefully regimented, CTS hoped to help overseas Chinese maintain ties with their ethnic roots in China.

While both the students studying abroad and the overseas Chinese visiting the "homeland" can be viewed as perfect examples of increasing international mobility among Chinese in the early twentieth century, these two groups also symbolized two different ideals of modern Chinese identity. On the one hand, Chinese students traveling abroad provided the perfect image of the Chinese modern traveler. Promoting modern travel practices by bringing the *liuxuesheng* (study-abroad students) into the limelight, *China Traveler* published a series of articles and photographs of their journeys. Besides advertising the CTS services, they also showcased what the Chinese modern traveler should look like. Young and well-educated, these students traveled across the globe via steam-powered means of transportation. Pictures in the magazine showed young men clad in Western-style suits and wearing bowties or neckties, with several women students standing confidently among the men. In a picture depicting a Chinese student group's visit to Yellowstone National Park, the Chinese students resemble American tourists. Indeed, it is the Native Americans in their traditional dress who stand out in the group photo, embodying "the Other" of the modern traveler (figure 3).

For overseas Chinese visiting their ancestral homeland or sending their children back for a Chinese education, CTS supplanted the traditional native-place or kinship-based organizations in providing travel services. This could potentially help transform their multiple social identities—rooted perhaps in native place, kinship or lineage, and secret societies—into a single national identity. While the direction of these journeys signaled the idea of "coming

FIGURE 3. "A group photo of Chinese student group and Native Americans at Yellowstone National Park." *Source:* "Shiwunian fumei xueshengtuan youlan huangshi gongyuan shi yu xi Mei turen ji hongren sheying," *Lüxing zazhi* [China traveler] 2, no. 2 (February 1928): 56–57.

home" and resuming a lost identity, a China-bound itinerary was not necessarily required to produce similar effects. Beginning in 1927, CTS helped Chinese Muslims make the hajj pilgrimage to Mecca. Generally, Chinese Muslims would travel from their homes in the northwestern provinces to Shanghai, and then journey together from Shanghai to Mecca, via Hong Kong, Singapore, and Colombo. They sometimes sought help from CTS to obtain proper travel documents and book tickets from Shanghai to Singapore. To make their journey more comfortable and less expensive, in 1933 CTS chartered a ship from the British Blue Funnel Line and sailed directly from Shanghai to Jeddah, the port city for Mecca.[99] In a pamphlet detailing the arrangements, CTS listed information about the visa application procedure, vaccinations, and the halal food provided on the ship. With this new arrangement, the number of Chinese Muslims seeking the CTS service to make the hajj to Mecca increased every year, and in 1936, 126 Chinese Muslims joined the CTS tour to Mecca.[100]

CTS's involvement in these religious pilgrimages broadened the outreach of its network and drew Chinese Muslims from the Northwest into its client base. These transnational pilgrims enjoyed a journey that was less cumbersome while incorporating themselves into a national consumer culture by purchasing a "national product." According to an article in the Chinese Islamic magazine *Yuehua* (*Crescent China*), the journey brought Muslim pilgrims from different provinces to Shanghai, transforming them into "Chinese pilgrims."[101] In this sense, travel helped enhance a feeling of national belonging while simultaneously reinforcing Chinese Muslims' Islamic identity.

Aside from organized travel for the demographics with unique travel needs, CTS also offered organized touring abroad for affluent Chinese citizens.

The most extravagant CTS tour of this kind was the Olympic Tour in 1936. Lasting for three months, the Olympic Tour left Shanghai on June 26, 1936, in order to reach Berlin in time for the opening of the Olympic Games. Differentiated by steamer accommodation, participants were divided into "first" and "tourist" classes, which cost £280 and £155 respectively. The costly fare also included tickets to museums, theaters, and most importantly, the Olympic Games.[102] Collaborating with Thomas Cook and Son in organizing the tour, CTS had to arrange a second, similar tour after the first one reached its maximum of twenty participants within a month.[103] According to the *Shenbao* sports journalist Jiang Huaiqing, tour participants would be accompanying the Chinese delegation to the premier global sporting event. More than a show of wealth, Jiang implied, the tour was also an act of national pride.[104]

Compared to the extravaganza of the Olympic Tours, the all-inclusive CTS tours to Japan during the cherry blossom season were less grandiose yet equally ambitious. An annual tour that commenced in 1926, this fortnight trip took Chinese visitors to see cherry blossoms at various "beauty spots" throughout Japan while setting aside some free time for other side trips and personal shopping.[105] Criticism of the tour began to surface in 1936–37, when Japan's continuous pressures on China triggered widespread resistance. The writer of an open letter published in an English newspaper in Shanghai stated that although he himself was a travel enthusiast, he could not understand CTS's "wisdom and reason in promoting such a tour" to Chinese tourists, whose "hard-earned money" spent in Japan would turn into funds supporting the Japanese military.[106] Despite the article's critical tone, it nonetheless demonstrates that the CTS tours caused the public to consciously link tourism consumption to the national economy and national pride.

From these cases, it is clear that CTS actively responded to diverse demands from its customers. As CTS took actions to build a national brand during the Nanjing Decade, purchasing their services not only helped Chinese travelers to assert their cultural, religious, and class identities but also allowed them to subsume these multifaceted identities under an emerging national identity. This in turn contributed to the rise of national tourism in China.

Touring the Nation

On October 10, 1936, the designated National Day of the Chinese republic, the *China Press* published an article written by Chen Xiangtao, the Yale-educated president of CTS. In the article, Chen showcased the tourism appeal of China by presenting three grand tours. The first tour strung together a series of destinations, ranging from Suzhou, Hangzhou, and Nanjing in the lower Yangzi region, to Qufu, Mount Tai, and Beijing in North China, and on to Hankou and Xi'an in the central and northwestern parts of the country. The second itinerary focused on the Yangzi Gorges in the southwestern province of Sichuan. The last tour went through South China, particularly Hong Kong, Macau, and Canton. These tours covered the major tourist attractions throughout China proper. In the concluding paragraph, Chen outlined the bright prospect of a national tourism in China:

> Persuing [*sic*] modern policy, China is not behind in developing her tourist industry. Thousands of miles of highways and railways are being built which have made touring through China much easier. Realizing the inadequacy of hotel accommodations inland the China Travel Service alone has during the past two years established about 20 hotels in both commercial centers and scenic spots wherever it is required. Unlike the China of yesteryears, both the Government and the Chinese people today welcome world tourists to China with open arms, for we fully realize that it is through tourists, international goodwill and understanding is made possible.[107]

It was not a coincidence that a speech broadcasting popular tourist routes in China was published on China's national day. On a day ritualized to celebrate the country, nothing is more appropriate than celebrating national spectacles. As Chen underlines, the development of a national transportation infrastructure and the modern lodging facilities built by CTS contributed to the expansion of China's tourism radius. He also pointed to the increasing awareness of the importance of the tourism industry among government officials, in the private business sector, and within the general public. The recommended tours mentioned in the article also suggested that by the mid-1930s tourism in China had shifted from a regional affair to a national endeavor. This development can be traced through the evolution of package and group tours.

Selling Package Tours

Compared with its tours to Japan and Europe, which attracted affluent Chinese travelers and generated much media attention, the CTS domestic package tours had a much broader reach. Like many of its other ventures, package tours as a product originated from CTS's collaborations with the state transportation sector. Coordinating with the "special tourism trains" run by the Shanghai-Hangzhou-Ningbo Railway, for example, CTS organized a package tour of the "Haining bore" beginning in 1927. Once the Hangzhou-Haining highway was completed in 1928, CTS Hangzhou teamed up with the provincial highway bus company to run a "special tourism bus" for tidal tourism. With spacious seating arrangements and catered meals, the tour became so popular that the agency added evening tours in 1928 and an extra day of tours in 1929.[108]

Another well-known CTS tour was the annual special tourism train to West Lake (*Xihu youlan zhuanche*). When the corresponding April holidays of Easter and the Qingming festival resulted in high demand for tourist services in Shanghai, CTS again joined forces with the Shanghai-Hangzhou-Ningbo Railway to offer all-inclusive tours for spring excursionists to Hangzhou. The tourism train not only allowed hundreds of urban residents to avoid the crowded coaches, but this highly orchestrated spring tour also offered catered meals on board, local transportation, and tour guides for sightseeing in Hangzhou, all courtesy of CTS.[109]

This business model spread to North China by the mid-1930s. The Tianjin branch of the CTS, for example, operated all-inclusive tours to Beijing, including a "maple leaf viewing tour" on the Fragrant Mountain in autumn and a "winter holiday tour to the old capital" during the New Year holiday season.[110] Between 1935 and 1937, the Qingdao-Jinan and Tianjin-Pukou Railways also collaborated with CTS to operate tours between Qingdao, Tianjin, and Beijing in April or May. Like the tours to Haining and Hangzhou, these railway package tours in northern China also included all transport, meals, accommodation, and guides.[111]

Steamship and highway companies also collaborated with CTS. When the opening of a steamship service allowed day trippers to visit Tonglu from Hangzhou, CTS Hangzhou began to operate spring tours to Fuchun River. A popular destination among literati travelers in imperial China, Fuchun River in central Zhejiang was famous for its storied history and natural scenery. Predictably, this CTS tour was a great hit. During its inaugural year in 1930, the agency arranged three trips that drew roughly seventy people each. To meet the demand, the capacity of each tour was increased to one hundred people in 1931.

Besides the natural scenery, the core of the tour was visiting the historical sites related to two local heroes: Yan Ziling and Xie Ao. Yan was a scholar during the Eastern Han dynasty (25–220) who refused to serve the corrupt court and chose to live a secluded life in the Fuchun Mountains; Xie Ao (1249–95) was known for having mourned the martyred Wen Tianxiang on the West Terrace near the Fuchun River and wrote a eulogy memorializing the Southern Song loyalist's death at the hands of the Mongols. Tour-goers busied themselves bustling from the shrine for Yan Ziling to the site supposed to be Yan's "Fishing Platform" (*diaotai*) to the West Terrace where Xie Ao spilled wine and wailed for his friend. While historical stories shaped the itinerary and added colorful details and cultural significance to the trip, tour-goers did not always passively accept the narratives offered by tour guides. At "Yan Ziling's Fishing Platform," some tourists questioned how the famed recluse could fish from a place more than two hundred meters above the water. Even though the CTS tour guide reasoned that geological changes probably elevated the original platform to its current height, the tour-goers' playful question reflected their sophistication.[112]

If the package tours for excursionists or vacationers adopted a seasonal rhythm, some of the early long-distance package tours organized by CTS were inspired by special events. In 1929, to "commemorate national unification and promote national products," the Nationalist government organized the West Lake Exposition (*xihu bolanhui*), which attracted more than seventeen million visitors from all over China.[113] CTS organized a special tour to assist Hong Kong visitors. The significance of tourists from colonial Hong Kong visiting the center of the newly established Nationalist state was self-evident. To maximize their enjoyment, CTS arranged for the tour to take place in September for optimal weather and assigned two staff members to accompany them throughout their trip. After a day of touring Shanghai, the group took an express train to Hangzhou and spent four days there visiting all the pavilions at the exposition site as well as nearby famous attractions.[114]

Informed by new travel trends, the CTS tours also influenced individual tourists or other travel groups in their own adventures. In 1930, Hong Kong YMCA arranged an "Eastern China Tour," bringing more than sixty members to visit tourist sites in Shanghai, Hangzhou, Nanjing, and Suzhou. Including more destinations than the CTS tour in the previous year, the YMCA tour nonetheless imitated many CTS-established conventions. For example, after visiting all the canonized sights surrounding the West Lake, the YMCA group also took a side trip to see the famous Haining bore. In addition to shopping for local products, they too visited iconic restaurants, including Tower beyond the Tower (*Louwailou*) in Hangzhou, where tour-goers sampled "traditional" local dishes like *cuyu* (sweet and sour fish) and *xihun chuncai geng* (West Lake brasenia soup). Although such dishes were not always palatable to Hong Kongers used to Cantonese cuisine, partaking in esoteric local dishes was still valued, as they provided a unique and authentic sense of the place.[115]

From 1934 to 1937, in the context of the Nationalist state's New Life movement, a number of municipal and provincial governments geared up their tourism development efforts. The most vigorous campaign was carried out in the southeastern provinces, the stronghold of the Nationalist state. In 1934, Zhejiang and four other neighboring provinces launched the Southeastern Scenic Tours project (*Dongnan jiaotong zhoulan hui*) to arouse public interest in the region's scenic beauty. To support this campaign, the Zhejiang provincial government established a Tourism Bureau (*mingsheng daoyou ju*). The neighboring Anhui Province declared Huangshan (the Yellow Mountains) to be a "scenic area" and established the Huangshan Construction Committee to facilitate tourism.[116] CTS responded to these regional movements by adding a Tour Department (*youlan bu*) in 1934, which offered crucial logistical support to the Southeastern Scenic Tours project.

Similar initiatives emerged in other parts of China too. After Beijing lost its status as the nation's capital and was renamed Beiping, its mayor, Yuan Liang, vowed to reconstruct the "old capital" into a new cultural and tourist center of China by developing tourism.[117] The municipal government of Qingdao founded a Development Association to support resort tourism in the beautiful seaside city.[118] Regulations for tour guides emerged in Guangzhou, suggesting that tourism there was rising.[119] With these state initiatives and CTS's expansion, domestic tourism in the second half of the Nanjing Decade moved well beyond Western-style resorts and a few tourist hot spots like Hangzhou.

Among the new offerings in package tours, two new destinations—Huangshan and Xuedoushan—became particular hits. Like Fuchun River, the glory of Huangshan was well-documented in classical travel literature. This legendary mountain range, however, only joined the ranks of popular tourist destinations in the mid-1930s, after highways to Huangshan were constructed and CTS opened a guesthouse there. The CTS organized its first Huangshan Spring Tour (*Huangshan chunyou tuan*) in 1935. The package included second-class train tickets, bus and sedan chair services, lodgings, meals, and tour guides. The itinerary was filled with more than forty must-see sights like iconic peaks, cliffs, waterfalls, and legendary natural phenomena like the "sea of clouds." The first Huangshan tour sold out so quickly that CTS organized a second twenty-person tour later in the same year. Meanwhile, another Huangshan tour departing from Nanjing was offered by CTS's Nanjing branch, which coordinated with the Jiangnan Railway Company to run a special tourism train to Huangshan from the nation's capital.[120] The Huangshan tours were so well-received that the Shanghai Bar Association and the Accountants' Association commissioned CTS to organize a group tour to Huangshan for their members in 1936.[121]

A scenic mountainous area made more accessible by new provincial highways, Xuedoushan became a desirable destination because it was the birthplace of the Nationalist leader Chiang Kai-shek. In coordination with the Southeastern Scenic Tours project, CTS constructed a guesthouse at Xuedoushan, the only modern lodge in the area and prominently located "at the center of all the famous sites of the entire mountain."[122] In 1935–36, CTS organized several tours to Xuedoushan, including a Christmas tour and a special tour celebrating Chiang Kai-shek's fiftieth birthday. The standard itinerary of CTS tours highlighted sites related to Chiang Kai-shek (such as his old residence, his family's ancestral shrine, and his mother's tomb), scenic spots like the waterfall at the "thousand-fathom precipice" (*qianzhangyan*), and activities like shooting rapids on rafts.[123] Although city dwellers appreciated the beautiful natural scenery and outdoor activities at Xuedoushan, ultimately, it was the celebrity appeal of being the hometown of Chiang Kai-shek, the national leader of China, that made it one of the sought-after destinations in the 1930s.

Anticipating the "red tourism" of China today, CTS tours in the Nanjing Decade promoted the tourism trend of visiting the hometowns of national leaders or pilgrimaging to the revolutionary sites associated with the origins of the republic. In December 1933, CTS organized a South China Tour, targeting more affluent travelers who were looking to splurge during the Christmas and New Year holidays. Accordingly, the tour took tourists to visit the casinos in Macau (a Portuguese colony reputed to be the "Monte Carlo of the Orient," as one tour-goer wrote), ride the Peak Tram up to Victoria Peak for a bird's-eye view of Hong Kong, and enjoy authentic dim sum in teahouses in Canton. Witnessing drugs, gambling, and prostitution—the epitome of capitalist decadence—prominently on display in these cities, the tour-goers were also ushered to revolutionary sites sacred to the Nationalist state. From Macau, the group made a side trip to Zhongshan, the birthplace of Sun Yat-sen, the founding father of the republic and the Nationalist Party. The sole tourist attraction there was Sun's former residence. The main hall of this two-story "old-style Western house" had been turned into a museum. Many photographs on display showcased images of Sun's family members and pictures from different stages of his life (including a picture of the 1929 interment ceremony at the Sun Yat-sen Mausoleum in Nanjing). When an elderly woman came out to greet the tour group in Cantonese, the tourists from Shanghai were surprised to learn that she was in fact Sun's older sister. The tour took a group photograph in front of the residence with Sun's sister standing among them.[124]

In Canton, in addition to the Sun Yat-sen Memorial Hall, the tour itinerary also included a visit to the Tomb of the Seventy-Two Martyrs at Huanghuagang. Erected in 1920 to commemorate the revolutionaries killed during a failed anti-Qing uprising, the tomb was accompanied by a castle-like memorial carved out of a huge piece of marble donated by overseas Chinese. Visitors could find donors' names permanently inscribed in the stone. It was clear that the tomb and memorial had cost an enormous amount of money. While most of the tour group did not seem to know much about the Huanghuagang uprising and the Nationalist revolutionaries commemorated there, the grandiosity of the tomb signaled its significance in the nation's founding story.[125]

After *China Traveler* published a travelogue by Chen Cunren, a famous Shanghai doctor who had gone on this South China tour, a reader originally from Guangdong wrote a letter to the travel magazine to make a few corrections. The reader understood the reasons behind the tour group's visit to Huanghuagang. "The seventy-two martyrs who died for the revolution were the founders of the Republic of China," he wrote, "and [the fact that] Mr. Cunren got to visit the site indeed made this entire trip worthwhile."[126] The reader makes an overt connection between Huanghuagang and the origins of the nation. His comment also suggests that tourism played a role in forging this connection. The appearance of revolutionary sites and the hometowns of national figures in CTS tours demonstrates the growing appeal of such sites during the Nanjing Decade. At the same time, these tours contributed to the national mythology of the Nationalist state. This was instrumental in the development of a nationalist travel culture in modern China.

By 1935, it was clear to the leadership of CTS that tourism could be a national business. Anticipating a surge in tourism nationwide, CTS hoped to bring its package tour products—which had been designed for the top echelon of the society—to average urban consumers. For this purpose, the travel agency inaugurated the CTS Travel Club (*Zhongguo lüxingshe youlantuan*) in August 1935.[127] By June 1937, nineteen CTS branches, including its Singapore office, offered weekly tours. These tours featured local destinations for excursionists, regional resorts and other "hot spots" for holidaymakers, and long-distance tours for travel enthusiasts (table 1). The diversification of CTS tours demonstrates how they catered to consumers wanting different levels of time and financial commitment. The increasing number of destinations also showed the broadening of the tourism radius in China proper.

These tours, numbering seventy-six altogether, also provides a snapshot of the scale of CTS's tourism operations right before the Japanese invasion of China. It is clear that along the eastern seaboard, tourism developments in three regions—the lower Yangzi region in the Southeast, the Beijing-Tianjin-Shandong triangle in the north, the Hong Kong–Macao-Canton hub in the south—had matured substantially. Just as Chen Xiangtao's 1936 speech

Table 1 Tour announcement of the CTS Travel Club

SHANGHAI	1	Yixing Jiaxing	SUZHOU	1	Jiaxing Hangzhou	WUXI	1	Huangshan	ZHENJIANG	1	Wuxi Yixing	NANJING	1	Tangshan
	2	Yandangshan Suzhou Fenghua Tiantong Yuwang		2	Wuxi Yixing		2	Nanjing		2	Danyang Maoshan		2	Danyang Maoshan
	3	Moganshan		3	Changshu		3	Shanghai		3	Nanjing Chuzhou		3	Chuzhou
	4	Zhenjiang Yangzhou		4	Zhenjiang Yangzhou		4	Changzhou Zhenjiang Yangzhou		4	Yangzhou		4	Qixiashan Zhenjiang Yangzhou
QINGDAO	1	Laoshan	JINAN	1	Laoshan Qingdao	TIANJIN	1	Fangshan	BEIPING	1	Fangshan	SINGAPORE	1	Malaysia
	2	Jinan Taishan Qufu		2	Taishan Qufu		2	Jinan Taishan Qufu		2	Tianjin Jinan Taishan Qufu		2	Hong Kong Guangzhou
	3	Dalian Lüshun		3	Tianjin Beiping Great Wall		3	Beiping Great Wall		3	Great Wall		3	Philippines
	4	Jinan Tianjin Beiping Great Wall		4	Dragon Cave		4	Beidaihe Qinhuangdao Shanhaiguan		4	Beidaihe Qinhuangdao Shanhaiguan		4	Hong Kong Shanghai
HANGZHOU	1	Shanghai	NANCHANG	1	Ganzhe highway Hangzhou	HANKOU	1	Nanjing	YICHANG	1	Sanyou Cave	CHANGSHA	1	Local districts Yuelushan
	2	Jiaxing Suzhou		2	Lushan Jiujiang Wuhan		2	Jiujiang Lushan		2	Wuhan Jiujiang Lushan		2	Hengshan
	3	Longmenshan Bailongtan		3	Local district		3	Changsha Hengshan		3	Wuhan Changsha Hengshan		3	Nanchang
	4	Huangshan		4	Changsha Hengshan		4	Yuehan Rail Guangzhou		4	Chongqing Beipei Chengdu Emei		4	Guilin

Table 1 (*continued*)

ZHENGZHOU			XI'AN			GUANGZHOU			HONGKONG			
	1	Kaifeng Shangqiu Xuzhou		1	Local districts Huaqingchi		1	Shanghai		1	Shanghai	
	2	Luoyang Tongguan Huashan Xi'an		2	Tongguan Huashan		2	Macao Zhongshan Hong Kong Kowloon		2	Macao Zhongshan Kowloon Guangzhou	
	3	Tongguan Fenglingdu Taiyuan		3	Zhongnan-shan Cuihuashan		3	Wuzhou Guilin		3	Singapore	
	4	Yanshi Songshan		4	Xianyang		4	Yuehan Rail-way Hankou		4	Guangzhou Yuehan Railway Hankou	

Source: Shenbao [Shanghai news], June 5, 1937.

indicated, a canon of tourist destinations had emerged in China, and travel businesses like CTS helped create them. Linking together different regional tourism hubs, CTS's vision of creating a national tourism market was close to becoming a reality.

The Unison Travel Club Group Tours and the Peripatetic Middle Class

As CTS gradually expanded, the Unison Travel Club became more ambitious in its group tours as well. As the club membership rose from roughly one thousand in 1927 to more than five thousand in 1937, its signature group tours became more frequent and sophisticated.[128] Suffice it to say that the Unison Travel Club represented the most active and committed tourism consumers in modern China. Rather than being orchestrated by industry professionals, the Unison group tours were organized by and for its members, who treated tourism as a central ritual for modern national citizens. Encapsulated in the practice and discourses of group tours, this shared view about tourism consumption contributed to the rise of national tourism and the emergence of a nationalist travel culture in China.

The Unison group tours can be divided into three types—Sunday day trips, multiday travels on public holidays, and long-distance journeys to North and South China. The first two types constituted the majority of the Unison group tours, and were largely inspired by the destinations reachable by the two main railway lines from Shanghai—the Shanghai-Nanjing and the Shanghai-Hangzhou Railways—or via the widely available coastal steamship routes and river routes in the water-laced lower Yangzi region. Towns located on the outskirts of Shanghai, such as Minhang, Gaoqiao, Jiading, and Nanxiang, were popular for Sunday day trips, although the club also arranged group outings within the city proper, such as summer evening cruises on the Huangpu River. Famous destinations such as Hangzhou, Suzhou, and Wuxi were easy to access from Shanghai, as they were the main stops on the two railway trunks from the city.[129] The cost for these tours was affordable for middle-income urbanites. A day trip to Mount Tianping in Suzhou, for example, cost five yuan in July 1932. Unlike the all-inclusive package tours offered by CTS, the Unison tours allowed the participants to opt out of the lodging and meal services. For example, those who joined the two-day Hangzhou tour in July 1932 were charged 3.5 yuan or 10 yuan depending on whether they needed the prearranged food and lodging.[130]

To further reduce transportation expenses, in 1933, the Unison Travel Club petitioned the railway authorities to extend group discounts to weekday express trains and all weekend trains. Acknowledging the club's contribution to tourism advocacy, the Ministry of Railways agreed to grant special group discounts as long as the club made arrangements with the railways in advance.[131] Unison was also able to reserve seats for its members, crucial for ensuring comfort during the peak travel seasons. This was considered one of the perks of joining Unison group tours.[132] The club also reached out to commercial steamship companies to charter entire vessels to popular destinations, enabling them to organize some of the largest group tours in the club's history.[133] Unison tour organizers were also meticulous in arranging logistics. During a tour to Wuxie, a mountainous area in central Zhejiang famous for its waterfalls, participants were puzzled when they found quilts on the rickshaws that picked them up at the train station. They later learned that the quilts were rented by the tour organizer for their overnight stay at a mountain temple. Although it was only early fall, the temperature in the mountains could drop precipitously overnight.[134]

As transportation, accommodation, and travel services improved in the 1930s, the ambitions of the Unison Travel Club members also expanded. With more disposable income, middle-class urbanites were particularly eager to travel farther afield.[135] This desire to explore more regions of China was also triggered by nationalism, especially anti-Japanese nationalism. From late 1931 to early 1932, the Japanese occupation of Manchuria and the attack on Shanghai temporarily shut down the leisure travel activities in North China and the lower Yangzi region. These armed conflicts also stirred up a widely shared anxiety about China's survival in the face of Japanese imperialism.

As the Nationalist state sped up its political consolidation by suppressing local opposition and encouraging economic construction in inland provinces, the urban bourgeoisie sought out tourism to connect with the greater nation. No longer satisfied to visit nearby attractions over a leisurely weekend, Unison members became enthralled by the idea of seeing the nation firsthand. To meet this demand, the Unison Travel Club took its most ambitious members out of their comfort zone and embarked on a few long-distance tours. These tours, especially the ones to North China, were a great success.[136]

Requiring significantly more time and expense, long-distance touring was only a burgeoning fad among travel enthusiasts in the early 1930s. Unison planned its first North China tour in 1931. A sixteen-day trip that would take tourists to Jinan, Tianjin, Beijing, Mount Tai, Qufu, and other major destinations in North China, the tour cost 160 yuan per person. Despite its high price, sixty-three people registered for it immediately. However, because of the crisis in Manchuria, the Unison North China tour did not become a reality until 1932.[137] But the volatile political situation also boosted the popularity of the tour, and to meet the high demand, the Unison Travel Club began to run two separate North China tours every year beginning in 1934. A second itinerary was developed, adding locations in Inner Mongolia such as Datong, Suiyuan, and Baotou to the original itinerary and increasing the cost to 235 yuan.[138] Later, Qingdao and Lianyungang were added to the route as well. In 1937, the number of participants on a single North China tour rose to 192.[139]

Two factors contributed to the success of the North China tours. First, the North China tour was packed with iconic sites, including the crown jewel of all Chinese tourist destinations—Beijing. Considered the "main course" of the tour, sights around the "old capital" constituted roughly half of the tour itinerary.[140] To promote the North China tour, *Unison* published a special issue in 1931 highlighting the city's splendor via a lengthy introduction to its attractions. Stressing the North-South differences, the special issue pointed out that the local architecture and customs in Beijing were entirely different from those near Shanghai, which would be of great interest to tourists from the lower Yangzi region. And there was a sense of urgency as well. Even though Beijing only ceased to be the capital in 1928, the 1931 special issue argued that the city had begun decaying quickly since the change of regimes, and a tour of Beijing could not be delayed.[141] This concern proved unfounded. On the contrary, the political decline prompted the Beijing government to champion tourism development, making it "the most interesting city in China."[142]

Travelogues and photographs by Unison tour members testified to the charm of the city. A physician and superlative photographer, Lu Shifu joined the Unison Travel Club's North China Tour in 1934 to "bring home the unique scenery of North China via the camera." He marveled at the great sights of the Great Wall, the imperial mausoleums, the Temple of Heaven, and the yellow-roofed palaces in the Forbidden City, and took many impressive images (figure 4). However, Lu also recognized that the package tours' fast pace could be a serious challenge for a photographer. He described the difficulties this way:

> Many spots within Beijing were indeed worthy subjects [to photograph]. But it is almost impossible [for the photographer] to capture their splendor and vividness when he is rushed and can only take a quick and superficial look. This was exactly the difficulty I faced as I was traveling [in North China] with the Unison Travel Club, which often scheduled visits to three to four sites every day. To photograph the camel caravans at Xizhimen gate, one should go there very early in the morning at five; the Temple of Heaven is best photographed with an array of clouds at three or four o'clock in the afternoon; if one wants to take pictures of the pagodas in Beihai Park, the optimal way is to shoot via long-focus lens from Scenery Hill the minute before sunset. And so on.[143]

Clearly, the number of impressive sites and sights included in the tour both delighted and frustrated Lu. The quick and superficial approach of the group tours prevented Lu from taking better photos. A typical middle-class consumer, Lu was drawn to the tour for its thorough coverage of the iconic sights. While the standardized itinerary ensured a consistency in travel experiences and canonized the sites for mass consumption, some individual needs had to be sacrificed in the process.

Nevertheless, for Lu and other tourists hoping to see as much as possible, the Unison tours offered a satisfying sightseeing experience. Many tourists from the lower Yangzi region also considered a trip to North China a cultural pilgrimage to the old and real China. In Confucius's birthplace of Qufu, tour-goers visited everything Confucian: the Confucian temple, Confucian family cemetery, Mencius's former residence, and other sites connected to Confucianism. Although Confucianism had ceased to be the state ideology since the abolishment of the civil service exam, it still had a profound impact on China's national identity. To Unison tour-goers, the connection between Confucius's hometown and China's cultural tradition and national spirit was self-evident. As they marveled at the majestic cypress trees at the Confucian temple, the grand architecture, and the steles and calligraphy from different dynasties commemorating the ancient sage, they felt a sense of respect

FIGURE 4. Lu Shifu's "Travel photographs of North China." *Source:* Lu Shifu, "Huabei lüxing sheying," *Shidai* [Modern miscellany] 7, no. 4 (1934): 16–17.

and pride. "It's inevitable [for us tourists] to offer a sincere bow to the sage made by Heaven and the saint of our culture," a Unison tour member commented.[144] Furthermore, tourists were excited to go to the Confucian mansion to meet the "Duke of Fulfilling the Sage," a teenager who was the seventy-seventh-generation descendant of Confucius.[145] Like Sun Yat-sen's elder sister, this young man represented a living link to a hero of the past. Becoming a fixture in Confucian tourism, his presence perhaps offered tourists a way to concretize an abstract connection to the Confucian cultural roots of China.

In 1931–37, the intensification of the national crisis in North China and a corresponding rise in nationalism aroused an unmatched interest in these northbound tours. After occupying Manchuria and part of Inner Mongolia in 1931–32, Japan forced the Nationalist government to withdraw their troops from the Beijing-Tianjin area and pushed to secure an "autonomous region" in North China. Of course, the political instability of the region could dampen tourism. However, the concern that North China might soon be carved out of China's national space also propelled many middle-class tourists from the lower Yangzi region to join the North China tours as a way to strengthen Chinese sovereignty. "The rallying cry of 'Heading to North China!' is so strong and powerful," a Unison tour-goer wrote, "how could it not evoke a resonance in you?!"[146]

Visiting North China gave tourism consumers a chance to perform their nationalistic self. For example, the same Unison tour-goer emphasized that besides the historic and touristic aspects, the opportunity to tour North China was appealing to him because it aroused deep nationalist sentiments.[147] Another tour-goer confessed that he joined the tour because he was interested in finding out "the extent of the crisis in Beijing and Tianjin."[148] He went on to observe the local society through the lens of the looming crisis as well. Witnessing Beijingers relaxing at a local teahouse, he commented, "[I]t's as though nothing will distract them from chatting and finishing their tea, even if bullets are flying over their heads."[149] Such benign scenes of everyday life in Beijing were interpreted as a heedlessness of the Japanese encroachment. If tourist sites like the Forbidden City and the Confucian Temple projected an expected national image, the peaceful local life did not quite fit the nationalist story of resistance that the North China tours had conjured up.

While criticizing their fellow countrymen in the North, some Unison tourists also asked themselves what responsibility they themselves had. During a visit to the Great Wall, a Unison tour-goer noticed that many parts were dilapidated. The Great Wall was worthy of its reputation as one of the greatest wonders in the world, yet, he commented, "do we, the younger generation, have the ability to protect it forever?" Besides a concern over the wall's physical soundness, he also worried whether they could "prevent it from being included in another people's map." Referring to the imminent threat from Japan, he then asked himself as well as his fellow travelers, "Have we fulfilled our responsibility simply by taking a glance at these sights?"[150] While seeing the physical landscape, natural scenery, and architecture of North China stimulated among tourists a national pride and a visceral reaction to the potential loss of these territories, this tourist's rhetorical question suggests that tourists like him also recognized the banality and triviality of the nationalism expressed through tourism consumption.

In the 1930s, long-distance tourism helped promote a sense of national unity. By familiarizing themselves with the iconic sites and landscapes of areas beyond the southeastern provinces, tour-goers found a tangible basis for a Chinese national identity and culture. Just as the emergence of CTS was triggered by the semicolonial condition of China, the passion for traveling beyond the southeastern provinces was fueled by the national crisis brought on by Japan's imperialist aggression. Together with the CTS package tours, the Unison Travel Club's group tours constituted an important part of travel as a business in modern China. These tours helped shape a nationalist travel culture, which increasingly associated tourism activities with the physical space and the political fate of the Chinese nation-state.

This chapter has traced the history of travel as a business from the early Republican period to the end of the Nanjing Decade by focusing on two key institutions: China Travel Service and the Unison Travel Club. Even though the tourism industry catered to foreign tourists rather than Chinese travelers when it first appeared in fin-de-siècle China under semicolonialism, Chinese urbanites embraced modern travel and tourism for their association with modernity. The foreign extraterritoriality and racial discrimination manifested in the foreign transportation and travel industries in China stimulated the Chinese bourgeois class to start their own tourism businesses and organizations. In this context, CTS and the Unison Travel Club emerged in Shanghai in the early 1920s with little support from the government. When the anticolonial Nationalist Party emerged onto the national stage, the trajectory of tourism development in China shifted.

Responding to the Nationalist state's efforts to establish a more centralized state, CTS played an important role in state- and nation-building agendas and became a central resource for Chinese travelers at home and overseas. Equating its services with a "national product," the CTS network became an adjunct to state operations, especially in areas newly added into the bureaucracy of the Nanjing government. As a voluntary civic association, the Unison Travel Club, in contrast, represented the consumer side of tourism development in modern China. Its growth during the Nanjing Decade, especially the expansive scale of its group tours, showcased how travel as a business was the product of a rising industrial modernity in urban China.

This history of travel as a business in China should also be situated within the development of national tourism globally. The early twentieth century witnessed the birth of national tourism around the world. In 1901, the New Zealand government created the Department of Tourist and Health Resorts, establishing the world's first government office for national tourism. The passage of the National Park Service Organic Act of 1916 charged the new bureau with promoting the "wonders" of the United States. After the creation of the Japan Tourist Bureau in 1912, the Ministry of Railways in Japan established another organization in 1923, the Japan Culture and Tour Association, to promote tourism nationally. During and shortly after the civil war in Ireland, several tourism organizations were founded with national ambitions. Creating Sovetskii turist (or "Sovtur") and Intourist in 1928 and 1929 respectively, Soviet Russia established its own state agencies to oversee tourism activities for its citizens and foreign tourists.[151] In different parts of the world, governments and voluntary organizations actively took part in tourism promotion and assumed an important place in branding the country's national identity.

In China, while neither the warlord government nor the Nationalist state facilitated official tourism organizations or created nationwide coalitions among transportation, hotel, and tourism service providers, CTS and the Unison Travel Club, with their comprehensive services, close ties to state agents, and broad reach to bourgeois consumers in coastal China, acted as surrogates for a national tourism organization. They coordinated services from different industries, including finance, ground and water transportation, accommodation, and restaurant and retail industries. They also bridged the gaps between the public and private sectors, making it possible for the tourism business to prosper.

Representing different economic actors, CTS and the Unison Travel Club both responded to the demand for national tourism and contributed to the rise of a nationalist travel culture in modern China. The popularity and diversification of their package tours demonstrate the spatial expansion of travel as a business in modern China. By extending the tourism radius to cross the whole of China proper, these tours helped generate an image of, and an experience of, an interconnected and unified national space.

Chapter 2

Travel as Narratives

Producing Quanguo *through Travel Print Media*

In an essay commemorating his ten-year editorship at *China Traveler*, the official magazine of China Travel Service, chief editor Zhao Junhao emphasized the role that the periodical had played in promoting tourism among urban consumers. "Our process is that first we attract [their] 'attention,'" Zhao wrote, "and before readers know it, they develop an 'interest'; with interest comes 'desire' and with desire comes 'decision' and 'action.'" To emphasize these steps, Zhao added the English translation to each keyword.[1]

A formula based on the famous AIDA (attention-interest-desire-action) model, a marketing model first developed by US advertising advocates, these steps captured the intimate connection between travel print media, travel narratives, and tourism promotion. By presenting leisure mobilities as an enjoyable, educational, and quintessentially rewarding modern activity, different kinds of travel-related publications became tools of seduction directed toward tourism consumers. As this chapter will demonstrate, guidebooks, journalistic travel literature, travel magazines, and other popular travel narratives were important rhetorical devices in tourism development. They also asserted a potent universalizing force in shaping the popular consciousness of China's national space.

Even though literary travel writing continued to be a serious genre in Chinese literary production, travel narratives intended for quick consumption took over the mainstream of print culture in the early twentieth century, which befitted an era of more mobility.[2] With the confluent arrival of modern travel and of print capitalism, the Republican-era print market was flooded with journalistic travelogues, photographic images of landscapes and scenery, and travel and tourist guidebooks with practical information. Reinforcing the allure of travel, these diverse travel narratives—textual and visual—commodified domestic tourism, shaped tourist culture and identities, and exhibited a spatial imagination of the Chinese nation-to-be.

In the Republican era, travel narratives appearing in book form and in periodicals became a mass-produced commodity. Heralding what we call "travel journalism" today, commercial guidebooks, travel magazines, and travel columns proliferated, becoming a ubiquitous part of print culture. Giant corporate publishing houses like the Commercial Press and Zhonghua Books began to add guidebooks, popular travel writings, and travel photo albums to their product lines. To promote its travel services and package tours, CTS published its own magazine, *China Traveler*. As the prime travel magazine in Republican China, *China Traveler* advocated for tourism development and consumption and promoted new travel trends among the Chinese urban middle class. In addition to travel magazines like *China Traveler*, many widely circulated magazines in modern China also carried regular travel columns.

This expansion of travel narratives in consumer culture helped popularize a panoptic view of China's national space. Large commercial publishers produced national guidebooks. *China Traveler* put out special issues focused on China as a whole, including one that featured scenic spots along a countrywide railway network. *Liangyou* pictorial, an iconic photojournalist magazine of the Republican period, serialized travel columns to report on the "Nationwide Photographic Tour" that they had organized. Through these ventures, publishers, tourism promoters, journalists, editors, and photographers evoked the notion of *quanguo*—the nation as a whole—to appeal to a wider national audience. This focus on the national market in turn allowed readers to view diverse forms of domestic travel from a national perspective. Providing seemingly authentic texts about and images of *quanguo*, travel narratives presented the oneness of China's national space as self-evident.

Corporate Publishers and Modern Guidebooks

Guidebook production in China did not begin in the Republican period. The lifting of restrictions on private travel in the mid-Ming dynasty (1368–1644) stimulated the production of geographical writings, among which the commercial route books could be seen as the precursor to modern-day travel guidebooks.[3] At the same time, as travel for pleasure became a serious business in a few popular destinations, guidebooks to specific places published by and for literati sightseers appeared in the print market, foreshadowing modern-day tourism guides.[4] Whereas route books were designed to guide the travel of merchants doing business across the empire, tour guides for literati sightseers were attuned to the performance of status and social networking. These traditions would continue into the Qing dynasty.

With the rise of mass print media in the second half of the nineteenth century, guidebook publishing in China entered a new age. As the center of the modern printing industry and the largest treaty port under semicolonialism, Shanghai was both the major site of production of and a popular subject in the new guidebooks that emerged in the late Qing. Offering a more objective selection of information and attractions, Ge Yuanxu's *Miscellaneous Notes on Visiting Shanghai* (*Hu you zaji*) published in 1876 and the visually rich *Illustrated Grand Sites of Shanghai* (*Shenjiang shengjingtu*) issued in 1884 by the Dianshizhai Studio were the harbingers of modern guidebook productions in China.[5] However, sophisticated Western-style guidebooks did not appear in China's book market until the first decade of the twentieth century.

Founded in Shanghai in 1897, Commercial Press, the corporate giant in modern Chinese publishing, was a pioneer in guidebook publishing. In 1909, Commercial Press issued its first Western-style guidebook: *Shanghai zhinan* (*Guide to Shanghai*). The prized product of the press's guidebook franchise, *Guide to Shanghai* was revised and updated throughout the Republican period. Akin to a city directory, it was printed with each page divided into two or three horizontal columns to allow for maximal text, covering diverse topics including "local administration," "public institutions," "industries," and "public utilities." A useful reference for newcomers to the city, the sections on "transportation" and "eateries, lodgings, and touring" were also useful for business travelers and other types of visitors.[6] Commercial Press also adopted the succinct and precise style often found in European guidebooks. In the 1925 edition of *Guide to Shanghai*, for example, the information on Chinese and Western hotels was organized into two different tables, and restaurants and eateries were divided into ten categories. The editors also organized tourist sites into seven genres, including "gardens," "amusement parks," "theaters," and "social clubs." The texts attached to individual sites were usually brief and practical, but readers were given the key information, such as address, noteworthy sights or activities, and whether Chinese were allowed inside the venue.[7]

Following this template, Commercial Press published *Practical Guide to Beijing* in 1919, and a few amplified editions appeared soon after.[8] Not only was Beijing the capital of the republic in 1911–27, but the city's rich history and culture also attracted the publisher's attention. In addition to the section on "eateries, lodgings, and touring," *Practical Guide to Beijing* devoted an entire chapter to the "scenic spots and historical sites" (*mingsheng guji*) and offered more detailed information about potential tourist attractions. For instance, in the 1926 edition, a short entry about the Temple of Heaven informed readers of where the site was located ("outside Zhengyang Gate and south of Tianqiao"), when it was originally built ("the eighteenth year of the Ming Yongle"), and what should be seen (such as the perfectly circular three-tiered altar and the dome in the Hall of Prayer). According to the same entry, the imperial ritual of offering homage to heaven and praying for a good harvest was still performed at the winter solstice in the postimperial years, and visitors were allowed to tour the site on National Day.[9] Although the guidebook was published before the grounds of the Temple of Heaven were turned into a public park in the 1930s, Commercial Press editors had already identified the tourist interest.

Besides its city guidebooks, another bestselling product in the Commercial Press guidebook franchise was *Zhongguo lüxing zhinan* (or *Guide for Travelers in China* in English), a guidebook to the entire country. Published in its first edition in 1912 when the Republic of China was founded, this national guidebook was updated regularly throughout the Republican period. Although various railroad companies began to publish their own travel guides during the warlord period, these guides were issued by individual companies and focused only on a single railway route.[10] When a travel guide covering all national railway routes finally appeared in the early 1920s, Commercial Press's all-China guidebook had already established itself as the most complete and authoritative up-to-date guidebook of the country.[11]

Each chapter of Commercial Press's national guidebook corresponds to a province and is divided into sections devoted to the main cities or towns in that province. The texts could be read as mini city guides, with many of the same categories found in Commercial Press's city guidebooks. Not every province got equal attention. Take its thirteenth edition, published in 1926, as an example. Whereas the chapter on Jiangsu Province was divided into thirteen sections (i.e., thirteen cities), under the chapter about Shanxi, only Taiyuan, the provincial capital, was listed. The varied chapter lengths reflect that travel activities were rather uneven across different regions in the early Republican era, with most travel concentrated on the eastern seaboard.

Aiming to produce a national guidebook, however, Commercial Press highlighted the increasing connectivity between different parts of China by including tables of railway and steamship routes, schedules, and fares.

Alongside the tables providing information about domestic passages on government railways and steamship routes was information about through travel on colonial railways and international steamship routes. Demonstrating how specific locales were networked nationwide via modern transportation, the guidebook also shows that China was integrated into the global travel network.

Besides these tables, another innovation of this national guidebook was the visual representations of China's national space. The same 1926 edition, for example, opened with a national map and two dozen scenic photographs representing every province in the country. Besides the major rivers and locations of cities and towns, the national map also marked national and provincial borders, finished and unfinished railway lines, and submarine cable lines. If the marked national borders aimed to define the geobody of the republic, other cartographical details gestured to an inner logic of an industrializing national space, which hinged on the expansion of transportation and communication networks. The spatial narrative of the nation was also reflected through the scenic photographs, which substantiated the national space with regional diversity. These aesthetic expressions of national space complemented the precise logic of industrial transportation represented by the tables of transportation routes, schedules, and fares.[12] Although, in the early Republican period, few travelers got to visit all provinces in China, this national guidebook became a useful reference for travelers, both real and armchair.

Commercial Press's formula was emulated by Zhonghua Books, Commercial Press's top competitor. In 1926, Zhonghua Books published a sweeping two-volume *Quanguo duhui shangbu lüxing zhinan* (*Travel Guide to Metropolitan Cities and Treaty Ports Nationwide*). Just like the Commercial Press national guidebook, it opens with scenic photographs and a national map and adopted similar categories to organize information for cities and towns. But unlike Commercial Press's dictionary-like organization and minimum prose, Zhonghua Books used a "pseudo-travelogue" style. Rather than being organized by provinces, *Travel Guide to Metropolitan Cities and Treaty Ports Nationwide* progressed through itineraries, combining cities and transportation routes. Whereas a major city could take up an entire chapter, other chapters might be devoted to several smaller locales on a main transportation line. As readers moved from chapter to chapter, it was as though they were following a travelogue or itinerary, which offered a system for how to "read" the national space via modern travel. Zhonghua Books also included dozens of city maps, a regional map along the Yangzi River, and even a map of newly constructed roads in the city of Guangzhou.[13] Encompassing more locations than Commercial Press's similar creation, this national guidebook had more elaborate descriptions of local scenery and sites of interest, signaling more attention to what might appeal to tourists. Compare these two paragraphs concerning Tiger Hill (*huqiu*), a famous tourist site in Suzhou, from the national guides produced by the Commercial Press (left) and Zhonghua Books (right).

Tiger Hill (7 li northwest of the city), the burial place of the King of Wu. There is Sword Pool, where, according to the legend, the King of Wu tested his sword. Adjacent is a cliff on which the four big characters "Tiger Hill, Sword Pool" were engraved by the Duke of Lu. Also, Thousand Man Rock, Sheng Gong Rostrum, Nodding Rock, White Lotus Pool, Hanhan Springs, and the Fifty-three Stone Steps. On top of the hill is Tiger Hill Temple. Behind the temple is Tiger Hill Pagoda.[14]

[Tiger Hill] Also known as Haiyong Peak. Located roughly 9 li northwest of the city. Visitors can hire a pleasure boat outside the Xu gate and reach the hill via the Xu River. Or they can hire a donkey and travel ten li along the embankment. At the entrance, there is an inscribed board that reads "Tiger Hill Buddhist Temple." Walking further, nearby are Ancient Mandarin Duck Pond, Zhengniang's Tomb, Hanhan Springs, and Sword-Testing Rock. Further north are Verdant Mountain Villa and Lengxiang Pavilion. Below are scenic spots such as Sword Pool, Two Immortals Pavilion, Kezhong Pavilion, Sheng Gong Rostrum, and Nodding Rock. According to Records of the Grand Historian, the tomb of King Helü of Wu was located outside the Chang Gate of Wu County. Built by ten thousand people, who dug near a lake [to form the hill]. Three days after King Helü's burial, a white tiger crouched on top of it. This is why it is called Tiger Hill. In the Wu region, the legend goes that when the first emperor of the Qin paid a visit to Tiger Hill to search for the King's sword, a tiger crouched on top of the tomb. This is how it got its name. The hill is not too high, but from it one can overlook the entire city.[15]

Whereas the Commercial Press's version presents a bare-bones introduction to the attractions at Tiger Hill, Zhonghua Books' treatment covers more useful information and fun tidbits, such as alternative visit routes and colorful histories and local lore.[16] The increase in tourism information in this general guidebook produced in the 1920s suggests that major publishing houses were responding to a growing appetite for such knowledge among urban readers.

Around the same time, guidebooks specifically geared toward tourists emerged as a popular subgenre. In 1922, Commercial Press issued a series of guidebooks featuring Western-style mountain and seaside resorts, including Beidaihe, Moganshan, Jigongshan, and Lushan.[17] As discussed in chapter 1, these Western-style resorts began to attract Chinese tourists when the Chinese transportation sector extended their services and Chinese businesses began to open hospitality facilities in the 1920s. From these early tourist guides emerged a new modern guidebook feature—a category called *youcheng*, which included step-by-step itineraries organized by time and area. While the majority of the itineraries were organized by number of days, some were organized by area, especially in the case of a large resort region.[18] Through these *youcheng*, guidebook authors suggested not only what ought to be seen but also how and in what order.

In *Guide to Moganshan*, for example, author Xu Ke designed a "three-day itinerary," covering transportation, lodging, and sightseeing. When discussing the lodging options for the first day of travel, Xu suggested that tourists could ask the steamship company to call the hotel to hold their rooms in the case of a late arrival. After visiting the main peak of a mountain on the second day, the itinerary suggests, tourists should descend from the north side to spot a giant rock resembling a frog. "Summering Westerners often bring wine there and sit and drink," Xu Ke added, hinting at the cosmopolitan charm of the resort. Similar fun facts mentioned in the itinerary included the mountain spring water, which "has high mineral content and can be bottled up to sell in Suzhou and Shanghai at a steep price."[19] Unlike the general guidebooks, tourist guidebooks targeted a specific niche—middle-class, fun-seeking leisure travelers. Not only did they help readers distinguish the must-sees from the maybes, but they also combined practical tips with fun trivia and insider's knowledge. As China's tourism industry blossomed during the Nanjing Decade, tourism-related information in general guidebooks and tourist guidebooks by corporate publishing companies increased accordingly.[20]

The publication of travelogues also picked up steam in the early Republican period. From 1912 to 1927, Commercial Press, Zhonghua Books, and other commercial presses published a series of "new travelogues" (*xin youji*), generating another boom in travel-related publications.[21] Among the new travelogue products, Zhonghua Books' eight-volume *Xin youji huikan* (*Collectanea of New Travelogues*) was so popular that it was soon followed by a sequel. Both collections were reprinted throughout the 1920s and 1930s.[22] Organized by the administrative units of the republic, these new travelogue collections started with travel writings about Beijing and its surrounds and progressed from provinces in China proper to the northwestern and southwestern frontier regions. It also encompassed Outer Mongolia and Tibet—regions that had declared independence shortly after the fall of the Qing. To accommodate the travelogues about journeys cutting across multiple provinces, both collections reserved a final section for "long-distance" travelogues. Like the national guidebooks, the collections' organization presented an idealized national space with increasing internal connectivity.

What exactly was new about these new travelogues? According to the historian Sun Qing's research, several new elements evolved in early Republican-era new travelogues. First, the authors of new travelogues mainly consisted of new-style intellectuals, who were either teaching or studying at Western-style schools or working as editors and contributors at modern publishing houses in treaty ports. New travelogues reflected their tastes, disciplinary training, and worldviews. Second, the selection and editing choices of the new travelogue collections added fresh spatial elements into the traditional literary genre of travel writing, accentuating the chronological development of China's national space. Third, these travelogues also added a different set of spaces into travel narratives, such as urban public spaces (parks, libraries, and telegraph offices) and traditional political and ritual spaces like palaces and temples. Moreover, new travelogues emphasized the types of mobility and travel experiences made available by modern transportation.[23]

Reflecting the rapid development of modern travel and tourism in Republican China, these new travelogues also served readers with diverse interests: literary, educational, and touristic. The usefulness of travelogues became more apparent during the Nanjing Decade. Unlike the large-scale multivolume travelogue collections published in the 1920s, travelogues published by Commercial Press and Zhonghua Books in the 1930s became smaller in size, with more updated contents, and therefore more suitable for tourists or everyday travelers. One example is a series published by Zhonghua Books in 1936 titled A Small Collection of Urban Geography.[24] Marketed as supplemental readings on Chinese geography for middle-school students and as tourist guidebooks for general readers, this series of travelogues was penned by the educator, children's author, and travel writer Ni Xiying, who had previously published two collections of travelogues with Zhonghua Books.[25]

Aside from the nationally important Shanghai and Nanjing, the series covered the regional centers and tourism-rich cities in the lower Yangzi region (Hangzhou), South China (Guangzhou), and North China (Beijing, Jinan, and Qingdao), as well as cities newly included in the tourism radius in central and western China (Luoyang and Xi'an). Aiming to convince readers that geography was not a dull subject, Ni Xiying's vivid and concrete travel descriptions were not only enjoyable and entertaining to read but also informative for tourists looking for information about the history, geography, social life, and most importantly, tourist attractions in various cities.

A closer examination of a few books from this series will help us see their effectiveness as tourist guides. Ni Xiying divided each book into ten chapters, and the sequence of the chapters followed a similar format. The first few chapters were devoted to background information, such as the history, geography, and transportation conditions of the given city. Each book concluded with a chapter on "impressions of everyday life" (*shenghuo yinxiang*). The middle chapters were reserved for travel essays about iconic scenery, historical sites, beaches and mountains, and other tourist attractions, which were often supplemented with photographs and simple maps. For example, in the book on Beijing, Ni devoted seven middle chapters to six main tourist sites in the "old capital" (*gudu*)—the Forbidden City, the three *hai* (lakes), and the scenic spots in the suburbs, namely the Summer Palace, the Great Wall, and West Mountain.[26] The chapters on tourist sites offered practical information, like the fact that the Forbidden City was divided into four routes that were open on different days. "On Mondays and Fridays, the West Route is open; on Tuesdays, the Central and the Outer West Routes; on Wednesdays and Saturdays, the Outer East Route; and on Thursdays and Sundays, the Central and the Inner East Routes." Ni added, "As it costs five jiao to visit each route, one will have to pay two yuan to see the entire inner quarter."[27] These details cast the historical sites within a new framework of tourism consumption, in which space and time were commodified.

Like many other popular travel writers in the Nanjing Decade, Ni Xiying was keen on constructing nationalist travel narratives through tourist sites and their symbolic meanings. In the book on Nanjing, the capital of Nationalist China, Ni Xiying devoted an entire chapter to the two mausoleums at Zijin Mountain. Ni framed the mausoleums in this way:

> On the west side [of Zijin Mountain], there is a glimpse of red walls within the mountain and among the shade of the trees. This is the mausoleum of Zhu Yuanzhang, who founded the Ming dynasty as a civilian. People from Nanjing call it "Ming xiao ling." In contrast,

> on the east side, there is a tall and expansive building made of white stones erected on the mountainside. All its structures are very clear in sight. This is the mausoleum of Sun Yat-sen, the premier (*zongli*) of the Nationalist Party and the founder of the republic. It is also called "Zongli mu." Among these two figures, one is the founding emperor of the Great Ming dynasty, and the other is the founding father of the Republic of China. When they were alive, they both worked for the revival of the Han people. And they were laid to rest on the same mountain. Although there are five hundred years between their endeavors, their contributions to the Han people are indelible.[28]

The most iconic travel sites in Nanjing, these two mausoleums cemented the sacredness of Zijin Mountain and Nanjing in China's national landscape. Zhu Yuanzhang had defeated the Mongols and restored the Han Chinese rule in the Ming dynasty, and Sun Yat-sen was considered to have led Han Chinese revolutionaries to expel the Manchus and establish the republic. The selection of Zijin Mountain as the burial ground for its founding father demonstrated the Nationalist Party's desire to make a conscious link between these two "revolutionary" figures. By underlining Han nationalism as a common ideal embodied by these two mausoleums, Ni Xiying guided his readers to absorb this nationalist myth constructed by the Nationalist state.

Besides the state-sanctioned founding myth, the past and present signs of foreign imperialism and Western modernity also contributed to the nationalist travel narratives. In a brief account of Qingdao's history, Ni Xiying traced the acquisition of Qingdao by the Germans in 1898, its occupation by the Japanese in 1914, and its return to China in 1922.[29] However, the history of foreign control was often depicted as a period of rapid modernization. In the case of Qingdao, as noted by Ni Xiying, the national and worldwide fame of the city was established during the seventeen years of German rule. "Upon its return, Qingdao had been transformed from a desolate fishing village into a prosperous city," Ni summarized, "just like a poor kid who had been sent to other family members to be fostered and returned home as a well-off and well-dressed adult."[30] Once the colonial pressure was securely locked away to the city's past, Qingdao's semicolonial history and its elevated status as a centrally administered city (*zhixia shi*) under the Nationalist state became a badge of honor that affirmed its place as one of the most modernized (i.e., Westernized) cities in the republic.

At the same time, Ni used his travel writing to bring out the unique qualities of the locales and the specific ways various cities fit into China's national history. Describing Beijing as a place of "high palaces and deep courtyards,"

Ni argued that the city had lost its position as China's capital because it had symbolized "corrupt" antirepublicanism and embodied the monarchical and restorationist forces of the past.[31] In the book on Xi'an, Ni argued that if Nanjing had a vibrant and youthful atmosphere and Beijing represented a grand and lofty quality, Xi'an and Luoyang—which were ancient capitals of imperial China—exuded classic beauty.[32] Even though these generalizations did not provide practical information for tourists or have a basis in objective facts, they nonetheless helped readers and travelers capture a sense of history and place. The moral discourse on Beijing, for instance, offered a model for interpretation and judgment of the city's past glory while allowing contemporary tourists to separate themselves from Beijing's antirevolutionary identity. At the same time, the varied stereotypes presented through travel narratives not only situated these cities within a national narrative but also branded each city for contemporary tourists.

This branding was not necessarily centered on history. The closing chapters in these travelogue books, which focused on everyday life, also guided readers in everything from local architectural styles and recent construction, to special cuisines and local pastimes, to bustling shopping and entertainment quarters. These snippets of information might be mundane to local residents, but to the main target audience of outsiders, depictions of unique local "flavor" allowed them to soak up the atmosphere of the particular city. These travelogues provided the emergent Republican national identity with local roots and outlined a national space with rich regional variations.

In his study of German and other European tourist guidebooks to Europe, Rudy Koshar argues that modern guidebooks taught tourists how to consume time and space, yet they also "pointed to a set of meanings and symbols that transcended the everyday life of commodification and consumption."[33] Although these Republican-era Chinese guidebooks and travelogue collections were similar, they were also embedded in a nationalist consumer culture. This close tie between tourism consumption and the nationalist imagination of China's national space is also evident in the travel magazine produced by CTS during the Nanjing decade.

China Travel Service's *China Traveler*

As the leading travel agency in modern China, CTS recognized the importance of travel publications as early as its days as the Travel Department of Shanghai Bank. In 1923–26, the Travel Department published a few guidebooks featuring Yunnan, Sichuan, and the eastern three provinces in Manchuria.[34] Located in China's Southwest and Northeast, these provinces were cumbersome destinations in the 1920s. Colonial railway and steamship services in Manchuria were not friendly to Chinese travelers, while the extra steps needed to travel between the east coast and the Southwest meant that careful planning was necessary. The Travel Department's guidebooks offered detailed information including steamship and railway timetables, must-have travel documents, local currencies, and recommended travel seasons. Although readers could also find local attractions and suggested itineraries, these guidebooks were designed for business travelers rather than tourists.

When the Travel Department was separated from the bank in 1927, the newly independent CTS also established a publishing department, promoting CTS's travel services and tourism business via print media. Besides publishing tourist guidebooks, starting in 1927, CTS published an official magazine, *China Traveler*.[35] A quarterly publication at first, it became a monthly magazine in January 1929 and continued to publish until the early 1950s.

Throughout its long run, *China Traveler* was mainly overseen by the accomplished newspaper editor Zhao Junhao, who fashioned the distinct style of the magazine. Because of its appealing content and high production value, readers in Republican China treated issues of *China Traveler* as collectables and treasured each issue, just like readers in the United States valued *National Geographic*. The circulation of *China Traveler* reached its peak at more than thirty thousand copies, putting it in the same league as the leading popular magazines in Republican China.[36] Flipping through the magazine, Republican-era readers could find updated railway timetables and steamship schedules, along with information about CTS facilities, services, and organized tours. Travel essays about trips to summer resorts, natural wonders, and historic sites were carefully paired with scenic photographs. Translations of foreign travelogues and discussions of new trends in tourism developments in the West also appeared regularly, keeping Chinese urban readers abreast of tourism practices around the world and helping readers imagine Chinese modernity through travel.

Discourses of National Tourism

Discourses advocating national tourism appeared in *China Traveler* when the tourism industry was burgeoning in the early 1930s. First, articles detailing tourism developments in Europe, North America, and Japan appeared in the magazine.[37] Soon after, *China Traveler* published opinion essays written by Chinese railway officials, in which they made direct calls for China to follow suit in tourism advocacy.[38] From 1935 to 1937, *China Traveler* also ran a column called Lectures on Travel (*lüxing jiangzuo*), featuring a series of interviews between editor Zhao Junhao and a celebrity traveler. Because the chosen interviewees were often prominent political and social figures, they had traveled throughout China and the world. Besides elaborating on their varied travel experiences, in such interviews, these elite members of society often emphasized the urgency of tourism development in China by pointing to successful examples in the West. They attributed the strength and economic vitality of Western nations to their advanced tourism industry, and argued that in order for China to strengthen, it needed to imitate Western efforts in developing tourism.

Chen Guangfu, for instance, was particularly impressed by hotels in the United States. Among a variety of amenities, customers could ask the hotel concierge to buy train tickets, reserve a bed in the sleeping car, arrange luggage transport, send telegrams, wire money, and even purchase theater tickets. "The front desk is just like a travel agency," Chen exclaimed, pointing out how Chinese-style lodgings lacked such efficiency. Different sectors in Europe, Chen noted, also contributed to tourism promotion. Whenever art conventions or other public events took place in Paris, the transportation authorities would sell tickets at half price to attract visitors. German steamship companies showed promotional movies on board to publicize famous scenery and resorts in Germany.[39]

In the same vein, Gu Shaochuan (known as Wellington Koo in the West), one of the most influential diplomats of the Republican era, detailed what he found to be an ingenious invention for tourism promotion in France. A fishing hobbyist, Gu was impressed by how French fishing yearbooks were designed as guidebooks. In addition to transportation information to various rivers and lakes, they also listed lodging and nearby attractions. Probably describing the *Michelin Red Guide*, Gu also talked about a French publication listing different road conditions and locations of gas stations and garages in France. "It even included the restaurants in each town and their famous dishes," Gu told Zhao Junhao in the interview, "so travelers could try the local delicacies."[40] The tourism industry in the West, according to these elites, had become a comprehensive operation that encompassed transportation, geography, cuisine, and organized leisure activities.

In these interviews, an admiration of the maturity of Western tourism was often paired with a sense of inferiority regarding the underdevelopment of tourism in China. Moreover, the lack of tourism infrastructure was often interpreted as potentially damaging to China's national image. A veteran newspaperman and nationalist partisan, Pan Gongzhan, for example, worried that foreign travelers often only saw the surface of China, since they seldom visited the *neidi*—the inland territories. Pan asserted that even though services were limited in the *neidi*, this was where the greatness and beauty of the country lay.[41] The diplomat Wang Rutang was concerned about the fact that aside from well-known tourist sites such as the Forbidden City in Beijing, West Lake in Hangzhou, and the few summer resorts developed by Western sojourners, scenic spots in other parts of China were not developed and might leave a negative impression on international tourists.[42] Their opinions regarding the areas of China that foreign tourists should or should not visit hinged on whether the place could project both a positive and an authentic image of China.

Through the elite voices it broadcast, *China Traveler* situated travel and tourism at the center of China's national image. While the celebrities interviewed for the column were from diverse backgrounds (ranging from businessmen to politicians, diplomats, educators, and newspaper editors), their shared views about tourism's place in the national economy and public diplomacy helped the magazine present a consensus on developing national tourism in China. Since robust national tourism was a defining feature of a country's economic vitality and a tool of projecting a positive image of the country to the outside world, the magazine argued, China needed to boost its own tourism development.

What's So Special about "Special Issues" and "Special Columns"?

Like the guidebooks and travelogue collections circulating in the Republican-era book market, the core content of *China Traveler* was devoted to informative pieces like "itineraries" (*youcheng*), "guides" (*daoyou*), and "travelogues" (*youji*). However, the magazine also welcomed readers' contributions; avid travelers and tourism consumers could submit short travel writings and travel photos based on their own journeys. Drawing from a broad spectrum of urban professionals, the magazine's authors included educators (ranging from university professors to middle and elementary school teachers), publishing industry professionals (such as newspaper editors and journalists), as well as doctors, lawyers, public servants, and business professionals. This general picture of *China Traveler* contributors matches the membership makeup of the Unison Travel Club; namely, white-collar urbanites with sufficient income and stable vacation days. *China Traveler* used its timely publication schedule and short-form format to set new social trends and even refashion nationhood. Such efforts are most visible in its special issues (*zhuanhao*) and special columns (*teji*).

Throughout the Nanjing Decade, *China Traveler* released a number of special issues in which the magazine promoted destinations or events that they deemed worthy of the public's attention. At the same time, readers were solicited to submit essays, and the resulting special columns give a glimpse into how Republican-era urban readers contributed to popular discussions regarding travel and tourism. Together, they showcase how travel culture in modern China was mediated through interactions among tourism promoters, travel writers and readers, and tourism consumers.

Between 1927 and 1937, *China Traveler* published ten special issues, each with a slightly different focus. One type of special issue was devoted to promoting specific CTS services and facilities. Dedicated to the Chinese study-abroad students (*liuxuesheng*), the special issue published in August 1929 showcased how the travel agency had been striving to meet the unique travel needs of *liuxuesheng*. An article in the special issue recalled the passage of the immigration act of 1924 in the United States, which changed the visa application procedure for Chinese students. Unprepared for such an abrupt change, students sought help from the newly founded Travel Department of Shanghai Bank, which petitioned on their behalf and successfully helped them acquire the proper paperwork.[43] Put together with a variety of information regarding travel and living abroad, this special issue demonstrated the agency's long-term commitment to and successful track record in providing much-needed service to geographically and socially mobile modern Chinese.[44]

In a 1935 special issue, *China Traveler* featured CTS lodging facilities. In the general introduction to the issue, Chen Xiangtao, the head of CTS, linked the lack of modern hotels in China's inland provinces to lackluster tourism development beyond the coastal regions. To encourage coastal tourists and international visitors to venture into China's inland regions, Chen emphasized, "nine out of ten CTS guesthouses, therefore, were located in inland provinces."[45] Articles detailed their modern design and equipment, the high standard of service, and rates of rooms, meals, and other services. These promotional articles also included information about local attractions and itineraries to scenic regions. For example, an article about the CTS guesthouse in Huangshan, a famous scenic site in southern Anhui, provided a guide to the mountains rather than focusing on the guesthouse itself. Not only did it offer a table that broke down the travel costs (see table 2), but it also provided several itineraries using the CTS guesthouse as the starting and ending point.[46] Photographs of craggy rocks and oddly shaped pine trees presented the iconic sights of Huangshan. The promotional contents about the guesthouses coupled with highly regimented itineraries presented the different scenic areas as standardized consumer products.

The second type, and the mainstay, of the *China Traveler* special issues was devoted to travel seasons and major tourism events. Dedicated to "spring

Table 2 Costs of visiting Huangshan from Shanghai

TRAINS						BUS (FROM XUANCHENG TO TANGKOU)		HUANGSHAN GUESTHOUSE ROOM RATES		HUANGSHAN GUESTHOUSE MEAL RATES			OTHER FEES	
SHANGHAI TO NANJING			NANJING TO XUANCHENG											
First class	Second class	Third class	First class	Second class	Third class	One way	Group	Single bed	Double bed	Breakfast	Lunch	Dinner	Sedan chair	Porter
11.85	7.5	3.75	7.05	4.7	2.35	7.3	5.15	1.75	3.5	0.2	0.5	0.5	1.6	0.8

Source: Lüxing zazhi [China traveler] 9, no. 9 (September 1935): 31.

travelers," for example, *China Traveler*'s first special issue was published in spring 1928, and centered around the "spring break" season.[47] A "Special Issue for Summer" (*xiaoxia zhuanhao*) appeared in July 1931.[48] For the peak travel season in spring, the magazine largely focused on popular destinations for Shanghai excursionists, such as Longhua, Changshu, and Yangzhou on the outskirts of Shanghai and in the lower Yangzi region. Summer resort areas, such as Moganshan, Guling, and Qingdao, were the highlights of the special summer issue.

In an essay about his family's trip to a summer beach destination in Qingdao, the author Ding Huikang, a Shanghai doctor who had studied at Universität Hamburg, detailed the transportation and lodging options, the famous beaches, and other travel sites in the city and in its surrounding areas, as well as the local cafés and delicacies.[49] Practical information aside, the author called attention to the German imprints on the city. Visitors strolling the tree-lined boulevards in the beachfront district could take in the grand mansions, while the tidiness and appealing aesthetics of the graceful streets, Ding pointed out, might make tourists forget they were in China. "You have to admire what the Germans arduously achieved," he commented.[50] An old site of a German fort on the waterfront had been turned into a tourist attraction, transforming a symbol of German military and scientific prowess into a location underlining the city's colonial past. A Western presence could also be directly observed. At one of the city's famous beaches, the author noted, one could easily spot foreigners, especially sunbathing women and playing children. "Their suntanned skin showcased the beneficial effects of natural sunbathing," Ding wrote, admitting that he had only attempted to sunbathe for a brief while before giving up when he got chilled. Similar to Ni Xiying's remarks on the colonial vestiges left on Qingdao's modern cityscape, Ding Huikang's travel essay about this beach resort built and frequented by Westerners also called attention to the link between colonialism and modernity.

However, by offering an account of his family's summering experiences in Qingdao, Ding Huikang, together with other authors in this *China Traveler* special issue, suggested that through tourism consumption Chinese could reaffiliate with these former colonial spaces and assert their own modern identity.

Some special issues of *China Traveler* featured a single destination explored from multiple angles. When CTS was actively involved in the 1929 West Lake Exposition, a large-scale industrial exposition organized by the Nationalist Government, *China Traveler* edited a special issue around the event.[51] Transforming the familiar tourist site into a glittering fairground, the West Lake Exposition spotlighted in *China Traveler* represented the emergent national industrial market and culture, through which a national tourism could thrive. In another special issue, *China Traveler* put cosmopolitan Shanghai into the limelight. While some articles provided information often found in commercial guidebooks, others zoomed out to project an overarching view of the city by introducing the Nationalist government's Greater Shanghai Plan or by comparing Shanghai to Paris and New York.[52] This combination of empirical snapshots and romanticized projections offered readers and travelers opportunities to view the city from local, national, and global perspectives.

Guling or Lushan, the mountain resort in Jiangxi Province, was the subject of another special issue published in June 1931. Appearing shortly after CTS had opened three branch offices in the region, this special issue featured photos of the CTS guesthouse and offices, the motorcar and sedan chair services, and the uniformed staff.[53] In a thorough account of Guling's natural scenic spots and noteworthy manmade structures, the author compared the underhanded practices of conventional travel service providers to the transparency of CTS branches. In the past, the author pointed out, the local conveyances were monopolized by a few companies, which set unreasonably high prices, while their sedan chair bearers routinely extorted extra fees by demanding tips at the halfway point of trips. In contrast, CTS set clear and reasonable prices for different local transportation options, which included standard amounts for tips.[54] The travel narratives in *China Traveler* conveyed a palpable sense of mistrust in local businesses. At the same time, the magazine underlined CTS's own service as the trustworthy brand for coastal tourists visiting this relatively remote resort area.

Finally, in July 1937, a collaboration between *China Traveler* and the Ministry of Railways gave rise to the special issue "Scenic Spots along the Railway Lines Nationwide." In 1934, the Ministry of Railways under the Nationalist government published an official railway travel guide. Covering all fourteen national railways, this handbook provided information such as tourist attractions, transportation connections, regulations, and timetables and fares while including hundreds of photographs of notable travel sites.[55] It was so exhaustive that it ended up being more than a thousand pages long, which made it unfeasible for average travelers to use. Describing their collaborative endeavor in 1937, the *China Traveler* editor explained that this special issue on railway travel was intended to provide a helpful index and systemic overview of China's domestic travel sites, so readers would not have to flip through the whole eleven years of *China Traveler* to look for certain tourist sites.[56] Whereas the lengthy travel guide produced by the Ministry of Railways did not consider the practical needs of tourists, *China Traveler*'s 1937 special issue aimed to offer a concise yet comprehensive companion to leisure travelers who increasingly chose the railway.

Almost a hundred pages longer than regular issues, this special issue included all the important attractions along the fourteen government railway lines. "Albeit not exhaustive, we have included all the major scenic spots," the editor wrote, "and by buying this one issue, readers will have a bird's-eye view [of all the tourist sites in China]."[57] Each of the fourteen sections was devoted to one railway line, and each section started with a few pages of photographs of iconic attractions along the route, followed by a textual introduction to every railway stop.

The texts were also standardized. Resembling the format in the guidebook issued by the Ministry of Railways, each section began with a brief account of a location, then traced its historical evolution from the time when the town first appeared in historical records to its present status within the republic. The concise background information was followed by a list of recommended tourist sites. The tourist attractions consisted of natural scenery and historical or revolutionary sites, along with hot springs and beach resorts, temples and gardens, and notable architecture representing recent developments. When some major cities, such as Shanghai and Jinan, appeared on more than one railway line, rather than repeating the same material, the editor referred the readers to the relevant discussion elsewhere.

A pronounced national outlook was also manifested through the magazine's deliberate choice to situate Nanjing as the starting point of the special issue, as though to emphasize that it was also the starting point of a national journey. Devoted to the scenic spots along the Shanghai-Nanjing Railway, the first section of the special issue opened with a few photographs of well-known travel sites in Nanjing, among which the Sun Yat-sen Mausoleum and the tomb of Tan Yankai (the former president of the Nationalist Government) were featured.[58] These memorial sites of the Nationalist Party's forefathers further affirmed Nanjing's status as the political center of the nation-state and as the symbolic site of China's national identity. Meanwhile, in the pictorial section on Shanghai, the familiar images of the iconic European-style landmarks were nowhere to be found. Instead, the magazine used photographs of new construction in Shanghai's municipal center. Located outside the International Settlement and the French Concession areas, the new municipal center of the Greater Shanghai Plan consisted of Chinese-style government buildings and public institutions (figure 5).

Carefully curated to characterize diverse regions in China, the images and text about these scenic spots were presented in a standardized format, which codified transportation routes and a modern travel network. Similar to the national guidebooks by commercial presses and the state agencies that portrayed the nation as a whole (*quanguo*), this special issue helped bring forth a national space that could be uniformly valorized.

FIGURE 5. "The scenic spots along the Shanghai-Nanjing line: Nanjing and Shanghai." *Source:* "Jinghu xian mingsheng: Nanjing and Shanghai," *Lüxing zazhi* [China traveler] 11, no. 7 (July 1927): n.p.

From 1934 to 1937, the magazine also offered special columns consisting of readers' contributions for its New Year issues. Although the wording of every year's "essay solicitation" theme was different, the magazine essentially asked readers to recount their memorable journeys and to express why they appreciated travel. Composed by writers, government and transportation administrators, newspapermen, professors, and average urban readers, the selected essays featured in these special columns demonstrated the expansion of the tourism radius in China and underlined CTS's vision of helping Chinese travelers tour ever farther.

In addition to accounts of personal travel experiences, tourism was also presented as a crucial part of public life and a modern citizen's civic duty. Yu Songhua, a contributor at the *Shenbao Monthly*, a supplement to *Shenbao*, for example, wrote a special column for *China Traveler* in which he reminisced about the group tours he participated in when attending the Chinese Economic Society's annual conference. Part of the activities arranged by the learned society, these group tours took conference attendees to popular tourist sites in Zhejiang and Shandong Provinces. Yu emphasized that these tours were a great opportunity to socialize with colleagues at the Economic Society and to see different provinces and the "laboring masses" as well.[59] In an essay responding to solicitations for the ten-year anniversary of *China Traveler*, Huang Boqiao, the general manager of the Shanghai-Nanjing and Shanghai-Hangzhou-Ningbo Railways, emphasized the role tourism could play in cultivating patriotic feelings among Chinese citizens. Huang attributed the indifference of many Chinese toward China's national interests to their lack of a thorough understanding of the country. Arguing that tourism could help remedy this, Huang wrote,

> Visiting famous mountains and rivers will allow one to recognize their scenic beauty; exploring the steep gullies and ancient caves will allow one to appreciate their remarkable structures; witnessing the historical relics connected to ancient sages will allow one to understand their outstanding words and deeds and sublime character; visiting the metropolitan cities will allow one to inspect the progress they have made in new construction; traveling to commercial ports will allow one to trace the ups and downs of their industry and commerce; touring towns and villages will allow one to learn about the spread of democracy and the local customs. Therefore, seasoned travelers will gain a thorough understanding of the geography, history, economy, and the prevailing customs of the nation. . . . With this thorough understanding, one's love and care for their country will arise spontaneously without reluctance.[60]

Echoing *China Traveler*'s main mission to "expound and propagate the scenic spots" in China, these New Year special columns supplemented the promotional messages of the magazine with a popular expression of civic nationalism.

Special contents such as this were part of the mission of the magazine, which aimed to direct urban readers to embrace commodified forms of tourist behaviors. Special issues allowed for more targeted promotional campaigns, and were particularly effective in commodifying particular scenic spots and destination cities. They also helped business interests and political powers from the national core impose order on the national space. At the same time, special columns provided a platform for average travelers and readers to employ tourism as a means to understand the nation.

Liangyou Travel Columns

Although quickly produced and ephemeral, journalistic travel writing and photography became commercially and culturally imperative to the success of any given magazine or newspaper in the Republican era.[61] Based on a nationwide tour led by its own journalists, travel columns in *Liangyou* in the mid-1930s showcased that the search for national and cultural authenticity, a quest at the center of the experience of tourism, also dominated travel narratives in popular pictorial magazines.

On September 15, 1932, a large crowd of journalists, publishers, and officials gathered at North Station in Shanghai to send off four journalists, who were about to embark on a Nationwide Photographic Tour of China (*quanguo sheying lüxing tuan*). Organized by *Liangyou*, this group of media professionals planned to tour and take photographs of "the Yellow River area, the Yangzi River area, and the Southwestern Provinces" in order to capture "real images" of the Chinese republic.[62] *Liangyou* magazine was the flagship publication of Liangyou Printing and Publishing Company, an enterprise with a focus on photography, literature, and fine arts. Striking a balance between highbrow and popular tastes, *Liangyou* was one of the most successful photojournalist magazines and the longest-running bilingual pictorial magazine in Republican China.[63] Rooted in a transnational commercial network formed by Cantonese merchant communities in Shanghai, Hong Kong, and the United States, Liangyou Company's distribution network covered not only major cities in China but also overseas Chinese communities.[64] Appealing to urbanites who became more drawn to visual content in the early twentieth century and an overseas Chinese audience with limited literacy in Chinese, *Liangyou* highlighted its photography-dominated format and provided its readers with captions in both Chinese and English. Juxtaposing popular cosmopolitan images from around the world alongside photographic depictions of a Chinese homeland, it was eagerly consumed by Chinese readers in semicolonial urban centers and those living in Chinese communities in diaspora. The *Liangyou* travel columns based on this Nationwide Photographic Tour demonstrate how visual representations of travel occupied a significant place in popular print media.

The popularization of photography in modern China coincided with the commercialization of tourism in the Republican period. Just as modern transportation technologies molded and stimulated modern tourism, photography not only created incentives for individuals to travel but also allowed distant parts of the world to be accessible to armchair travelers.[65] The application of photography to the representation of travel was not new in China in the 1930s. In 1910, Commercial Press produced a photo album titled *Zhongguo mingsheng* (*Views of China*), consisting of 182 photographs of iconic sites from all the provinces.[66] The popularity of this photographic travel album led Commercial Press to issue more than twenty other *Views of China* albums between 1914 and 1926. Unlike the comprehensive framework of the original version, these later additions were based on real journeys by avid travelers in the early Republican era. Through these photographic albums, urban consumers could gain familiarity with different tourist sites across the country, ranging from Huangshan in Anhui to the West Lake in Hangzhou, and from Mount Taishan and the Shrine of Confucius in Shandong to Wuyi Mountain in Fujian.[67] During the Nanjing Decade, commercial photography magazines, such as *Keda zazhi*, the Chinese version of *Kodakery* issued by the Kodak Company in Shanghai, also featured travel columns as a way to promote cameras, film, and other products related to photography.[68] While continuing to endorse photography's unique place in travel narratives, *Liangyou*'s 1932 nationwide tour and the subsequent travel columns provided a more balanced act between textual and visual elements, and refashioned the journalistic style of travel narratives.

The purpose of *Liangyou*'s Nationwide Photographic Tour, as Liangyou Company owner Wu Liande pointed out, was to present China as a modern nation to both Chinese and international readers by familiarizing them with "new construction" and eradicating the obsolete image of the Chinese people as those with bound feet or queues.[69] The magazine launched the tour at a moment of national crisis. In September 1931, Japan invaded and occupied Manchuria, fomenting a public outcry in China. The heightened anxiety over China's territorial integrity had reinforced calls for national strength and unity. Just as tourism promoters and transportation administrators viewed domestic tourism as a way to shore up the country politically and economically, journalists and intellectuals considered the production and circulation of photographs of different parts of China as a way to assert the oneness of the nation.

The valedictions given by government officials and elite intellectuals at the *Liangyou* journalists' departure reflected this sentiment. Ye Gongchuo, the former minister of transportation and communication under the warlord era government, for example, praised *Liangyou*'s "nationwide tour" as a timely effort. Ye was concerned with the surveys and tourism activities carried out by Westerners in China and emphasized that China had already been "X-rayed" by foreigners while the Chinese themselves were "still sleeping." He expected that the *Liangyou* photographic tour would engage urban readers with texts and images of the "real" China.[70] The choice of the word "X-ray" here also underlined the significance of visualizing technology in this endeavor to understand and represent the nation.

Liangyou magazine ran two travel columns in 1932–34 based on the tour. Telling their readers to "sit back at home and travel the nation," these columns allowed average urban readers to follow the journalists' arduous journey. After the tour concluded in May 1933, the more than ten thousand pictures captured during the journey were edited into a collectable photographic album by the Liangyou Company.[71] Before that, while the tour was still ongoing, the photos were serialized in *Liangyou*. "Reports from the Nationwide Photographic Tour," the first column of seven installments, was penned by Liang Desuo, the chief editor of *Liangyou* and a member of the photographic tour. Based on Liang's travel journal, this initial column traced the journey from Nanjing to Shandong, from North China to the Northwest, and from the central plains to South China.

Like the organization of many nationwide guides, collections of travelogues, and *China Traveler*'s special issue "Scenic Spots along the Railway Lines Nationwide," *Liangyou*'s tour of *quanguo* also started with the nation's capital. This time it was the Nationalist capital of Nanjing. Hyperbolically calling Nanjing "the largest city in the world," Liang Desuo described everything they saw with grandiosity to demonstrate the promising future of the nation. However, the *Liangyou* group's arrival in Nanjing coincided with the Mid-Autumn Festival and the news of Japan's recognition of Manchukuo, the puppet regime Japan had set up in Manchuria. Invited to a moon-watching dinner party—a traditional celebration of the festival—at the Sun Yat-sen Mausoleum, Liang Desuo recounted his dark mood and complicated feelings in the travel column:

> It is almost midnight when the bright moon finally breaks through the clouds. The pure bright light shines upon the mausoleums of Sun

> [Yat-sen] and the Ming [emperor], where our national leaders are buried. Today when we are facing internal hardship and foreign aggression, under this dismal moon, how can we bear to listen to the sighs from the deceased [leaders] buried in Zijin Mountain; how can we bear to listen to the songs by Shangnü near the Qinhuai River?[72]

Unlike Ni Xiying's guidebook-style travelogue highlighting the timeless value represented by the mausoleums, Liang's journalistic approach underlined the timeliness of travel accounts. By framing his discussion around a specific holiday that celebrates family reunion, Liang drew readers' attention to the contrasting tragic event of the complete severance of Manchuria from China. While celebrating Nanjing's youthful energy as the new center of state power, Liang underlined the long shadow of the national crises cast on Nanjing's most symbolic site, since the situation of Manchuria was a fresh wound in the long history of China's national humiliation at the hands of imperialist powers.[73]

Unlike the promotional and sometimes didactic tone of the travel narratives in *China Traveler*, travel narratives in *Liangyou*, as this first installment indicates, were less about the practicalities of visiting different destinations. In their tour of *quanguo*, travel was presented as a patriotic act, and there was a journalistic duty to report the "real" conditions of the entire country. Travel writing and photography were no longer reserved for the beauty of tourist sites. They were also interpretive tools that could shed light on and dissect different problems within China. Emerging from the late Qing reformers' travel narratives about their overseas journeys, this reportage approach to travel narratives carved out an important space between consumerist and literary travel writing. Just as the crisis of Manchuria prompted coastal urbanites to join package tours to the northern provinces and other parts of China, the widely shared concern about China's fate in the early 1930s also drove readers to consume reportage-style travel narratives.[74]

The *quanguo* showcased in *Liangyou*'s travel narratives, however, did not present any less of a complex image. After Nanjing, the *Liangyou* photographic tour traveled to Qufu and Zouxian, the birthplaces of Confucius and Mencius in Shandong Province. Liang Desuo's travel journal from Shandong was paired with a two-page photographic spread (figure 6). Filled with photographs of pavilions, stelae, cypress trees in the temples, and the monuments commemorating famous historical stories related to Confucius and Mencius, the spread highlighted the "monumentality" of these sacred sites by intentionally excluding humans from most of the frames. The only exception was a portrait of Kong Decheng, the seventy-seventh-generation patrilineal descendant of Confucius. Surrounded by pictures of sculptures, "beautifully carved stone columns," and "finely engraved work in stone," Kong Decheng's portrait looked more like a "monument" than a human being. Liang Desuo's account described his meeting with Kong Decheng. Not impressed by this thirteen-year-old boy's lineage, Liang was surprised by his pale appearance and lack of expression during their meeting. Liang wrote, "For a thirteen-year-old, he is a silent scholar afraid of the sunlight. He meets with travelers mechanically. [The fact of] travelers coming to visit him is just as meaningless as seeing Confucius's old house, its ancient well, or the tree Confucius planted. He shows his boredom toward life in his expression. What a pity that he buries his life in this historical site of Confucius's house. Prolonging its history has become his responsibility."[75] Contrary to the enthusiasm from tourists in Unison tours, Liang Desuo's text raised doubts about the cultural meaning of touring the hometown of Confucius and Mencius in the twentieth century. Using Kong Decheng as symbol of Confucian cultural heritage in modern China, Liang questioned the practice of having a teenager meet visitors as his daily job, and, by extension, his critique revealed the anachronistic ideals embedded in these sites. Here, the pictorial and the textual parts of the column are bifurcated: while the photographs presented a travel narrative that celebrated quasireligious scenes in

FIGURE 6. "A visit to the land of ancient sages." *Source:* "Shengdi xunli," *Liangyou* [Young companion] 71 (November 1932): 8–9.

order to show the unchanging cultural value of Confucian heritage, Liang's text called into question the practice of touring these sites, which had come to embody a bygone era.

Nonetheless, Liang Desuo understood the value of historical sites and cultural relics. In an installment published in February 1933, he showed particular concern about the protection of heritage sites in remote areas in central and northwestern China. Recording their journey through the "central plain," Liang and his colleagues visited the South Grotto Temple (*Nan shiku si*) in Shaanxi, a site whose history could be traced back to the Six Dynasties (220–589), according to Liang's travelogue. Besides lamenting the damage to some of the masterpieces caused by natural decay, Liang also pointed out that the culprit responsible for some of the headless statues was in fact theft by foreigners. At the Longmen Grottoes outside Luoyang, Liang underlined that theft was also a threat to this famous Buddhist heritage site.[76] As will be discussed in chapter 3, this concern over foreign plunder of Chinese artifacts was particularly strong when Chinese coastal elites traveled to and wrote about inland and northwestern China.

Even though Liang Desuo questioned the practice of preserving outdated values from the feudal past, the *Liangyou* tour group considered protecting and maintaining "history" through travel as equivalent to maintaining the cultural identity of China. To showcase the huge scale of the statues carved on cliffs at medieval Buddhist sites like the Longmen Grottoes, *Liangyou* photographer Zhang Yuanhen used human figures for contrast in the images. Whereas the statues represented the material size of the grand project and symbolized the pursuit of permanence by the ruling class in history, the small human forms in the photographs also gave scale to the moment of tourist encounter in the present tense. Similar to the pursuit of antiquity among travelers in imperial China, capturing these encounters via camera "affirmed various modes of continuity with the past."[77] A tactic to maintain "history," travel photography in popular print media also transformed historical sites and travel destinations into iconic symbols of the nation. Alongside the wide circulation of travel texts among middle-class urbanites, the dissemination of photographic images of landscapes, monuments, and other sites of interest in tourism also enriched the collective perception of a national identity at a time when China's territorial sovereignty was under serious threat.

Paralleling the cultural heritage and historical relics, the *Liangyou* photographic tour also paid attention to new construction and local reforms. One such example was a photographic report of their visit to Ding County in North China, where Yan Yangchu (known as James Yen in the West), a Yale-trained educator, carried out his "mass education" experiment. A reformist experiment aiming to "improve the nation" and "eradicate the illness of foolishness, poverty, weakness, and selfishness" among the Chinese people, Yan's movement was described as "the outlet to illuminate our nation" in the *Liangyou* column, which highlighted the national importance of this local reform attempt. Treating this experimental site as an exemplary place worth visiting, *Liangyou* photographers captured the reformist ambience on film. By juxtaposing images of improved modern farm life with those of educational programs, the column conveyed the vision of a rural China undergoing modernization.[78]

The *Liangyou* travel column on Ding County was a rare example of a positive portrayal of new developments in the "real" China. More often than not, disastrous infrastructure and primitive means of transportation became the symbols of the inland provinces. As Madeleine Yue Dong has pointed out, Republican travelers from Shanghai often complained about the "unmodern aspects" of their travel destinations, confirming their cosmopolitan and superior identity through journeys to the rest of China.[79] In the same essay about the visit to the hometown of Confucius and Mencius, Liang Desuo described the mule cart as the one means of transportation prevalent in north and central China.[80] While Liang's criticism was similar to the grievance voiced by Shanghai tourists in *China Traveler*, he castigated the lack of nation-building projects in outlying regions. In another installment titled "Transportation in Hunan and Guangdong and the Problems of the Nation," Liang described in detail how the touring journalists traveled south from Hankou over newly constructed highways and railways. With this new infrastructure, Liang pointed out, people from Guangdong who wanted to visit Hubei and Hunan would no longer need to take a detour through Shanghai. However, according to Liang, this did not save a lot of time or cost less money. The problem was that the roads had not been completed, and the missing links between different sections delayed travelers.[81] The *Liangyou* travel column did not simply applaud the achievements in transportation like some of the travel essays in *China Traveler*. Instead, it presented a more realistic picture of travel conditions in inland China.

Similarly, in an installment titled "Impressions of the Northwest," Liang Desuo complained that they spent more than a month traveling a distance of three thousand *li* (about one thousand miles) from Baotou to Lanzhou. "If there were trains," he wrote, "it would take only two or three days." Instead, long-distance buses covered only four hundred *li* of the distance. After a one-day bus trip, the group spent thirty-five days traveling by mule cart. Liang said he could not regard the routes in northwest China as proper roads, claiming that they "were carved out by mule cart wheels and human steps." When

Liang's colleague complained that "they should have built a good bridge," another passenger replied with bitter sarcasm: "Should have? There are so many '*should haves*' in China. If this place was given to the foreigners, the railways would have been built long ago." Seemingly contradictory to complaints about the depredations of colonial modernity, this offhand comment from a fellow traveler pointed to the uneven developments between different regions of China. With the national core located in China's eastern seaboard, successive governments from the late Qing to the Republican period made plans to build railways in central and northwestern China but were slow in implementing them. The worse the transportation was, Liang argued, the harder the development of this part of China would be.[82]

While he emphasized the construction of new infrastructure in the interior as integral to China's survival, Liang also worried that such development would bring problems. One issue would be energy sources. Liang argued that more roads meant more vehicles; more vehicles meant a need for more gasoline. But most of the gasoline supply came from foreign countries and was controlled by foreign companies. Thus, Liang called for a search for energy substitutes. A second problem was people's livelihoods. Liang noted that although most people did not object to developing the transportation system, those who made their living by rowing ferry boats or carrying luggage for travelers worried about losing their sources of income. Liang observed that these new construction projects did not bring benefits to most ordinary people but rather aggravated their burdens. "People's anxieties about their livelihood will be the main obstacle in the way of construction," Liang argued, "but if construction does not benefit most average people, what is the value of it?"[83]

After Liang Desuo left *Liangyou* to launch *Dazhong Huabao* in 1933, the "Reports from the Nationwide Photographic Tour" in *Liangyou* ended abruptly. However, in his new pictorial magazine, Liang picked up where he had left off by inviting Wang Xiaoting to serialize Travelogues of China (*Zhongguo youji*), whose last installment was the example with which this book opened. Highlighting the vastness of China, Wang's column of twelve installments foregrounded photographs of less-traveled destinations. Wang featured images of the lost territories of Jehol in Inner Mongolia and different locales in the ill-defined Sino-Tibetan frontiers, located on the edge of China's national space.[84] Perhaps in response to *Dazhong Pictorial*'s travel column, in August 1934, *Liangyou* began to serialize a second travel column dedicated to domestic destinations, which was also more photograph-centric. Titled *Liangyou* Readers' Travel Train, it would run from 1934 to 1936, even though only the first few installments were based on material gained from the Nationwide Photographic Tour.[85]

As the sole photographer in the Nationwide Photographic Tour, Zhang Yuanhen became the key author of the early installments of *Liangyou's* second travel column. Centering on the presence of a Belgian Catholic Mission in the Hetao region in the remote corner where Shanxi, Ningxia, and Inner Mongolia met, for example, Zhang used his photographs in the column's first installment to trace how Western missionaries had managed to establish a "Catholic City" in China's deep interior. Like Liang Desuo, Zhang warned readers there was no time to waste in developing the northwestern region, because "outsiders have already seeded evil influences here, so we must eliminate this huge obstacle first." Here, the discourse of developing the Northwest, a political and social campaign that emerged shortly after 1931 (see chapter 3), was directly related to "a remedy for the economic bankruptcy of inland China" and the national crisis of foreign expansion.[86] Combined with the text, the visual narratives also conveyed this sense of urgency. Photographs of the Catholic church, the fortress, the city wall and gates suggested that the Catholic community had built a wide-ranging infrastructure. Images of the foreign bishop in Chinese attire and Chinese followers holding a Bible study group gave readers a jarring impression of the deep penetration of a Western presence in China (figure 7).

FIGURE 7. "The secret country in the Hetao region." *Source:* "Hetao zhong zhi mimi guo," *Liangyou* [Young companion] 92 (August 1934): 6–7.

Producing travel narratives in a different commercial context than that of *China Traveler*, Liangyou Company focused its printing business on commercial publishing and photojournalism. As its flagship magazine, *Liangyou* represented this mission in the 1932 tour by producing gripping images and compelling narratives of the nation. Unlike the travel narratives in *China Traveler*, *Liangyou* travel columns combined a celebratory tone around the mobility enjoyed by coastal urban Chinese with a heightened social critique. Juxtaposing texts with photographs, *Liangyou* presented different visual angles from which to observe China's condition. Compared to the travel narratives featured in *China Traveler*, which maintained a focus on tourism, *Liangyou* travel writings provided more critical comments on China's situation, both related to and beyond leisure travel. Even though the gravity of the obstacles China faced during the Nanjing Decade was present in the discourse of tourism in *China Traveler*, *Liangyou* journalists voiced a more consistent anxiety over the presence of imperialist powers and their overt or latent threat to China. However, both *China Traveler* and *Liangyou*—as well as their parent companies—recognized the significance of travel as narratives. Boosting tourism consumption and magazine circulation, travel as narratives represented and substantiated *quanguo* at a time of anti-Japanese nationalism.

Modern travel is rooted in mass print media. Modern travel and mass print media also transformed travel as narratives in modern China. In the Republican period, the growth in tourism and travel prompted the production of Western-style guidebooks and other travel-related popular print media in an increasingly competitive national book market. Guidebooks and periodicals produced by major commercial publishing companies in turn helped stimulate public interest in travel and tourism. Popular travel narratives, both textual and visual, also served as critical tools to scrutinize a nation in the making. By studying the commercial guidebooks featuring China's national whole and the travel magazines featuring diverse destinations across the nation, this chapter also traced the rising usage of the term *quanguo* in travel narratives. Indeed, much of the travel print media examined in this chapter bore the term *quanguo* in their titles, from the national guidebooks to a special issue of *China Traveler* to *Liangyou*'s Nationwide Photographic Tour. Products of modern Chinese bourgeois travel culture, these travel narratives targeted tourism consumers, business travelers, and general readers, and presented a national landscape that could be surveyed with a panoptic view.

At the same time, however, they also reinforced the division between the core and the periphery within China's national space, which resembled a similar order between the metropoles and colonies. Despite the different emphases of commercial guidebooks and travel narratives in *China Traveler* and *Liangyou*, one common thread unified Republican travel narratives: a constant reference to the nation from the perspective of the national core. Through these commercial travel narratives, we can capture the zeitgeist of the Nanjing Decade, which conjoined tourism, popular culture, and a nation-building agenda. The *Liangyou* travel columns embraced a more journalistic and visual focus to report on the national space, while the discrepancy between the coastal core and the inland periphery undergirded the authors' desire to expand travel and modern development inland. While struggling with the nation's survival and dreaming of modernity, these cultural workers in urban centers, like their counterparts in the business world, were formulating a similar notion of modern China's national space.

The separate approaches of the print media discussed here suggest that there was no predestined itinerary to an idealized national space. Particularly explicit in the *Liangyou* travel columns was the message that beyond the tourism radius, China's national space was fraught with peril. The following three chapters of this book are devoted to travel outside that safe radius: travel as business and travel as narratives.

Chapter 3

"Head to the Northwest"

Modern China's Movement Westward

In August 1925, on his way back to China after studying at the Sorbonne for five years, Liu Bannong wrote a rambling letter to his friend Zhou Zuoren, who had been his comrade in arms during the May Fourth Movement in China in the previous decade. In the letter, Liu complained about a US archaeological expedition that was setting out to "excavate antiques" in Xinjiang, a frontier province in China's Northwest (*xibei*). "How can we let the Americans dig at will in our Xinjiang?" he exclaimed. "Could Chinese go to America and explore if antiques were found in their land?"[1] A sardonic line laced with dismay, Liu's rhetorical question reflected Chinese intellectuals' growing concern with China's Northwest in the 1920s and 1930s. On a geographic periphery originating in the Central Asian frontier of the Qing empire, the northwestern frontier area of China, Xinjiang included, was also a political fringe of the young republic. Nevertheless, as Liu's epistolary complaints indicate, the northwestern "hinterland" was far from obscure.

Although tourism development in the Northwest was not on their minds, Chinese intellectuals in the early Republican period were aware of the allure of the Northwest to Western explorers and travelers. Becoming known for its commercial, mining, and railway prospects, the Northwest attracted international adventurers who arrived to conduct surveys and gather data and artifacts beginning in the late nineteenth century. Consequently, when the new-style Chinese intellectuals found their way to Western metropoles, they encountered these Westerners' field reports and collected objects, along with a novel concept—the "Silk Road"—which prompted them to pay attention to the geopolitical margins of their own country.[2] Their encounters with these "Chinese" artifacts outside China also triggered a visceral sense of anxiety about China's impotence in the face of foreign encroachment. Encompassing both the multiethnic borderlands and traditional Chinese heartland, *xibei* was reconfigured into a prominent site of Chinese civilization, as well as the geopolitical locus of China's limited sovereignty. In other words, for these May Fourth intellectuals and the educated public on the east coast, the marginalized northwestern region was positioned at the center of their reimagining a Chinese nation-state and refashioning a Chinese national space.

In this historical context, travel and tourism become an important means for coastal Chinese to engage with the northwestern inland and frontier provinces. Chinese intellectuals, for one, dreamed of going to the Northwest themselves. Attracted though they were to *xibei* in the 1920s, however, Chinese intellectuals lacked material support from the state and academy. They were forced to team up with the very rivals they feared—foreign explorers. Nonetheless, in the tumultuous warlord era, Chinese academics based in Beijing spearheaded a westward movement through Sino-foreign joint explorations.

During the Nationalist rule, especially in the early 1930s, this inchoate westward movement gained new urgency. Not only did the Nationalists became more committed to incorporating the frontier provinces into their orbit of direct control, but the uncertain fate of the Northwest was brought to the forefront of Chinese elites' concerns when the Japanese occupation of Manchuria and advancing encroachment in Inner Mongolia and North China ushered in an era of deepening frontier crisis. In this new climate, railway companies, commercial institutions, publishing companies, and ordinary urbanites were also drawn to this "westward movement." Calling for their fellow countrymen to "head to the Northwest" (*dao Xibei qu*), coastal intellectuals and urbanites channeled their yearning for national strength into a fantasy of traveling to the nation's margins.

This chapter sketches out this westward movement in modern China via three specific journeys: a lone Chinese scholar joining a US mission to Dunhuang in 1925; the negotiation and execution of a large-scale Sino-Swedish scientific expedition to Xinjiang in the early years of the Nanjing Decade; and the commercial travel writings on the northwestern frontier facilitated by the Ping-Sui and the Longhai Railways, the two major railway lines connecting China's eastern seaboard and the Northwest, in the mid-1930s. In their northwestward journeys and travel narratives of the Northwest, east coast–based Chinese travelers echoed old imperial spatial imaginations while incorporating new ideas of science, colonialism, and modernity. In the process, they formulated a spatial imagination of the Northwest as an integral part of the Chinese nation-state.

Dunhuang, the Silk Road, and the Northwest

In the Republican era, the Northwest usually included Shaanxi, Gansu, Ningxia, Qinghai, and Xinjiang.[3] Associating the region with the upper and middle reaches of the Yellow River, people often referred to Shanxi and western Inner Mongolia as part of the region as well. A broad swath of territory with a culturally diverse population, part of it—namely, Mongolia, Xinjiang, and most of Qinghai—had been brought into the Qing empire only in the eighteenth century. While the Qing policy of divide and rule had maintained the multiethnic and multireligious status quo of the region for two centuries, the Northwest became a key concern of the Qing frontier defense in the second half of the nineteenth century, leading to heavy-handed military campaigns and state-building projects. Despite these efforts, the Northwest turned chaotic soon after the fall of the Qing in 1911. Connecting China proper to Central Asia, the Northwest, especially Xinjiang, was also a venue for international rivalry at the turn of the twentieth century. The conflicts between British, Russian, and other powers for supremacy in Central Asia—and the international scramble for China—made the Northwest a volatile region within the fragile new Republic of China.

At this historical juncture, the Northwest also occupied a unique position in knowledge production after the concept of Silk Road was invented in Europe in the 1870s. As a historical narrative and developmental vision of East-West exchange, the conceptualization of the Silk Road signaled both the spatial and temporal significance of China's Central Asian frontier.[4] For example, because ancient artifacts and religious and secular manuscripts were excavated from a few Silk Road sites in China's Northwest that "enabled a link to be made between philology, the study of comparative religions . . . and antiquities," the region had attracted foreign explorers and archaeologists since the late Qing period.[5] At the same time, the Euro-American obsession with the search for the origins of races and civilizations—including the origin of Chinese civilization—moved eastward to Central Asia, which was viewed as a crucial historical corridor of Eurasian exchange.[6] From the Hungarian-born British archaeologist Aurel Stein to the French Sinologist

Paul Pelliot, and from the Swedish explorer Sven Hedin to the Russian Orientalist Sergei Oldenburg, foreign explorers surveyed deserts and valleys and pillaged temples, tombs, and ruins, carrying away priceless documents, fossils, and artifacts which would later be acquired by museums and libraries in Western metropoles. Among all the sites they explored in the Northwest, none could trigger more bitter feelings among Chinese intellectuals than Dunhuang, a place later regarded as a poignant chapter in "the sorrowful history of Chinese academics."[7]

In 1899, Wang Yuanlu, a Chinese Taoist priest and the abbot of the Mogao grottoes in Dunhuang, stumbled upon a sealed cave. Full of manuscripts and paintings enclosed shortly after the end of the first millennium CE, the cave made Dunhuang "the time capsule of Silk Road history."[8] Recognizing the value of the discovery, foreign explorers, especially Aurel Stein and Paul Pelliot, managed to cajole the priest into selling them the majority of the collection and the art objects.[9] Stein's collection ended up in London; Pelliot shipped his acquisitions to Paris. As a result, while Western scholars could study the documents and artifacts in their home institutions, Chinese academics had to travel to London or Paris in order to copy them out by hand. Writing his dissertation in Paris, for example, Liu Bannong encountered the Pelliot collection of the Dunhuang manuscripts in the Bibliothèque nationale de France. These documents, dating from the fourth to the eleventh century and written in Chinese and multiple Central Asian languages, offered unparalleled resources on the religious and secular history of the areas along the old Silk Road.[10] Interested in the folk literature, vernacular language, and everyday socioeconomy recorded in the Dunhuang collection, Liu Bannong painstakingly selected and copied more than a hundred items in Paris, which were later published in China.[11] A nationalistic reaction to China's humiliation at the loss of the Dunhuang collections to Western institutions, Liu's endeavor to recover these materials from the West should also be positioned against the backdrop of the National Studies Movement (*guoxue yundong*), which was getting underway in 1920s Beijing.

The National Studies Movement was led by prominent Chinese thinkers like Hu Shi, who aimed to establish a discipline dedicated to studying Chinese traditional culture in a scientific way. Rather than a New Culturalist crusade, the mission was in fact shared by so-called cultural conservatives. Despite the ostensible differences between the two groups, both sides, as Lydia Liu has pointed out, shared a similar intellectual background and "were centrally preoccupied with the problem of Chinese national identity."[12] The New Culturalists in the movement, based at the National

Studies Department (*guoxue men*) at Beijing University, aspired to introduce Western scientific disciplines such as archaeology, geology, and paleontology into China's classical studies. Loosely defining "national studies" as an umbrella term for all disciplines dealing with China's past, they also inherited the May Fourth spirit of giving attention to the vernacular traditions in China.[13] These *guoxue* scholars, Liu Bannong included, shifted their attention from annotated classics to local dialects, customs, and folklore. Meanwhile, the supposedly more conservative camp of the movement, including Liang Qichao, Wang Guowei, and Chen Yinke—who were the founding professors in the Research Institute of National Learning (*guoxue yanjiu yuan*) at Qinghua University—engaged in the revival of China's traditional culture after they became disillusioned with Western civilization in the post–World War I era.[14] While they rejected the notion of writing in vernacular Chinese, these Qinghua scholars were not more traditionalist than their counterparts at Beijing University in academic background and research methods. The politically conservative Chen Yinke, for instance, had spent his teens in Japan and later in Europe and the United States studying with famous Western Sinologists, making him a rare polyglot among Chinese academics. Wang Guowei, a classicist, was among the first group of Chinese scholars to recognize the value of the Dunhuang manuscripts.[15] An advocate for the introduction of archaeology into China, Wang had also celebrated the field as a new scientific approach to discover China's past.[16] As their shared interests in the Dunhuang manuscripts have indicated, both the "radical" May Fourth intellectuals and the "conservative" classical scholars harbored a similar view. They believed that national studies should be understood as an indispensable part of the reorganization of China's national heritage, which was crucial to the nation-building process. They also had in common the belief that China's national studies should be as scientific as Western Sinology.[17]

These intellectual developments in the early Republican era placed the Northwest at a center of knowledge production about China's past. Epitomized by Dunhuang and the Silk Road, the significance of the Northwest in China's historical connection with Central Asia and Europe marked a new direction in studying Chinese history and traditional culture. Moreover, two seemingly separate projects—a Western project of expanding knowledge of the "Orient" through exploration in Central Asia and a Chinese movement to establish a modernized national studies—converged at the Northwest. Recognized internationally as the prime site for studying a variety of disciplines, the Northwest was also a symbolic space pivotal to China's national history and national identity for many Chinese intellectuals.[18] They were not entirely contradictory projects. As Fa-ti Fan has pointed out, the international cosmopolitanism embedded in modern science partly gave rise to the attention given to the issue of national sovereignty within scientific explorations.[19] It was in this context that Chinese academics actively participated in explorative travels to the Northwest. Among all the possible destinations, Dunhuang was naturally the place they most coveted.

In 1925, an opportunity knocked at the doors of Beijing University. Although not many manuscripts were left in Dunhuang by the early 1920s, its spectacular caves with pre-eleventh-century murals and sculptures beckoned leading Western museums to send archaeologists and art collectors to China's Northwest. A year after his first trip to Dunhuang, Langdon Warner, a US art historian and archaeologist, returned to China and prepared to set out on a second trip to the Mogao caves. Employed by the Fogg Art Museum at Harvard, Warner had first found his way to Dunhuang in the autumn of 1923 with Horace Jayne of the University Museum of Pennsylvania. Both Warner and Jayne took great interest in the ancient wall paintings decorating the grottoes there, most of which had been painted during the Tang dynasty. After bribing the abbot Wang to allow them to remove some of the murals, Warner tried to transfer them using cloth soaked in strong chemical adhesives.[20] Warner and Jayne then returned to China a year later with a bigger team equipped with better recipes both for the fixative and for adhesives.[21] This time, they also sought out a Chinese expert to help translate the inscriptions of the stone tablets and were introduced to the National Studies Department at Beijing University. As a research institute, the National Studies Department did not have the resources to fund field trips to the Northwest, so when Warner's request arrived, the department welcomed the invitation and sent Chen Wanli to take part in the expedition.

A versatile young intellectual, Chen Wanli came from a medical background but taught himself photography and founded several photographers' groups in Beijing.[22] He was also an experienced traveler who studied local gazetteers and travelogues and was well versed in the knowledge of Buddhist statues, tablets, and other historical artifacts. Chen was very enthusiastic about the opportunity of travel to the Northwest, which he called "a dream I have not been able to forget for more than a decade."[23] Separate from the main group, Warner remained in Beijing a little longer and traveled alone, while the rest of the mission, including Chen Wanli, journeyed northwestward in February 1925. After a short train ride from Beijing to Taiyuan in Shanxi Province, the group moved on by means of horseback and cattle-drawn carts. Sharing rooms in mountain inns and sometimes traveling on foot to explore local villages, Chen and his US colleagues worked side by side along the journey. Chen not only helped translate the scripts of the steles and tombstones they encountered but also took pictures of and notes on the Buddhist cave temples they visited in Shaanxi and Gansu Provinces. Shocked by how high the travel costs were, Chen acknowledged that without the Americans, it would be hard for a Chinese scholar to visit Dunhuang.[24]

Nevertheless, even a collaboration that seemed mutually beneficial could turn sour. Approaching Dunhuang, Chen started to hear rumors about the Americans' last expedition. Local resentment about Warner's "theft" of the murals in 1923–24 made the local officials extremely wary about the Americans' reappearance. At first, the provincial government rejected the group's request to remove more murals, even though Jayne claimed that they would be displayed in Beijing so both Chinese and foreign experts could study them.[25] With little hope of gaining permission, Jayne asked to photograph the statues and murals inside the caves. Suspicious of the foreigners and concerned about the reaction from locals, officials refused the Americans' request to stay for three months, and only allowed them to be on-site for two weeks. Moreover, they were not allowed to camp overnight inside the caves. Perhaps because they were forced to abandon the original removal plan, or because they were intimidated by the local hostility, Warner and Jayne resolved not to stay long. As a result, the team spent only three days in the Dunhuang caves, which frustrated Chen Wanli a great deal. Chen complained in his travel journal:

> The members of this expedition had all hoped to stay in the caves of the thousand Buddhas for three months. However, because of Warner, [we] could only stay for three days. On top of that, [we had to] travel back and forth eighty *li* every day [between the caves and our lodging inside the town], which limited the actual time spent inside the caves to only five hours a day, a total of fifteen hours over three days. Getting only this [amount of time in the caves] after a journey lasting more than a hundred days was really not what I had expected.[26]

For Chen, Warner's irresponsible actions of the previous year had robbed a Chinese scholar like himself of a chance to study the site thoroughly. Unable to figure out why the local authorities and villagers had turned hostile seven months after his last trip to Dunhuang, Warner suspected Chen of being the "fly in the ointment," and thought it was Chen who had informed the locals of the true nature of their expedition.[27] Perhaps perceiving the anger directed at him, Chen Wanli did not follow the expedition onward but returned to Beijing alone.[28]

As it turns out, it was not Chen Wanli who had derailed the expedition. According to the memoirs of William Hung, a professor and dean at Yenching University, it was Wang Jinren, the translator for both of Warner's Dunhuang expeditions, who had caused the second visit to backfire. After his first journey to Dunhuang, Wang Jinren told his professor William Hung about Warner's plunder in 1923–24. Despite his position as a dean at Yenching, a US missionary university with close ties to Harvard, Hung paid a secret visit to the vice-minister of education in the Beijing government. This was likely why both provincial and local governments forbade Warner to touch any historical relics in Dunhuang in 1925.[29] Although it is not clear whether Chen Wanli or any Beijing University professors were aware of Professor Hung's efforts and whether they were part of an attempt to stop Warner, in 1950, during the height of the Chinese Communist Party's anti-US propaganda, Chen Wanli published an article hinting that the purpose of dispatching him to join the Dunhuang expedition was solely so that he could monitor the US imperialists and prevent them from stealing national treasures.[30]

Irrespective of what had stirred the antagonistic attitude among locals, the premature ending of this cooperation should not be understood simply as a conflict of national interests between Chen and Warner. Rather, it can be seen as an instance of the predicament Chinese scientific travelers faced in international ventures: the irreconcilable conflict between the pursuit of scientific knowledge on the one hand, and the claim of sovereignty on the other. Chen Wanli, although disheartened by the wretched condition of the Dunhuang caves, seemed to be more disgruntled by his lost opportunity to conduct research in the Northwest. As a Chinese, he could not side with the Americans and object to local officials' decisions. But as the first Chinese academic traveler to visit Dunhuang, he yearned for the liberty to conduct research without interference, a liberty many Western and Japanese explorers had previously enjoyed. His nationalistic commitment to survey China's hinterland and his professional desire to pursue scientific projects in cooperation with international explorers—both of which had propelled this westward journey—were tightly interlocked. But for Chen, like many Chinese intellectuals of the time, the sense of national belonging could collide with a belief in the universal value of science.[31]

Another incident that occurred on the way to Dunhuang also illustrated this implicit discordance. In Jingzhou, Shaanxi Province, the expedition stumbled across a cave temple from the Northern Wei period (386–534) called the South Grotto Temple (*Nan shiku si*), the same temple the *Liangyou* Nationwide Photographic Tour encountered in 1932–33. In 1925, the sculptures and other historical artifacts in the temple had gone missing or been damaged. One of Warner's justifications for removing the frescoes from Dunhuang was the damage he witnessed in the Mogao grottoes in 1924. Although this did not explain why he had prepared cloth and strong adhesives even before witnessing the damage, Warner stated that the only way to save the Dunhuang grottoes was to remove the artifacts from China. Echoing his US colleagues' logic, Chen Wanli believed that the only way to safeguard the valuable treasures at the South Grotto Temple was to carry them to Beijing. Jayne and other US members concurred and suggested that Chen demand that local authorities keep some of the sculptures from another site for him to carry back east to Beijing University.[32]

But the episode at the South Grotto Temple was fraught. Just as Chen was assisting the Americans in using axes to chop away the contoured mud covering the original sculptures, twenty or so local villagers showed up at the caves. Furious, they blocked the caves and demanded that the Americans restore the statues they had "destroyed." With their limited language skills, Jayne and the other Americans struggled to respond. Seeing that the confrontation was escalating, Chen became afraid that it would turn into "an international incident." He came forward and acted as a mediator, and later resolved the dispute by promising a small indemnity for each statue they had uncovered. At the same time, he managed to sneak out a stone tablet.[33] Commenting on this confrontation in his travel journal, Chen found himself in a very uncomfortable position, as he considered the Americans' approach to the statues to be an "appropriate research method." "I agreed [with their method] wholeheartedly," Chen wrote, "and I even helped them." But he regretted that the confrontation had happened, lamenting that "people from a friendly country" had become embroiled in a heated dispute. Nevertheless, Chen was hesitant to blame the villagers for their "ignorance." Trying to attend to their concerns, Chen further pinpointed the villagers' misstep as lying in the extreme methods of protest they used: besieging the caves and attempting to search the foreigners. Chen emphasized that the central issue was not the question of sovereignty. "It is one thing to claim [the possession of the temple caves]," Chen concluded, "but the misuse of methods would instantly result in disputes." Identifying this incident as the biggest lesson people should learn "when traveling in inland regions," Chen blamed the local villagers' xenophobia for the unexpected obstruction.[34]

Chen's 1925 journey to Dunhuang reflects the complex self-positioning Chinese scientific travelers had to engage in when they joined Sino-foreign joint expeditions to the Northwest. The use of Western-inspired scientific methods to study China had universal appeal to both Chinese and foreign explorers. As shown in Chen Wanli's assistance to the Americans, as well as the Americans' suggestion that they help Beijing University obtain some of the historical artifacts they encountered in Shaanxi, Chinese and foreign members of such expeditions shared a sense of obligation to rescue historical objects from backward local populations. Gu Jiegang, a historian at Beijing University, touched on the incident at the South Grotto Temple in the preface he wrote for Chen's travel journal. Gu simply pointed out different attitudes toward antiquities, attitudes bifurcated between "us" and "others." He wrote,

> As to the antiquities, I truly feel that ourselves and others stand in two worlds. Statues, murals, and suchlike, in our opinion, are only regarded as antiquities, art, and history, and we always want to study them; however, in other people's eyes, they simply represent the concept of deity, and thus should be protected for worship. . . . There is no way to reconcile these absolutely different views. Because their purpose was "use," [they could only] save [the antiquities] or destroy [them], or even steal and sell them for profit. . . . [T]he abbot Wang from the Thousand Buddha Temple, who had been stealing and selling antiquities for ten years, is one good example.[35]

Gu viewed the local villagers in Shaanxi and the abbot Wang in Dunhuang as similarly backward and ignorant, even though the villagers prevented the foreigners from carrying away any statues, while Wang had allowed Warner to peel away murals. For Gu and his fellow Beijing scholars, the demarcation between "us" and "others" was more complicated than whether one protected Chinese antiquities from foreigners. The division lay in whether one viewed these artifacts as objects for scientific research. Chen Wanli and his colleagues in Beijing aligned themselves with their Western colleagues and considered the locals as "others" who had hindered their scientific research. This otherness of the locals, in their opinion, was also the framing condition of China's lagging modernization vis-à-vis the unceasing penetration by imperialists.

Although far from becoming a popular tourist destination, Dunhuang, especially the mural art in the Mogao caves, gained wider recognition after Chen Wanli published his travel writing and photographs.[36] In the 1930s, government officials on inspection tours and journalists conducting reporting trips in the Northwest made concerted efforts to visit Dunhuang.[37] During the War of Resistance against Japan (1937–45), coastal artists, who had relocated to western provinces because of the war, also found their way to the Thousand Buddha caves. Among them, the preeminent Chinese painter

Zhang Daqian spent almost three years studying and hand-copying hundreds of Dunhuang murals. Zhang's copies were later exhibited in the wartime capital, Chongqing, and the Dunhuang mural art once again astounded the public audience.[38] Introducing the Chinese public to the history and art of Dunhuang, Chen Wanli's 1925 trip also helped establish the notion of "backward locals," which would shape tourist experiences in the Northwest in the mid-1930s. Whereas urbanites from coastal regions searched for national myths and histories by touring the Chinese heartland, tourism in the Northwest would also operate on the assumption that traveling to these remote provinces helped middle-class tourists assert their autonomy, cosmopolitanism, and superiority over the "backward locals," who were less educated, less mobile, and unable to fully appreciate China's national essence.

Influenced by the nationalist movements in the early twentieth century, Chinese intellectuals were acutely aware of the controversies involved in foreign-led explorations on Chinese soil. Whereas the Chinese travelers to the Northwest recognized that the value behind such scientific missions was not geographically bound and their discoveries should be appreciated by all modern-minded people globally, it was also self-evidently important for them to assert Chinese sovereignty over the Northwest by participating in these joint projects and bringing the discoveries back to Beijing or Shanghai, rather than to Paris or London.

This inner conflict between a bifurcated self-positioning as a "scientist" and as a "Chinese" was not uncommon. As discussed at the beginning of this chapter, Liu Bannong voiced his objections to a rumored US expedition to the Northwest. In fact, Liu's letter was triggered by the exact news of Chen Wanli's participation in Warner's expedition. Although Chen and his colleagues at Beijing University considered it an unprecedented opportunity for Chinese to break the barrier of foreign monopoly on expeditions to the Northwest, Liu was concerned about the possible harm that could be caused by imperialist explorations. In the late 1920s and the early 1930s, the Chinese intellectuals who had made possible the joint explorations of the Northwest usually had to play the dual role of universal scientists and nationalistic patriots. Upon his return to Beijing from Paris, Liu Bannong himself was soon to become a representative figure in such stories.

Negotiating a Common Ground

As demonstrated by Chen Wanli's 1925 trip to Dunhuang, Chinese academics in Beijing became more involved in the exploratory travels to the Northwest in the 1920s. Several factors contributed to this development. Beginning in the 1910s, a community of international elite scientists had made Beijing their home base. Preparing for their impending expeditions to the Northwest and Central Asia or serving alongside Chinese colleagues in scientific associations and institutions of higher education, they engaged many Western-trained Chinese intellectuals in popularizing the practice of fieldwork in disciplines such as geology and paleoanthropology.[39] At the same time, Chinese intellectuals who had been trained in Euro-American and Japanese metropoles returned to China and took over the professionalization of Chinese academia. Like Liu Bannong, they were exposed to European science and Sinology as well as library and museum collections of Chinese and Central Asian historical documents and artifacts during their study abroad. Modern disciplines such as archaeology, geology, and paleontology entered Chinese academia with the return of these professionalized scholars. With the deepening of the National Studies Movement in the late 1920s, the need for scientific methods to study China's past were broadly discussed and became accepted in academia. The mission of the movement also entered the popular consciousness. Lastly, in the late 1910s and 1920s, there was a rising tide of Chinese nationalistic movements in urban areas, and intellectuals also became more vocal about their objections to explorations in the Northwest led wholly by foreigners.

Two years after Chen Wanli's groundbreaking trip, a larger-scale Sino-Swedish expedition departed from Beijing for Xinjiang in May 1927. This Scientific Mission to the Northwestern Provinces (*Xibei kexue kaochatuan*) was led by Sven Hedin. A prominent traveler who had made three monumental journeys crossing the Gobi desert in Xinjiang and the Himalayan range in Tibet at the turn of the twentieth century, Hedin returned to China in the winter of 1926.[40] Commissioned by Lufthansa German Airlines, Hedin convened a team of scientific explorers and technicians to revisit China's Central Asia and investigate prospects of opening an air route for Lufthansa between Berlin and Beijing via Urumqi (known as Dihua in Chinese at the time). Hedin's team planned to set up meteorological stations and survey the terrain for landing sites, which required comprehensive surveys not only on the ground but also from the air.

However, just as Hedin's plan was green-lighted by the warlord government, a newspaper in Beijing reported his impending expedition. Questioning Hedin's intentions, the news media soon painted this mission as another barely disguised foreign looting of Chinese antiquities, which would be conveniently shipped abroad by aircraft.[41] Beijing academics reacted immediately. A meeting was called at Beijing University, to which twelve academic organizations in Beijing, including Beijing University, Qinghua University, the History Museum, and the Beijing Library, sent representatives. As a result, the Association of Chinese Learned Societies (*Zhongguo xueshu tuanti xiehui*) was formed, and Liu Bannong, who was then teaching at Beijing University, led a committee to renegotiate with Hedin.

In Hedin's memoire, Liu Bannong was the "real soul" of the Beijing committee overseeing the negotiation, during which Liu and his Chinese colleagues emphasized the issue of sovereignty. As Hedin soon found out, Liu "considered Western interference as a danger and a threat to the hegemony of Chinese science in China."[42] Liu insisted that not only should Chinese participate in the expedition, but they should also have a Chinese field director who would be equal to Hedin in the decision-making process. Every artifact collected in China would have to remain in China. Hedin and other European members of this expedition would not be allowed to publish meteorological data, maps, and other scientific statistics without China's permission, a rule that also applied to photographs and films. Furthermore, the travel routes, once determined and approved by the Chinese committee, could not be altered. Hedin, although hesitant, finally agreed to the proposals Liu and his colleagues made.

As we can see, the demands of the Chinese committee exceeded the realm of scientific research per se. Some of the terms were directly linked to their concerns over Chinese territoriality. For example, they prohibited Hedin from drawing maps to a scale more detailed than 1:300,000; nor should meteorological observations be used for any military purposes. Even the word "expedition" (*yuanzheng*, or *tanxian*) was disallowed and replaced by "mission," because "expeditions," as Liu Bannong and his Chinese colleagues understood them, were carried out among the "blacks and savages," and it would be insulting to use the term in China, a country with an ancient culture.[43] The hard negotiation resulted in a nineteen-article agreement, detailing the responsibilities and rights of both Chinese and foreign team members.[44] This contract marked a triumph for Chinese academia in forging an equal collaboration with foreign explorers.

What made this collaboration possible, however, was not simply the unyielding demands made by the Chinese. The negotiation happened in the middle of the Nationalists' United Front campaign against the northern warlords, which, as Hedin observed, had stirred a wave of nationalistic and antiforeign sentiments among urbanites in China. Because of the Chinese press's attack on Hedin's expedition, another expedition led by Roy Chapman Andrews, an explorer dispatched by the American Museum of Natural History, came under attack, and Hedin's fellow Swedish geologist Johan Gunnar Andersson, the discoverer of Peking Man, was also subject to criticism.

Sensing a possible renewal of the Chinese state through the Nationalist campaign, Chinese academics protested pending foreign expeditions in the firmest terms. The uncompromising attitude even prompted Hedin to question Liu Bannong at one point as to whether the Chinese scholars were imposing "a Treaty of Versailles" on him.[45] Even though Hedin had obtained permission from the warlord government, he discerned that without support from and affiliation with Chinese institutions, his costly expedition would be truncated if regime change occurred in China.

Aside from the political climate, his Chinese opponents also impressed Hedin. Both Liu Bannong and his colleague Xu Xusheng (also known as Xu Bingchang), a history professor and philosopher from Beijing University who later served as the Chinese field director of the expedition from 1927 to 1928, spoke fluent French after years of study at the Sorbonne.[46] Yuan Fuli, another US-educated Chinese scholar from Qinghua University who later joined Hedin's expedition, was "an extremely learned and skillful geologist, paleontologist, archaeologist and topographer who spoke perfect English."[47] The fact that these Chinese opponents "had been in touch with the modern methods of Western science" was considered proof of their legitimacy.[48] Furthermore, as a great admirer of Wilhelmian Germany and its statist nationalism, Hedin sympathized with the "nationalistic attitude" of the Chinese scholars. He told Liu Bannong that he himself was a nationalist in his own country.[49] He argued that the Chinese claim to all archaeological objects found during the expedition "was in complete conformity with the legal regulations in all civilized countries."[50] He explained further to his Western audience not only that "knowledge transcends political frontiers and the prejudices of the different races," but also that the "Chinese would enjoy the same rights as the Europeans," because they "were in their own country, at home; we, on the other hand, were guests."[51]

On the Chinese side, Xu Xusheng, the head of the Chinese contingent of the expedition team, expressed a slightly different opinion when detailing the motivations that had driven Chinese intellectuals to be involved in Hedin's expedition. While Xu concurred with Hedin that "knowledge and science" should be "international" and "borderless," he highlighted that the material objects and individual researchers involved in knowledge production processes belonged to their nation. He added,

> As for the foreigners, utilizing their superior financial resources, they take away scientific materials from our country as they please and without limit. With respect to invaluable sources that are hard to come by, they extort them by trickery or by force, and carry them out

> of China secretly! If we cannot redeem this situation, the academic future of our country will suffer great losses. . . . As to the foreigners themselves, we will welcome cooperators with friendship but think up a method to resist those who want to invade us culturally and plunder our invaluable scientific materials, so that they will not harm our country.[52]

If for Hedin the universal value of science and the shared belief in nationalism formed the common ground on which he and his Chinese opponents could reach mutual understanding, Chinese academics like Xu Xusheng did not romanticize the seemingly egalitarian impulses behind joint expeditions. Hedin believed that Chinese academics' embrace of Western scientific methods lent them an equal footing in their collaboration and had elevated China to the rank and order of civilized nations. While not questioning this Eurocentric view of science and knowledge production, Xu Xusheng was still deeply aware of the imbalanced power dynamics between foreign explorers and Chinese academics. The Chinese scholars' insistence on the issue of territoriality and strictly equal status was their effort to balance the power dynamic in an uneven relationship. In the same paragraph, Xu Xusheng used the word "friendship"—which was also mentioned by Chen Wanli and Hedin—to evoke an amicable spirit. But it could also be interpreted as a euphemism for mutual interests and advantages.[53] Chinese researchers needed foreign funding, equipment, and expertise to practice professional techniques and carry out explorations in the hinterlands.[54] At the same time, Western explorers could only pursue their ambitious missions by avoiding direct conflict with the increasingly nationalistic educated class in China.

The emphasis on friendship also underlined the shared politics animating these joint ventures. At the outset, this negotiation between Hedin and Beijing academics seemed to be a mission impossible, as it was between two irreconcilable interests: an imperialist industrial project energized by global capitalism, and a struggle for national salvation that was intrinsically anti-imperialist. However, the Han Chinese resistance against foreign exploration was also informed by their own desire to penetrate the multiethnic frontier provinces in the Northwest. After the westward conquest in the eighteenth century, the Manchus had fashioned themselves as a fellow Central Asian steppe people to exert a more personalized rather than centralized rule over the Mongols, Tibetans, Uyghurs, and other nomadic peoples in the region. Yet shortly after the Qing empire collapsed in 1911, the Republican regime faced separatist movements not only in Outer Mongolia and

Tibet but also among the multiethnic peoples in Xinjiang and Inner Mongolia. The central government in Beijing had only nominal control over the local warlords in the Shaanxi-Gansu-Ningxia area within larger *xibei*, some of whom were from or had close relationships with the Chinese-speaking Muslim communities in the Northwest. As the Nationalist revolutionaries and Han Chinese intellectuals further linked these "separatist" movements to foreign imperialist influences, they justified the Qing borders as vital to the legitimacy of a Nationalist republic and saw the formal incorporation of the Northwest into the nation-state as a pressing issue.[55] The nation-building effort embedded in the Chinese participation in joint explorations of the Northwest reflected both Chinese resistance against foreign imperialism and their rejection of the prospect of relinquishing the Qing colonial frontiers.[56]

Han Chinese intellectuals did not view their desire to shore up the nation by strengthening Han control over the frontiers and their intense anti-imperialist sentiment as contradictory. Their stance was in tune with the development of colonialism and nationalism globally. The US westward expansion showcased how the majority nation formation could include incorporating lands and peoples in their close proximity. In Germany, Britain, France, and Russia, colonial, racial, and nationalistic policies were often spearheaded by explorers and geographers. While late Qing and Republican Chinese seldom equated the Qing westward expansion with the European overseas colonial possessions, some did liken Chinese rule over Xinjiang to Japan's colonization of Hokkaido and Taiwan.[57] Further, colonial expansion was not considered a bad idea if Chinese were conducting the colonizing.[58]

When Xu Xusheng and his colleagues joined Hedin's exploration in 1927, the idea of *zhibian* (frontier colonization) had also been circulating among Han Chinese officials and intellectuals, a school of thought that would gain a broad resonance nationally in the early 1930s.[59] In short, in the 1920s and 1930s, Chinese discourses of the Northwest were informed by the Qing expansion movement, Western and Japanese colonialism, as well as a nation-building vision largely based on a majority nationality group's design.

From this angle, the Han Chinese desire to participate in the exploration of the Northwest may bear resemblance to Hedin's urges to explore Central Asia. In both cases, the region was viewed as an "empty" and wide-open space, waiting to be explored and developed. Both groups considered themselves civilizing forces vis-à-vis the backward local populations. By professing a common faith in modern (European) science and a high regard for nationalism, the two sides recognized that science had major political implications for constructing nationalistic and imperialist sets of knowledge. Even as Chinese academics decried the term "expedition" (*yuanzheng*) and removed it from the mission's formal title, it was not the idea of racial hierarchy per se they were protesting. The perceived insult lay in where the Westerners had located Chinese on the racial hierarchy. Whereas Chinese academics eventually convinced Hedin that they were intellectually—and perhaps racially—equal to Western explorers, these Beijing scholars had no qualm about considering the non-Han peoples in the Northwest as less civilized. In the same vein, what set the two sides apart was the question of which geopolitical center and whose imperialist or nationalist orbit the Northwest should revolve around. As a colonial technique, the practice of exploration and travel played a key role in aiding colonial and imperialist powers. For the Chinese intellectuals striving to unite a national space based on its imperial legacy, this colonial technique fit neatly into their nation-building endeavor.

It was because of this common ground that this hastily formed joint exploration initiated by Hedin turned out to be a lasting mission that continued on and off from 1927 to 1935. When the original plan to survey Xinjiang from the air was denied by the Xinjiang authorities in 1928, the main sponsor of the mission, Lufthansa, pulled their funding and called back their experts. Hedin and the Association of Chinese Learned Societies appealed to the Swedish and Chinese (by then led by the Nationalists) governments respectively, which granted extra funding to extend the exploration and sent scientists from China and Sweden to join the second phase of the mission. After this second stage concluded in late 1933, Hedin gained the ear of top figures in the Nationalist government, including Chiang Kai-shek, who appointed him to be an advisor to the Ministry of Railways and outfitted him for an expedition to investigate the potential for modern transport routes between Inner Mongolia and Xinjiang in 1934–36.[60] The fact that an imperialist exploration inspired by European scientific ambitions and industrial capitalist interests in Asia could eventually morph into a frontier expedition endorsed by the Chinese Nationalist government underlines how intimately the travel circuits of colonialism and nationalism overlapped in the Northwest.

Locating the Northwest in China

On May 9, 1927, families and friends gathered at the Xizhimen train station in Beijing to bid farewell to the departing Scientific Mission to the

Northwestern Provinces. They were heading to Baotou via the Ping-Sui Railway, where the joint team, consisting of eighteen Europeans and ten Chinese researchers, along with twenty-two Mongolian and twelve Chinese servants, and more than two hundred camels, would proceed by caravan through western Inner Mongolia to Xinjiang.[61] The geologist Yuan Fuli, the archaeologist Huang Wenbi, the geologist and paleoanthropologist Ding Daoheng, the cartographer Zhan Fanxun, and the photographer Gong Yuanzhong were among the Chinese contingent. The Europeans, consisting of Hedin's old associates and aviation and meteorology experts sent by Lufthansa, were from Sweden, Denmark, and Germany. Xu Xusheng, a Sorbonne-trained philosopher who had never done any fieldwork or traveled to the Northwest, volunteered to be the Chinese field director. Xu was enthusiastic despite his lack of experience. In a photograph taken at the train station, he wore a Sun Yat-sen suit with a telescope or water bottle strap over one shoulder, along with puttees and leather ankle boots. The outfit was completed by a fedora hat and a walking stick, mirroring his senior colleague Sven Hedin. Posing awkwardly alongside the three-piece-suit-clad Hedin and the Swedish ambassador, Xu looked both overprepared and ill-suited for the exploration.[62]

As a nonscientist, Xu played more a symbolic than a practical role in the expedition. Unlike his scientist colleagues, he had no responsibility for the surveys, excavations, or experiments. Nor did he engage in any day-to-day tasks such as managing the camels, searching for fresh water, or cooking meals, with which the many unnamed helpers were tasked. However, as a field director and an eyewitness to the first stage of the mission, Xu wrote in his journal every day. After his return in 1929, Xu published his travel diary with encouragement from *Eastern Miscellany* magazine and his friend Lu Xun. Rather than editing his travel journal into a "popular form," as Sven Hedin did in his rendition for the general reader, Xu kept the original style and presented a day-to-day account of their journey: from their departure from Beijing toward Baotou to the difficult journey to the oasis of Etsin-gol in the Gobi Desert, and from their entry into Xinjiang at the town of Hami to their eventual advance to Urumqi in late 1928.[63] Xu's field notes offered readers a sense of what it was like to be a modern explorer on the northwestern frontier.[64]

In the West, Central Asia evoked a fabled image because of Marco Polo's legendary traversing of its oases and deserts in order to reach China.[65] Similarly, Han Chinese elites had associated the Northwest with the vague notion of *xiyu* (the western regions). A loosely defined western domain that could be found in both dynastic histories and popular literature, it referred to a region that lay beyond the pale of civilization. Paradoxically, in the 1920s and 1930s, the name Central Asia or the Northwest connoted a sense of the "unknown" to Westerners and Chinese alike. Its extreme natural conditions, diverse ethnic and religious makeup, and political chaos perpetuated a mysterious image, at least when viewed from the outside. The locations of many sites mentioned in historical travel accounts were lost after being buried under the desert, while at the same time the native knowledge of the region was largely discredited as hearsay by Western and later Han Chinese explorers and disregarded by the educated public in the metropoles. In short, in the opinion of Western explorers and Chinese academics, because the Northwest had yet to be scrutinized through the lens of modern science, its largely uncharted territories still needed to be discovered or rediscovered.

Aside from setting up temporary meteorological stations to collect atmospheric data for Lufthansa, the main focus of this joint expedition was to do topographical, geological, paleontological, and archaeological surveys while collecting ethnological and ethnographical information by measuring, photographing, and filming the local peoples. Though not a scientist, Xu Xusheng clearly viewed himself as a fellow explorer. He consciously maintained a "scientific" style in his travel journal. For example, as their caravan meandered through the Gobi Desert, Xu recorded the names of places they encountered. Since many of the oases, trading spots, and mountain passes were only locally known and often referred to in Mongolian, Uyghur, or other non-Han languages, Xu painstakingly transliterated the names into phonetic symbols with their possible meanings in Chinese, instead of just using Chinese characters.[66] Scientific data, such as altitudes and the highest and lowest temperatures of the day, were often included in his daily entries. Sometimes, Xu also put down a code consisting of numbers and letters. One of the earliest such code reads "3,988; N. 76, 30 W."[67] In order for readers to decipher it, Xu explained that the first number indicated the distance the caravan had progressed that day, and the rest pointed to the exact direction in which they traveled. "All based on Mr. Hedin's calculations," Xu added.[68]

Xu also described some of the experiments carried out by Hedin and others in his travel journal. To investigate upper air currents, the expedition conducted balloon observations between June 1927 and October 1929. Xu watched the initial series of balloon releases with great interest. He even copied in his diary what was written on the note attached to each of the released balloons. It asked the finders of the balloons to record "the date, location, nearby population density and distance to main roads of the find" and to return them to Qinghua University in Beijing in order to receive a reward.[69] Even though Xu was a layman where natural science is concerned, he still could discern the cutting-edge methodology used in this large-scale atmospheric experiment and probably hoped his descriptions would be elucidating to Chinese meteorologists and the general public.

After arriving at the Etsin-gol River in late September 1927, Hedin decided to map the river and the two connected salt lakes—the Socho-nor and the Gachun-nor—from the water rather than from the land. Their Mongolian helpers built a few boats, and after a few test launches, Hedin and his team settled on a raft-like vessel built by joining two hollowed-out tree trunks together with wood planks.[70] Hedin and Haslund—the Danish geologist in the group—took the raft to the river, while Xu and the rest of the team were divided into two columns to follow them on land. As Xu described in the travelogue, Hedin and Haslund undertook grave risks. The raft was crudely made, equipped only with some makeshift paddles and a small surface for Hedin's mapping gear. While some parts of the water were so shallow that the boat often ran aground, other parts were dangerously deep. The weather was also turning wintry in October in the Gobi Desert, and the water was cold enough to be dangerous were the boat to become submerged far from the shore. Yet Hedin persisted. Xu noted that the sexagenarian Hedin was more athletic than he was himself, despite being only in his forties. Perhaps ashamed of his own lack of physical strength, Xu exclaimed that after witnessing Hedin's herculean navigation, he was convinced that China needed a "Spartan-style" school to cultivate athleticism and perseverance among Chinese youth, as these qualities were crucial in the pursuit of science.[71]

As the joint exploration's field director for the Chinese contingent, Xu was keen to report the discoveries made by Chinese researchers. For example, after hearing about geologist Ding Daoheng's discovery of a "massive iron ore deposit" at Bayan Obo (which means "mount of riches" in Mongolian), Xu excitedly predicted that the mine would surely become "one of China's major sources of wealth."[72] At the northern foot of the Tianshan mountains in Xinjiang, Yuan Fuli found dinosaur fossils from the Jurassic period, which Xu emphasized as the first such discovery in Asia.[73] Whereas Xu had little expertise in geology and paleoanthropology and made few comments on Ding and Yuan's breakthroughs, he was invested in the archeological sites along the Silk Road and was particularly vocal about the looting and damage that had been done by Western explorers. At Khara-Khoto ("black city"; *heicheng* in Chinese), a ruined frontier town of the Tanguts from the twelfth and thirteenth centuries, Xu noticed missing statues and the traces of earlier excavations, which, he suggested, had probably been carried out by Western adventurers such as "Kozlov, Stein, or Warner."[74] Similarly, at the Bezeklik Caves, a Buddhist site near Turfan, Xu mentioned all the missing murals and sculptures in the grottoes, which he attributed to German explorer Albert von Le Coq, who had led several German expeditions in Xinjiang in the early twentieth century.[75] In fact, even before entering Xinjiang, Xu had written about Le Coq in his travelogue. At Etsin-gol in Inner Mongolia, Xu borrowed a copy of *Auf Hellas Shuren in Osttukistan* (*On the Trail of Hellas in Eastern Turkestan*)—Le Coq's account of German expeditions to Turfan—from Hedin to prepare for their exploration of Xinjiang. Learning about German activities in Turfan and the hundreds of cases of treasures they shipped back to Berlin had filled Xu with indignation. This is clear from his travel journal: "I am not a Nationalist (*guojia zhuyi zhe*) and have always advocated that science—knowledge—is the public property of mankind. But the fact that outsiders could grab any old thing from our own home at will was like an old tree with all its branches and leaves stripped away. Even if the old trunk is not dead, it is miserable and colorless. How could you not feel angry and heartbroken in the face of such wretchedness!"[76] Hedin argued that the German seizure of the Silk Road artifacts near Turfan was actually "fortunate," because shortly after the departure of the Germans, the region was hit by an earthquake, and these materials would have been destroyed if they had not been removed. Xu, however, was not convinced, pointing out that an earthquake was an accidental event, and did not ameliorate the foreigners' original intent.[77]

Agonized by these open wounds along the Silk Road and wary of foreigners' insatiable appetite for "Chinese" treasures, Xu Xusheng felt protective of the archaeological findings in the Northwest. Huang Wenbi, an archaeologist from Beijing University, heard about the relics of an ancient city near Olon Sume from locals and set out to look for it. What he found was the ruins of a rectangular walled city. At the site, Huang saw the remains of the foundations of houses and temples, and also discovered a Chinese tablet and a Mongolian tablet, from which he made rubbings before leaving.[78] Afterward, Xu discussed these findings with Huang enthusiastically. On the basis of their knowledge of the official dynastic histories from China's imperial era, which Xu Xusheng had carried with him to the Northwest, both men suspected that this ruined city was the historical town of Jingzhou from the Jin (1115–1234) and Yuan dynasties (1271–1368).[79] However, Xu objected to Huang's idea of relocating the tablets from the ruins to the nearby Bat Khaalga monastery (known as *Bailingmiao* in Chinese) and entrusting them to the Mongol aristocrats there. Afraid that the Mongols would sell the tablets to foreigners, Xu insisted on writing to their colleagues at Beijing University to request that they send a team to retrieve the tablets.[80] What Huang and Xu did not know was that two months before Huang's "discovery," Henning Haslund had already encountered this ghost town. The American scholar Owen Lattimore then reached the same place in 1932 under the guidance of a Mongol guide named Arash, who had worked for Huang and Haslund in 1927.[81] Unlike Haslund and other Western explorers, Huang Wenbi did not pay attention to the Nestorian dimension of the site. Instead, like Xu Xusheng, Huang was preoccupied with locating lost frontier towns established by central regimes that were mentioned in Chinese dynastic histories.[82] This reflected the academic trend of national studies, which emphasized the application of archaeology in corroborating Chinese historical texts, especially the official dynastic histories. It also met the political need to include the Northwest within the geohistorical realm of the Chinese nation-state.

This Chinese endeavor centering on how far the Middle Kingdom extended westward also mirrored the pursuits taken up by some Western explorers in Central Asia, which sought to discover how far east the influences of Western civilization—especially those of the Greco-Roman culture and Christianity—had traveled.[83] Albeit with opposite orientations, both of these quests embodied the kind of worlding projects inspired by the Silk Road, in which the logic of world-making was determined by where the perceived center lay. As Tamara Chin has pointed out, "the Silk Road offers . . . a condition or strategy for geopolitical thought and action."[84] These worlding projects had implications beyond historical narratives. Just as articulating the West's historical links to Central Asia reinforced the West's modern-day imperialist ambitions in these distant lands, underlining China's long-lasting influence in the Northwest provided much-needed legitimacy for a Chinese nation-state that encompassed the contested former Qing frontiers.

In the 1920s and 1930s, when Chinese intellectual travelers began to familiarize readers with the geography and history along the Silk Road, their travel writings also unveiled the present-day conditions of the region, especially the complex ethnic composition and the volatile political reality. At the time, anthropometrics had become the standard practice of Western explorations of non-Western societies. Within the joint mission, David Hummel, the physician of the group, for example, made 60 anthropometric measurements and blood tests in Inner Mongolia, while Henning Haslund conducted 170 anthropometric measurements of "East Turks, Lopliks, Dede Mongols, Torguts, and Tibetans" in Xinjiang in 1928.[85] Seeking to correlate physical and biological traits with racial identity, Western members of the joint team often highlighted the non-Chinese features among the local pastoral peoples in their more casual travel writings. Haslund, for example, described the native Uyghurs in Xinjiang as people "with Aryan facial traits" whose women were "very beautiful" and "could pass for southern Europeans."[86] Hedin once depicted one of the Mongol camel handlers as "a magnificent type resembling an American Indian."[87] Rather than highlighting physical appearance, Xu Xusheng's travel journal painted a subtle image of a linguistically, culturally, religiously, and politically diverse society in the Northwest. In Inner Mongolia, Xu recorded how their caravan encountered varied native peoples and other fellow travelers on daily basis. From Mongolian tax collectors sent by Mongol princes to Han Chinese farmers and shopkeepers relocated in the late Qing era, and from Tibetan lamas in local monasteries to multiethnic merchant caravans engaging in long-distance trade between Outer Mongolia and North China, the various groups he met in desert oases and yurt communities evinced a northwestern frontier that was far from empty.[88]

Although Xu spent the majority of his time in Xinjiang communicating and negotiating with the Han Chinese governor, Yang Zengxin, and had very little free time left in the field, he was still constantly reminded of the complicated ethnic composition of the Xinjiang society. For example, Xu learned that Burhan (*Bao'erhan*), the interpreter and main liaison sent by Yang, was a Tatar expatriate originally from Russia.[89] Aside from the Tatars (*dada*), Xu also recorded the "Turban heads" (*chantou*, a Chinese term of the day for Uyghurs), Kazaks (*hasake*), the Torgut Mongols (*tu'erhute*), the Hui (Chinese-speaking Muslims), and other non-Han groups in Xinjiang. These diverse peoples also molded their identities in the crucible of interethnic alliances, conquests, displacements, and migrations. Once Xu Xusheng met a fourteen-year-old coach driver who told him that he was also originally from abroad. Confused by the teenager's muddled description of his origin, Xu later figured out that the boy was descended from the Hui Chinese who had fled to Russian Central Asia from Shaanxi and Gansu during the Muslim rebellion in the 1870s. Just like the Torgut Mongols who returned to Inner Mongolia during the High Qing after decades of exile in Russia, these Hui Chinese—also known as the Dungans among Russians and Uyghurs—resettled again in Xinjiang after the Russian revolution. These complicated histories prompted Xu to comment that the ethnic differences in Xinjiang were extremely hard to untangle.[90]

As Xu's travel journal demonstrated, across the steppes and deserts in the Northwest, multiethnic peoples formed communities and maintained cultural, ethnic, economic, and political norms that were strikingly different from those in the Han heartland. This anomaly was shaped by the legacy of the Qing. One of the noticeable differences of the frontier society came from the unusual configurations of its ruling elites. Instead of being governed by civil servants appointed by the provincial and central governments, on the frontier, military strongmen, non-Han aristocrats who held hereditary titles, and religious leaders of Tibetan Buddhism and Islamic faiths exerted power. This tactical system of control on the non-Han periphery was not easy to grasp, even for highly educated Han Chinese travelers.[91] Despite his matter-of-fact presentation, Xu Xusheng's travel notes reflected a certain level of puzzlement over the nomenclature of the titles for frontier elites. When Huang Wenbi hoped to entrust the tablets he had found at Olon Sume to the *beile wang* presiding over the Etsin-gol area, Xu explained that this Mongolian nobleman, commonly known as the *beile wang* by locals, was in fact a *zhasake* who had been "demoted from the junwang rank to the beile rank."[92] For Han Chinese readers in the early Republican era, the terms *junwang* and *beile* might sound familiar, as they were top-ranking titles used largely within the Manchu imperial clan. The term *zhasake*, on the other hand, was not widely known. It is a Chinese transliteration of the Mongol term *jasagh*, which was used by the Manchus for the leaders of Inner Mongol banners.

Similarly, when the joint mission entered Hami in Xinjiang, Xu Xusheng went to pay his respects to the *huiwang* of Hami. Literally meaning the "King of Hui," the term *huiwang* here referred to the Muslim prince Shah Maqsut, whose family was among the few native Muslim Khanates who had received hereditary fiefdoms granted by the Manchu court for their support of the initial Qing conquest in the 1750s.[93] "The huiwang is seventy-one years old, grey-haired but in good health," Xu wrote after their first meeting, continuing that "he dresses in Chinese style clothes and speaks fluent Chinese. If one didn't know the truth, it would probably be hard to tell that he's of another ethnicity."[94] Despite being impressed by his high level of Sinicization, Xu was not thrilled by Shah Maqsut's praise of Yuan Shikai, the first president of the republic, who had been denounced by the New Culturalist intellectuals for his cultural conservatism, autocratic rule, and attempt to restore monarchy. As the Khanate prince reminisced about the good old "peaceful" days under Yuan and lamented the current chaos caused by the civil war between the northern warlords and Nationalists, Xu expressed his mild annoyance in his travel journal, reminding readers that these Muslim Khanates—along with other non-Han aristocrats with hereditary fiefdoms on the northwestern frontiers—were anachronistic relics from the Qing era. Out of their own self-interest, autocratic warlords such as Yuan Shikai had preserved this outdated system even after the fall of the Qing.[95]

If the antiquated hereditary fiefdoms clung to by non-Han elites signaled a challenge to China's nation formation, competition among Han Chinese warlords further complicated the volatile political reality. In the late 1920s, Chinese intellectual travelers located the unrest in the Northwest against the backdrop of the political chaos in the rest of the republic. As the joint team traversed politically unstable regions in Inner Mongolia and Xinjiang in the midst of the Nationalist military campaign against northern warlords, their caravans encountered suspicion and interrogation from various local authorities. As mentioned above, the caravan was a multinational team traveling with more than two hundred camels. The mission also carried some sizable meteorological equipment—some of which resembled cannons—as well as real weapons for self-defense. Perhaps because of the size of the group or the noticeable Chinese presence on the team, different warlords suspected that they had been dispatched by neighboring competing warlords on a hostile military mission. Nor did the suspiciously shaped pieces of research equipment and firearms help quell the speculations. The "Christian General" Feng Yuxiang in Gansu, for example, thought the team had been sent by warlord Zhang Zuolin, who had controlled the Beijing government in 1927 after ousting Feng from the capital. Feng jailed two Chinese members of the mission when they were sent to Lanzhou for supplies. It was not until Cai Yuanpei, who was informed by Xu Xusheng of the incident, intervened and asked the Nationalist government to explain the purpose of the joint mission to Feng that they were released. Similarly, when Yang Zengxin, the governor of Xinjiang, who had succeeded the last Qing governor in 1911, learned about the joint mission heading from Beijing to Xinjiang, he conjectured that Feng Yuxiang had sent the group to attack him.[96] More speculation emerged in Xinjiang when a censor submitted to Yang a letter addressed to a Chinese student participating in the mission. The author of the letter, a friend of the student in Beijing, joked about the camels under their command, writing, "I congratulate you on having two hundred soldiers in your caravan." A simple joke was misconstrued as indicating a suspicious troop of two hundred people, and caused Yang to send thousands of his soldiers to the border. Once the joint team had arrived at Hami, Yang's subordinates closely examined their equipment and luggage before allowing them to move forward to Dihua (Urumqi).[97]

Compared with this vigilant attitude toward a scientific mission from Beijing, the frontier authorities seemed to be rather nonchalant about foreign explorers entering their jurisdiction. After arriving in Dihua, Xu Xusheng heard from Hedin that two German explorers were at Kucha, a Silk Road site along the northern edge of the Taklamakan Desert, and planned to go on to the Lop Nur.[98] Enraged by the news, Xu blamed the "inconsistent government decrees" in the Northwest, which allowed foreign explorers to "sneak across the Chinese border at will" but obstructed "legitimate investigations of scientific materials" such as theirs.[99] The phrase "inconsistent government decrees" here was perhaps Xu's polite way to describe the many different layers of administrative and military units with which he and his fellow travelers had to negotiate during their journey. With varying degrees of independence from the national government, as Xu's travel narrative suggested, these multilayered units underlined the patchwork nature of the frontier ethnopolitical reality, which resulted in a porous borderland vulnerable to foreign penetration and threating to the integrity of the national space.

While Han Chinese travelers like Xu Xusheng largely identified with the Nationalists' antiwarlord campaign, they were not blind to the value of upholding the Qing frontier legacy. For the warlords and Han officials in the Northwest, existing practices provided continuity, allowing them to rule through different native middlemen. Honoring the Qing promises regarding the privileges of indigenous rule also prevented the local non-Han elites from seeking outside support and from fomenting their own national independence movement. When Hedin relayed the rumor that a large number of the Torgut Mongols at Etsin-gol were "sympathetic towards the Mongolian Republic and might declare allegiance to it at any moment," Xu Xusheng disputed the likelihood of that by pointing out that the local "Torgut prince"—whose tribe had gained their fiefdom since the Kangxi reign of the Qing dynasty—would lose his title and position if he were to let that happen.[100] Inheriting the Qing frontier legacy also lent legitimacy to the Chinese republic's territorial claim over the Northwest and other frontier regions. For Han Chinese travelers, the key was to rearticulate and ameliorate the relationship between the Han Chinese and their ethnocultural other. Differentiating themselves from the warlords who were, in their view, simply taking advantage of the indigenous peoples during the chaotic postimperial transition, Chinese intellectual travelers envisioned themselves to be scientific and nationalistic agents who could inform the central state, academia, and a broad learned public about the frontier society through their travels and travel writing. Exactly who got to depict the frontier society and how it should be represented, however, were not always clear in Sino-foreign joint missions.

Within the Sino-Swedish scientific mission, Paul Lieberenz, a German photographer and cinematographer, was enthusiastic about filming the expedition as well as the local peoples. For example, he went to the Festival of Maidar at the Bat Khaalga monastery and shot "1,100 metres of film," documenting local Mongolians carrying out a multiday celebration with rituals, music, and dances.[101] The Chinese team members were not always pleased with Lieberenz's filming. When Xu Xusheng learned that the film Lieberenz was making was his own personal production, he was disturbed. Aside from not being thrilled that 20 percent of the profit generated from the film screening would go to Lieberenz, Xu was also alarmed that the film would need to be sent to Berlin to be developed, which meant that it could be shown to foreign audiences without proper censorship from Beijing. After lengthy discussions with Hedin—who in turn talked to Lieberenz—Xu Xusheng managed to amend the joint mission's agreement to ensure that Beijing had final say over the content and circulation of any photographs or film.[102] Xu's insistence on the Beijing committee's right to censor visual content showed his strong belief that Chinese should have control over representations of the Northwest and, by extension, of China.

What should be represented was not always clear-cut, even among the Chinese members of the team. Early in the expedition, the international caravan was crossing a prairie near Baotou when an itinerant opera troupe showed up at their camp. The foreigners invited them to perform for Lieberenz's camera, and the performance proved too vulgar for the Chinese intellectuals' taste. According to Xu Xusheng's travel journal, the stories were all about "flirting men and women," and he found the performance offensive.[103] Enraged by the off-color jokes, Huang Wenbi objected to the filming. Protesting to Xu, Huang argued that such "folk" performances were so indecent that the "Southern government" (by which he meant the Nationalist government) had already banned them in order to "reform social customs."[104] To Huang's surprise, Xu had little objection to filming and argued that although this kind of performance should be banned in the future, because of its huge regional popularity folklorists would want to study it and therefore it should be filmed.[105] For Huang, such footage would defame China and present the backward side of the country to the rest of the world. However, for Xu Xusheng, vulgar or not, the performance was an intrinsic part of local society. Collecting and studying every element of the social life and culture of the borderland was precisely the reason they had set out on the expedition in the first place.

Furthermore, Xu Xusheng did not think there was any difference between the vulgar and the refined in terms of their scientific value. He viewed the local opera the same way that Huang, an archaeologist, would think of an artifact excavated from underground: as an object of study. However, Xu's proposal to ban this kind of local opera after letting folklorists study it also exemplified an intriguing interrelation between the administrative apparatus and social scientific study in the Republican era. As Tong Lam has argued in his study of the social survey movement in the 1920s and 1930s, the discourses of science and reason permeating professional social surveys and amateur writings about China's social world allowed such research to become a novel political technology that contributed to China's nation formation.[106] For Xu, the backwardness of the northwestern society did not lie in the vulgar performance itself but rather in its influence as indicated by its huge local following. Without proper research, Xu seemed to believe, the backward social mentality behind its popularity would not be eradicated even after the performance was banned. This logic of thorough scientific research leading to effective control illuminates the perceived link between culture making and nation building. Just like social surveyors, travel writers like Xu hoped to bring forth interventions to eliminate anomalies and impose sameness across the national space.

Besides being categorized as an anomaly, non-Han indigenous peoples were often consigned to the past in travel writings by Western explorers and Han Chinese travelers. This practice of the "denial of coevalness," to use Johannes Fabian's well-known term, was often exemplified in the depictions of "superstitious" religious practices.[107] Once in Inner Mongolia, Xu Xusheng accompanied Huang Wenbi and a few Chinese students and Mongol helpers to excavate a few old tombs. Soon after their arrival, the local Mongol chief sent someone with his calling card to invite the Chinese to his nearby yurt. Sensing the possibility of an objection, Huang Wenbi went to the meeting with a Mongol assistant. The Mongol chief begged them to stop digging in his land. He told Huang Wenbi that just a few days prior, a few foreigners took away some rocks from their *obo*—the stone heaps piled on the top of mountains that serve as altars in local Buddhist practice—which had already caused the death of three sheep and made one person ill.[108] Although feeling guilty about disturbing their peaceful lives, Xu Xusheng pitied how superstitious the Mongols were.[109]

If Xu was only mildly disturbed by the Mongols' plea based on their religious beliefs, he dished out harsh criticism of Islam in Xinjiang when they visited the Bezeklik Caves near Turfan. At this Silk Road site, Xu noticed many empty niches where statues of the Buddha once stood. On the basis of the damage he found, he concluded that it was the result of deliberate destruction. He blamed the great loss of historical artifacts on imperialist archaeologists for their theft, on the Chinese for failing to protect their objects of scientific interest, and on the Muslims for their anti-idol fanaticism.[110] On the surface, Xu seemed to have identified three equal culprits and implied their simultaneity. However, there are subtle differences in these groups' attitudes toward the Buddhist statues. Both the greedy foreign plunderers and the failed Chinese protectionists probably shared a similar modern view that the Buddhist artifacts were valuable objects for scientific study. In contrast, with their view of the Buddhist statues as idols, Muslims were depicted as religious zealots whose aggressive actions seemed to emanate from a bygone era.

This deliberate denial of coevalness is even more obvious when we consider how much the joint mission had relied on the local Mongols, Uyghurs, and other indigenous peoples for service and information. It was evident that many "lost" sites were hardly lost to the native peoples, without whose guidance foreign explorers and Chinese scholars could not make their "discoveries." However, even as Han Chinese travelers acknowledged that the inclusion of indigenous non-Han peoples within the Chinese nation-state was essential to the country's postimperial transition, the agency of these non-Han peoples was either omitted or downplayed in travel narratives. When discussing the future of Xinjiang with Hedin, for example, Xu Xusheng agreed that education should play a crucial role. They both thought that it would be beneficial to the indigenous population to learn about local weather, geography, vegetation, fauna, and ethnic formation. Xu argued that learning these modern disciplines would help solve many practical problems in Xinjiang, such as road construction, forest cultivation, taming and utilizing rivers, and ore extraction. The greatest issue in the administration of Xinjiang, Xu emphasized, was "racial difference" (*renzhong fenqi*), which made it impossible for the Muslims to be assimilated by the Han.[111]

Xu criticized the existing approach to assimilating Muslims in Xinjiang, which forced them to give up Islam and to study Confucianism instead. He argued first of all that Confucianism was not a religion, so one could not expect it to replace the religious beliefs of the local population. Second, this approach was ignorant and disrespectful toward the history, religion, and culture of the Muslim people. Xu warned that simply using "old morals" (Confucianism) and state power to consolidate authority over the Muslims and other minorities was bound to fail. "The Russians could not succeed in Poland. Nor could the English succeed in Ireland," Xu wrote. "China is even weaker than those two countries; how can we expect [a similar tactic] to be fruitful in Xinjiang?"[112] He then concluded that Hedin's suggestion to build a modern school in Xinjiang was a practical approach. Teaching Muslims the "precise science of modern times" and the "cultures of Arabia and other ancient nations," Xu concluded that Han Chinese would be able to tell the local Muslim population,

> You have your own religion and your own history. We recognize all that, and we do not want to interfere. However, your innate culture is an ancient relic, and not a precise form of knowledge in modern times. It is fine if you do not want to live in the present era. But if you do, you cannot ignore modern science. The school we set up is a place to provide you with modern science, which will give you methods to use in real life. It does not conflict with your religion. You can pursue your studies with ease. While it is good of you to maintain respect for and pride in the Arabian culture, do you really understand it? It is fine if you do not want to know. But if you do, there is no other way than to come to our school to study.[113]

Considering the weapon of "Confucianism" to be just as archaic as "Islam," Xu Xusheng held "modern science" as an olive branch as well as a conquering device to captivate the indigenous non-Han population. It was clear to him that rather than assimilating to Han cultural values, the Muslims and other local non-Han peoples would be won over by preaching modern science, a line of reasoning inspired by the actions of foreign explorers. While pointing to unsuccessful cases in Poland and Ireland, Xu conceived an idea altogether not that different from the Western and Japanese colonizers' approaches in their colonies: building authority in the name of science and modernity.

As Tani Barlow has asserted in her analysis of China's colonial modernity, modernity is intrinsically colonial since one cannot possibly exist without the other in the nineteenth and twentieth centuries.[114] To reimagine the Qing frontiers under the rubric of a Chinese nation-state, Xu and his fellow travelers portrayed the Northwest as a distant, empty land with savage indigenous peoples that should be better integrated through urban civil power (i.e., science and industrial capitalism). This settler-colonist-style argument resembled US tropes about "empty lands" and "Indian savages" in their western frontiers. Just as that trope had turned the colonization of indigenous land by white settlers into a benign westward movement and American exceptionalism, the push to use science to modernize the northwestern frontier in China helped repaint the history of Qing continental expansion as an inspirational narrative of strengthening the nation and resisting Western and Japanese imperialism.[115] Yet simultaneously, Xu and his colleagues also presented the Northwest as a familiar and indispensable part of the national space, given its long history of affiliation with various Chinese imperial states. By presenting the Northwest as always a part of China throughout history, Han intellectuals reformulated Chinese national history to incorporate the non-Han peoples. The travels of Han Chinese to the Northwest during the Republican era occupied a crucial position in China's transition from empire to nation-state. This would inspire the mainstream narrative of the Northwest in journalistic and tourist narratives in the 1930s.

Popularizing the Westward Movement

Encapsulated by Chen Wanli, Xu Xusheng, and other Beijing academics' participation in Sino-foreign joint explorations of the Northwest, this intellectual "westward movement" in the 1920s was embedded in China's postimperial transition, the multilateral competition for influence and profits in Central Asia, and the transnational networks of knowledge production of Eurasia. It was also during the early Republican period that Chinese writings about "the Northwest" and "the frontier" (*bianjiang*) moved beyond the text-based evidential research valorized in Qing statecraft writings to a more scientific and journalistic style with an emphasis on fieldwork (*shidi kaocha*) and scientific facts.[116] The Sino-Swedish joint mission played a significant role in this transition. After its initial negotiation, the mission captured public attention. Upon their departure, print media, especially those with significant visual components, reported on their journey with great enthusiasm.[117] Later, Chinese members of the joint team gave public lectures about their travel experiences, popularizing the notion of scientific travel in the Northwest among urban audiences.[118] Like Xu Xusheng, Chinese scientists who had participated in other Sino-foreign joint expeditions to the Northwest also published their travel journals about the Northwest in the early 1930s.[119] At the epicenter of this movement, university faculty and students and other members of the learned public in and beyond Beijing became attuned to associating scientific travel with the Northwest.

Continuing as an intellectual endeavor throughout the 1920s, this westward movement exploded into a popular national campaign in the 1930s. After seizing power, the Nationalists soon faced the blunt aggression of Japanese imperialism and China suffered a string of military attacks from Japanese forces. Among them, the loss of Manchuria in 1931 not only dealt a bitter blow to the Nationalist government but also injected into the public consciousness an unprecedented sense of national crisis. Taking refuge in Luoyang in early 1932, the Nationalist central government looked further west for areas of possible retreat and selected Xi'an, a city in the northwestern province of Shaanxi, as an "alternate capital." Soon a Planning Committee for the Western Capital (*xijing choubei weiyuanhui*) was formed to develop Xi'an and the surrounding region into a strategic base from which to resist Japan and assert firmer control over frontier provinces.[120] Espoused by right-wing Nationalist Party politicians and thinkers as an opportunity to transform the "empty" frontier into what Jeremy Tai underscores as "a vital place in restructuring China's political economy,"[121] this broadening of the westward movement also coincided with the Chinese Communists' retreat to northern Shaanxi, which thrust the movement to the top of the priority list for the Nationalist government.

Throughout the rest of the Nanjing Decade, the number of government initiatives, research societies, and publications dedicated to the Northwest mushroomed.[122] "Head to the Northwest" (*dao Xibei qu*)—a slogan popularized by political speeches, popular print media, and even a movie—became a mantra reverberating in the zeitgeist of the 1930s.[123] From top politicians to intellectuals to average urbanites, becoming an explorer-tourist to the Northwest became a key patriotic act. Shen Songqiao and Zhihong Chen have identified a broad spectrum of travel narratives of the Northwest that emerged in the 1930s and pinpointed their role in constructing the imagination of China's national space and refashioning Chinese identity.[124] According to their studies, the most active creators of travel narratives of the Northwest in the 1930s were Nationalist government officials and journalists from news agencies all across the political spectrum. Whereas the former group of travelers emphasized the pivotal place of the Northwest in China's territorial integrity and its deep connection with China proper, the latter faction, like the *Liangyou* journalists in the previous chapter, brought more critical scrutiny to the local society. Dispatched by *Dagong bao*, *Shibao*, *Shenbao*, and other successful commercial newspapers, renowned journalists like Fan Changjiang, Chen Gengya, and Gu Zhizhong journeyed from coastal cities like Shanghai and Tianjin to Shaanxi, Qinghai, Gansu, and other northwestern provinces. Like the *Liangyou* travel columns about China's interior, their travel reports about the Northwest called attention to harsh realities: an underdeveloped transportation infrastructure and rampant corruption among local politicians, as well as the persistent famine and other disasters that plagued the rural hinterland.[125] Despite their different messages, both the political and the journalistic scrutiny boosted public interest in the Northwest. Government agencies in cities like Xi'an made efforts to improve transportation, beautify the urban environment, build tourism facilities, and publish guidebooks publicizing scenic spots and tourist sites in their cities and the surrounding areas.[126] They also solicited assistance from tourism businesses like CTS, which not only opened new branches and built lodging facilities in the Northwest but also issued a guidebook to Shaanxi in 1935.[127]

While the remote Buddhist cave temples in Dunhuang and the exotic scenes of non-Han peoples roaming the deserts in Xinjiang and Inner Mongolia fascinated urban readers, travel accounts by Nationalist officials, journalists, and intellectuals also highlighted historic relics and natural scenery in the Northwest that had become more accessible via modern transportation. A politically charged popular movement, the Head to the Northwest campaign also created commercial opportunities for transportation institutions and the tourism industry. Catering to this skyrocketing public attention to the Northwest, the Ping-Sui and Longhai Railways—the two major trunk lines connecting China's eastern seaboard to western China—expanded their services and designed new tours to attract tourists.

Curating the Northwest

Connecting Beijing to Baotou in Inner Mongolia, the Ping-Sui Railway was a major achievement in China's modern railway development. Its symbolic significance lies in the fact that the first section of the railway—between Beijing and Zhangjiakou (Kalgan)—was the first railway designed, built, and managed by Chinese. The only steam-powered transportation line linking China's east coast to the Northwest until 1932, the Ping-Sui Railway played a crucial role in the economic development of the Northwest. However, because of the volatile political climate and the underdeveloped tourism infrastructure in the Northwest, unlike other major railway lines in coastal China, the Ping-Sui Railway did not attract many nonbusiness travelers in the 1920s. That changed dramatically in the 1930s, when academic travelers from Beijing began to explore the Northwest using the Ping-Sui line. Responding to the increasing demand, the Ping-Sui Railway started a weekly express service between Beijing and Baotou in 1932; by 1934, it was running every day.[128] To promote sales in this new climate, the railway management also began to issue "tourism round-trip tickets" (*laihui youlan piao*).[129] A common feature on major railways in coastal China, the tourism round-trip tickets gave passengers a discount and allowed them to travel within a certain period of time instead of in a fixed timeslot. This feature met tourists' needs for a flexible schedule and lower cost.

Attention to scenic spots and historical sites along the Ping-Sui Railway also grew in the 1930s. In a 1932 article appearing in the *China Weekly Review*—an English newspaper in Shanghai—Dickson Tong, an author based at Yenching University, argued that tourism developers should stop neglecting the Northwest. Along the Ping-Sui Railway, Tong emphasized, one could find many places of scenic beauty and historical and cultural interest, including "the outer part of the Great Wall, one of the four wonders of the world" and "the grave of Wang Chao-chun, a favorite concubine of Emperor Yuan Ti of the Han Dynasty." Among all the tourist sites one could visit by train, Tong was most excited about the Yungang Grottoes, a Buddhist site near Datong in Shanxi Province. "The Stone Buddha Temple, with thousands of buddhas," he wrote, was "some sixty or more feet high, carved on the rocks of a mountain stretching two *li* long and built in the Wei Dynasty."[130] Tong was not alone in recognizing the appeal of the Yungang Grottoes. Aside from the Beijing-Suiyuan and the Beijing-Baotou round-trip tickets, the Ping-Sui Railway designed a weekend- and holiday-only package tour to the Yungang Grottoes. Leaving Beijing on Saturday and returning on Monday on express trains, the tour participants would have an entire day to explore the famous cave temples and visit the historical capital of the Northern Wei dynasty (460–524).[131]

On top of these discounts and offers, the railway also issued group tickets for school students on field trips to the Northwest. With a minimum of thirty people, a group could purchase discounted tickets to five different stops on the Ping-Sui route.[132] These promotions were the railway's response to the rapid growth in academic and adventure tourism to the Northwest. In an essay about his travel experience along the Ping-Sui Railway in 1934, the writer Xiao Qian—then a student at Yenching University—used the voice of an old local resident to describe the influx of students to the Northwest every summer:

> Every year, droves of students clad in Western clothes come to investigate [the Northwest], inquiring about how many "worn-out shoes" [referring to prostitutes] or opium pipes we have, so you can use the figures to impress your teachers or readers. . . . Those who have good taste go to see historical sites. You visit the Zhaojun tomb, just a desolate mound, to ponder the past, or go kowtow at the Temple of Yu. Then, you take the same train back to Beiping and tell your fellow countrymen with your chest out: "I have been to the Northwest and have first-hand material!"[133]

Xiao's satirical depiction exposed the hypocrisy behind urban youths' social surveys in the Northwest. Instead of effecting real change, many simply went to the Northwest to participate in a popular trend. However, Xiao's travel essay, as well as other similar travel writings of the time, sheds light on one of the effects of this movement: an increasing familiarization with the Northwest among coastal populations. Curated by railway companies and the tourism industry, travels to the Northwest became increasingly standardized with common itineraries, which helped make the frontier region accessible to tourists from the core regions of China.

This kind of curation could take many different forms. The Ping-Sui Railway itself organized a group tour in 1934 to publicize its tourism appeal. Hoping to draw on the momentum of the Head to the Northwest movement, Shen Chang—the MIT- and Cornell-educated director of the railway—decided to invite a few prominent Beijing academics to travel to the Northwest. Even though the region still lacked basic tourism services, the railway administration had special connections with military authorities in the Northwest, which allowed them to provide all the necessary travel services for their guests, from lodging and meals to local transportation and guides.[134] In return, the participants would publish their travel writings as promotional materials for the railway. Shen Chang knew exactly whom he should invite: Bing Xin (also known as Xie Bingxin), the renowned writer and professor, and, incidentally, his sister's classmate at Wellesley College. After earning a master's degree in English literature at Wellesley in 1926, Bing Xin returned to Beijing to teach at her alma mater, Yenching University, where her husband, Wu Wenzao, a Columbia-trained socioanthropologist, also held a position. Through Bing Xin and Wu Wenzao, seven other Yenching professors joined this Ping-Sui Railway Travel Group (*Pingsui yanxian lüxingtuan*) in the summer of 1934.[135]

After the conclusion of the group tour, the Ping-Sui Railway produced ten different publications, including two bilingual pamphlets on the Great Wall and the Yungang Grottoes.[136] Within these publications, the railway also singled out five titles as "reading materials for traveling along the Ping-Sui Railway" (*Pingsui tielu yanxian lüxing duwu*) for tourists and ordinary readers interested in the Northwest. To highlight the intellectual appeal to academic tourists, the Ping-Sui Railway's travel reading materials included three short social-scientific studies, ranging from an anthropological description of Mongolian yurts to a historical account of a Han land merchant's struggle in Inner Mongolia to a social survey of the Catholic missions along the Ping-Sui Railway.[137] Different from guidebooks or travelogues, they offered social reportage and popular folk knowledge, and also conveyed the political implications of the Head to the Northwest campaign. For example, on the basis of what we might today call oral history, historian Gu Jiegang published a study on Wang Tongchun, a Han militiaman and land merchant who was successful in "reclaiming" agricultural lands in the region of the Great Bend of the Yellow River in the late Qing period. By recovering the story of a folk legend in Inner Mongolia, Gu's account set Wang Tongchun as an exemplary figure—a "national hero" (*minzu yingxiong*)—from an earlier episode of Han Chinese development of the Northwest and called for his compatriots to imitate Wang in the Head to the Northwest movement.[138]

Indeed, travel to the Northwest, even in its most leisurely form, conveyed the excitement of scientific discovery and the urgency of nation-building—two sides closely intertwined as articulated by the Chinese members in the Sino-Swedish joint scientific mission. Echoing Gu Jiegang's message, Bing Xin emphasized travel as a crucial way to incorporate the frontier region into the nation-state and stressed the link between resisting imperialism and going to the Northwest:

> Since the fall of the Northeast, the whole country has suddenly become aware of the significance of frontier defense. There is a public clamor to develop the Northwest. But where on earth is the Northwest? What is the frontier condition of China's Northwest? People feel completely in the dark on these questions. And with the loss of the Northeast, animal husbandry and land cultivation in the Northwest have become a source of wealth for the whole country. But we lack any understanding of the land, products, commerce, and other conditions of the Northwest. The Ping-Sui line is the gateway for people to go to the Northwest and for the products in the Northwest to get out. It is a route every compatriot should travel and investigate.[139]

The purpose of their 1934 tour, Bing Xin emphasized, was "to pay attention to the scenery, historic sites, elegant architecture, customs, and religion as well as the local economy and products along the railway line."[140] Here, tourism became an organic part of the 1930s campaign to develop the Northwest. Just like the social studies, the popular travel writings penned by Bing Xin and other Beijing academics enticed intellectuals and average readers to partake in the broad political movement through travel.[141]

The travel writings published by the Ping-Sui Railway Travel Group also successfully standardized the tourism itineraries along the railway line. The Yenching professors focused on three main towns during their tour: Datong, Suiyuan, and Bailingmiao, which became the main destinations for other coastal travelers to the Northwest in the 1930s. Datong was the ancient capital of the Northern Wei dynasty and the coal capital of modern China. The tourism appeal of the town lay in its historical sites and modern coal industry. Its most famous site, the Yungang Grottoes in Datong's western suburb, were at the center of the Ping-Sui Railway's tourism promotion campaign. First excavated in the fifth century when Buddhism was patronized by the Tuoba Xianbei—a non-Han people of nomadic origin who unified northern China and established the Northern Wei dynasty, the Yungang Grottoes consisted of thousands of Buddhist statues that had been carved into the cliffs and caves of the Wuzhou mountains over several centuries. The Beijing academics marveled at their historical and artistic value. In addition to bringing readers on a virtual tour of the site by detailing their three-day visit to the Yungang Grottoes, Bing Xin and Zheng Zhenduo, a fellow writer in the tour, stressed the profundity of the site by comparing it to other wonders of ancient civilizations, such as the Egyptian pyramids, the famous artworks at the Parthenon in Athens, and the Venus de Milo.[142] By situating the Yungang Grottoes on par with the world's greatest ancient ruins, the writers cast the Northwest not as a backwater of the republic but rather as a symbol of China's early civilization, reinforcing its indispensable place within the nation.

At Datong, another recommended attraction was the coal mine in the nearby Kouquan county. Just as tourists from the Japanese main islands were encouraged to visit sites of production in Japan's overseas colonies and semicolonies, coastal Chinese travelers viewed sites like modern coal mines as examples of achievements in modernization and state-building in the Northwest.[143] Both Bing Xin and the aforementioned Xiao Qian described the unique sensory experience of descending underground to the coal pits of the Kouquan mine: the darkness, the stuffy air, the narrow space crammed with blackened workers, and the eerie sounds made by shaky lifts and coal cars.[144] The palpably hellish conditions they captured in their travelogues also demonstrated the creeping advance of capitalist modernization.

Known as Hohhot in Mongolian, Suiyuan was the second main destination on the Ping-Sui line that coastal travelers frequented in the 1930s. Located in western Inner Mongolia, it was the capital of Suiyuan Province, which was established as a full-fledged province by the Nationalist government only in 1929.[145] Divided between old and new towns, Suiyuan was famous for a number of Tibetan Buddhist monasteries inside the city and the Zhaojun tomb on its outskirts. Bing Xin's travel journal recorded visiting four different monasteries. The history of these monasteries, she pointed out, could be traced back to the High Qing period when the Manchu's western conquest reached the region. For example, the monastery called *Xiao zhao* in the old town of Guisui was rebuilt in 1698, and renamed *Yanshou* temple by the Kangxi emperor himself. Whereas the structure of the monastery was of both Chinese and Tibetan styles, a stele erected at the site bore texts in four languages: Chinese, Manchu, Mongolian, and Tibetan. From the stele, Bing Xin learned that the Kangxi emperor resided there after he suppressed the Dzungar Mongols.[146] In his travel essay, Xiao Qian ranked the Shiretu Juu (*shelitu zhao*) as his favorite monastery in Suiyuan. Charmed by its "hybrid Eurasian style," Xiao and his travel companions even convinced the lama there to show them the armor that was believed to have been left there by the Kangxi emperor.[147] While the mixtures of architectural styles and the juxtaposition of multiple linguistic markings demonstrated this frontier town's complicated religious and cultural history, the stories of the Kangxi emperor's sojourn cast light on Suiyuan's unique position in a central regime's westward expansion.

Situated outside Suiyuan, the Zhaojun tomb, which was also referred to as the "green tumulus" (*qingzhong*) in many tourism promotion materials about Suiyuan, was believed to be the grave of Wang Zhaojun.[148] Wang was a Han Chinese court lady who was sent to marry the Xiongnu Khan by a Chinese emperor in the Western Han dynasty (206 BCE–8 CE) in exchange for peace. In an essay about his visit to the tomb, Zheng Zhenduo pointed out that although it might not be the real location of the famous beauty's final resting place, this "poetic ancient site" was still worth visiting to ponder on the past.[149] Although these historical sites turned tourist attractions came from different time periods, they collectively represented Suiyuan as a frontier town bearing witness to the interactions between China's central regime and its northwestern frontier. These interactions, from military conquests to religious patronage to marriage alliances, not only formed the unique cultural and religious institutions in Inner Mongolia but also shaped contemporary narratives of the connections and ruptures between the center and the periphery.

Located north of Suiyuan on the other side of the Daqing mountains, Bailingmiao (Bat Khaalga sume in Mongolian) was not technically on the Ping-Sui Railway route. Nor was it as accessible as other tourist destinations in the Northwest. However, as the home of an important Tibetan Buddhist monastery complex and the center of the Mongolian autonomous movement in the early 1930s, it attracted many Han Chinese travelers who were intellectually curious and politically minded.[150] Accompanied by an interpreter and an associate of the Ninth Panchen Lama, who had been visiting the monastery at Bailingmiao since 1932, Bing Xin and her Yenching colleagues headed to Bailingmiao from Suiyuan in several all-terrain vehicles. Even with modern means of transportation, it took them two days to reach Bailingmiao, where they were received by the Mongol Prince Demchugdongrub (known in Chinese as *De wang*), the prince of the Sunid Right Banner in the Shilin-gol League and the leader of the Mongolian autonomous movement.[151] In perhaps the most exciting and exotic part of their journey, the group stayed in Mongolian yurts and were treated to performances of horsemanship and wrestling by Mongolian elites.[152]

Aside from a guided tour of the Bailingmiao monastery complex, Bing Xin was also fascinated by the nomadic lifestyle of the Mongols, taking pictures of their dwellings, clothes, and other scenes of their daily life. In her travel journal, Bing Xin called the Mongols "our brothers of the same mother," who were also under threat from "powerful neighbors." She appealed to her readers, presumably Han Chinese in coastal regions, to support the Mongols so they would know that they were not alone in their fight.[153] Here, "powerful neighbors" referred to Japan and Soviet Russia, which had successfully instigated the separation of Manchuria and Outer Mongolia from the Chinese republic. Nevertheless, Bing Xin's own wishful thinking on behalf of the local Mongols belied the reality of the tricky situation unfolding at Bailingmiao. Gaining new momentum in the aftermath of the Manchurian incident, the call from Mongol elites for a higher level of Inner Mongol autonomy was particularly worrisome for the Chinese nation, considering the growing Japanese encroachment in Inner Mongolia and North China.

In another travel essay about Bailingmiao published in *China Traveler*, author Huang Yubin, while praising the hospitality of Prince Demchugdongrub, directly acknowledged the existing tensions between the Chinese and the Mongols. Huang attributed such tensions to the Mongols' deep-rooted superstitions, a comment that resonates with the opinions voiced by Xu Xusheng in the late 1920s. Echoing Gu Jiegang's praise of Wang Tongchun's endeavor to expand Han settlements in Inner Mongolia, Huang Yubin argued for Han migration and land reclamation as an ultimate solution to ethnic tensions in the region.[154] Unlike Datong and Suiyuan, which attracted travelers with their easy accessibility and iconic tourist attractions, the hard-to-reach Bailingmiao lured visitors with its authentic Mongolian flare. Its astonishing rise to become the power center of the Inner Mongol autonomy movement also attracted high-profile visitors from the Nanjing government and intellectual explorers like university faculty and students. Although Han travel writers depicted the Mongol elites as friendly hosts and urged their readers to work harder to shore up ties with them, their travel writings nonetheless evoked an underlying anxiety about the political loyalties of ethnic non-Han peoples to the Chinese republic.

Traveling to Inner Mongolia in 1935, Huang Fensheng, one of the chief editors of *Weekly on Mongolia and Tibet*, spent more than two weeks at Bailingmiao observing the council meeting of the local Mongol autonomous political committee. Aside from his official capacity, Huang was also a tourist. Like Bing Xin, Huang was eager to act like a "local." Treated to a feast of roasted whole lamb (a Mongolian custom to welcome special guests), Huang learned to use his left hand to hold down the meat on the platter and cut with his right hand using a Mongolian knife. According to him, the local Mongol lamb tasted less gamey than that in China proper. Local Mongolians used cow dung as fuel to roast the lamb, which was rumored to reduce the muttony flavor. While recording this tidbit in his travelogue, Huang argued that instead of the cow dung, it was probably the local environment that gave the lamb its unique flavor.[155] Although only an outsider, Huang presented himself as an authority on the Northwest for his coastal readers. In another piece, Huang pointed out that the Bat Khaalga monastery—from which Bailingmiao got its name—was considered sacred by the Mongolian believers in Tibetan Buddhism, and the town was not to be entered by women. "Modern girls" were nowhere to be found here, Huang noted. However, he also pointed out that "the precepts of Buddhist lands are often broken by our modern times." Since Bing Xin's and Lei Jieqiong's visit in the previous summer, at least two other groups of women travelers had also penetrated the sacred ground of Bailingmiao.[156] Even as he acted as a conveyer of the local authenticity in Bailingmiao, Huang admitted that the power of modernity—the fast marching of tourism in this case—would inevitably change this remote religious site in the Northwest.

Indeed, the success of the tourism promotion campaign by the railway and other institutions was manifested in the popularity of Northwest travel photography in the mass print media. *Kodak Magazine* featured popular destinations in the Northwest in its column Kodak Travelogue (*keda youji*). A travel column for amateur photographers to showcase their adventures, it was designed to broadcast the easy-to-use Kodak hand cameras and roll films and to demonstrate their great performance in outdoor environments. The appearance of the Yungang Grottoes and Suiyuan in the column in 1934 and 1936 indicated their increasing accessibility and appeal, attracting not only student and academic travelers from Beijing but also tourists from Shanghai and other cities.[157] Connecting two urban middle-class recreational hobbies—leisure travel and photography—the Kodak Travelogue also presented the Northwest as a playground for their targeted consumers. Just as using the latest model of Kodak camera would indicate a consumer's purchasing power and status as a more advanced amateur photographer, touring the Northwest and taking snapshots of magnificent Buddhist statues and remote ancient tombs could showcase one's adventurous spirit and free-spending modern lifestyle. In tourism to the Northwest, it is impossible to separate middle-class citizens' sense of civic pride and responsibility from their desire to consume new tourism destinations.

Hardship and Adventure

Just as the Ping-Sui Railway took full advantage of the Head to the Northwest movement to promote its tourism appeal, the Longhai line also helped advocate for travel to the Northwest, especially in Shaanxi Province. Shortly after the central government extended the Longhai line to Tongguan, railway administrators invited experts from government agencies, the business and banking sectors, and universities to participate in exploratory trips to Shaanxi.[158] Once the railway reached Xi'an in December 1934, tourists could purchase tourism round-trip tickets to multiple destinations in the province to visit the famous Huaqingchi hot springs and to hike Mount Huashan.[159] From Shaanxi, more adventurous travelers could also venture further into Gansu via newly constructed motorways, or even go on as far as Ningxia and Qinghai.[160] Intersecting with two main north-south trunk lines at Xuzhou and Zhengzhou, the Longhai Railway also allowed more access to the Northwest. In short, the extension of the Longhai line to Shaanxi helped broaden the geographical scale of this westward travel and shifted the center of movement beyond the intellectual hub of Beijing to include political and economic centers such as Nanjing and Shanghai.

Focusing on Xi'an and Shaanxi Province, the CTS guidebook to the Northwest argued that Xi'an should be a must-visit city for tourists from major cities like Shanghai, Beijing, and Tianjin. Not only did it become more accessible via train, but Xi'an was also worth spending time in because "as an ancient capital . . . for roughly nine hundred years, it is filled with historic relics, artifacts, and scenic spots."[161] Indeed, the guidebook's suggested itinerary contains twelve days of activities, including touring sites within the city limits and those in the surrounding areas.[162] Leading a tour of thirty people from Shanghai to Shaanxi in 1934, Shen Xingchu, a core figure in the Unison Travel Club, designed a classic itinerary for tourists originating from the lower Yangzi region. While Shaanxi's Mount Huashan, one of China's five sacred mountains, was the main destination, Shen's tour itinerary also included twelve hours in Kaifeng, the historic capital of Northern Song (960–1127), and twelve hours in Luoyang, another ancient capital in Henan Province.[163] Since both Kaifeng and Luoyang were also on the Longhai line, a tour to the Northwest via the Longhai Railway also meant a cultural tour of multiple historic capitals in China.

While the Longhai line allowed many coastal urbanites to sample Chinese history through tourism, the harsh conditions travelers underwent became a trope in commercial travel writing, mirroring journalists' travel accounts of the Northwest. In the mid-1930s, *China Traveler* published a string of travelogues on long-distance journeys to the Northwest. Joining a 1932 exploratory trip in Shaanxi organized by the Longhai Railway, Bian Wenjian, a manager at the Shanghai Commercial and Savings Bank, detailed his journey in a lengthy travelogue.[164] Even though the group traveled with a large staff and were offered the best amenities available in the Northwest, the journey turned dreadful once they left Xi'an. The lack of modern roads, subpar food, and appalling rural poverty quickly diminished the travelers' optimism for the future of the Northwest.[165] Shortly after publishing Bian's article, *China Traveler* serialized Shu Yongkang's travel journal about his journey from Shanghai to Qinghai, in which he painted an image of the Northwest that was even more jarring.[166] On their way from Xi'an to Lanzhou, for example, Shu and his travel companions, including Shanghai journalists Gu Zhizhong and Lu Yi, were stranded for a week when their long-distance bus broke down in rural Shaanxi. Terrorized by bandits, the authorities could barely protect local communities and travelers who were passing through. Local militias were prone to abuses of power. Shu also noted that in rural Shaanxi "9.5 people out of 10 smoke opium."[167] The poor hygiene also turned the most mundane acts of daily life, such as washing one's face and relieving oneself, into stressful tasks. Published at a time when coastal urbanites were growing more interested in the Northwest, these anecdotes preached caution when setting out to visit less developed regions of the country.[168]

Less flattering depictions of travel destinations were not always intended to deter potential travelers. Modern guidebooks and popular travel writings often included information about inadequate facilities or mentioned the lack of cleanliness at certain destinations, so readers could be fully prepared for their journeys.[169] Descriptions of hardship were also found in popular travel writings about the Northwest, adding a sense of adventure and uncertainty into otherwise curated and safe tourism experiences. In 1934, Zhang Henshui, a best-selling fiction writer in the Republican era and a regular contributor to *China Traveler*, embarked on a three-month journey to the Northwest. In a piece serialized in eleven parts in *China Traveler* in 1934–35, Zhang seldom mentioned the significance of developing the Northwest.[170] To him, the various sites he visited in Luoyang, Xi'an, and Lanzhou were representations of a Chinese tradition not that different from those former imperial spaces turned tourist sites in Republican Beijing.[171]

Zheng Henshui's 1934 tour disembarked from Beijing, and his first stop was Zhengzhou via the Beijing-Hankou Railway. Sightseeing aside, the main task in Zhengzhou was shopping. Zhang reminded readers that Zhengzhou was the last city where one could purchase, at reasonable prices, all the necessary items for a northwestward journey. For example, Zhang emphasized that one should prepare canned food and crackers if planning to climb Mount Huashan in Shaanxi. Other required items included a camp bed, gas stove, salt, water filters, and so on. "When you get to Gansu Province, the water there is muddy," Zhang warned readers, "and you will want to filter the water and boil it on your stove."[172] From Zhengzhou, Zhang transferred to the Longhai Railway to continue on to Luoyang, where he visited two famous Buddhist sites: White Horse Temple (*Baima si*), the first Buddhist temple in China, and the Longmen Grottoes, a Buddhist site dating from the Northern Wei dynasty like the Yungang Grottoes. The Longmen Grottoes, Zhang noted, were disappointing. Many of the small statues were either headless or completely missing as a result of theft, while the bigger statues were intact but hard to view without hiking up the cliff.[173] In Shaanxi, one of the main items on Zhang's itinerary was hiking up Mount Huashan. Although he enjoyed the natural scenery, he complained that some of the paths up Huashan were so narrow that the sedan-chair bearers needed to dismantle the chairs at certain points and reassemble them further on in order to carry travelers and their luggage up the mountain.[174]

At Lintong, another popular tourist destination in Shaanxi, Zhang bathed in the hot springs at Huaqingchi, a site believed to be the place where the famous beauty Yang Yuhuan, a consort of the Tang dynasty, had enjoyed the water.[175] In Xi'an, Zhang used a guidebook to direct himself through a long list of must-see attractions, such as the Forest of Steles (*Beilin*) and the Great Goose Pagodas (*Dayan ta*).[176] Zhang paid attention to local customs as well. As he pointed out, it would be prudent to learn about the local customs and habits to avoid offending local residents. Compared with travelers from the eastern seaboard, Zhang noted, the locals in the Northwest were used to simpler clothing. Strict distinctions between men and women were still deeply ingrained in the Northwest, so Zhang warned his Shanghai readers not to simply ask any woman on the street for directions. Above all, Zhang emphasized that "people from the Southeast"—the lower Yangzi region and other urban centers on the east coast—needed to respect the strong patriotism and native pride in the Northwest. He even instructed visitors about what to say to locals: "[You] should say that 'we have come back to our ancestral home' . . . or 'people in the Northwest are so hardworking that we from the Southeast cannot compare.'"[177] Like Ni Xiying's travel writings about inland cities discussed in chapter 2, Zhang's depiction of local customs alerted his readers to regional differences but also indicated his concern that tourists might appear arrogant to locals.

With help from local officials, Zhang journeyed from Xi'an to Lanzhou via the Xi'an-Lanzhou motorway, which was still under construction. Traveling with two engineers inspecting the motorway project, Zhang spent nine days on the road, during which he encountered abject poverty, rampant banditry, and faint traces of progress. Having received a classical education, Zhang found himself recognizing the names of counties outside Xi'an from the classics he had once memorized. This was not entirely a surprise, as he knew that the Guanzhong plain was the cradle of early Chinese civilization. Although struck by the surreal feeling of visiting places where venerable saints and legendary kings once resided millennia ago, Zhang's meditations on the past were often interrupted by disbelief at how unimpressive many famous locales looked in person. When the group was forced to spend a night in a small county because of heavy rain, Zhang and his traveling companions were offered lodging at the local elementary school. Recently built, the building's windows had no glass or paper to block the wind and rain. One of the engineers had stayed there before, and told everyone that there was no need to worry about bandits, because the county was so poor that even the bandits had long abandoned it. But he also added that during the night he had spent there, two wolves had tried to get in. Zhang was so shocked by this anecdote that he barely slept that night.[178]

Further on at Lanzhou, Zhang deemed the Yellow River and the iron bridge over the river the sights worth seeing, especially for travelers from the Southeast. Before the iron bridge was built by the Germans in the late nineteenth century, people in Lanzhou had maintained a floating bridge at the same location. That floating bridge and subsequent iron bridge were for years the only bridge spanning the Yellow River. While that was no longer the case, the bridge was nonetheless a rare place where travelers could see the huge waterwheels propelled by the rapid currents and ox-skin rafts floating downstream with passengers and cargo.[179] Although Lanzhou was harder to reach than Xi'an, with the upcoming completion of the Xi'an-Lanzhou motorway, Zhang saw the city's potential to become a popular destination in the Northwest, and he added information about transportation, lodging, and special local products to his travelogue.[180]

"Compared to traveling in the southeastern provinces, traveling in the Northwest is arduous," Zhang wrote in his conclusion to the series. Despite the hardship, Zhang emphasized that it was worth the trip.[181] Unlike Bian

Wenjian, who had traveled to the Northwest on an official tour organized by the Longhai Railway administration and the Shaanxi government, or Shu Yongkang and his journalist colleagues, who had set out from Shanghai to Qinghai to conduct sociological investigations, Zhang Henshui was a successful commercial writer without any official, professional, or academic titles who had acted on his own as a leisure traveler. His essays in *China Traveler* normalized leisure travel to the Northwest. Cities like Luoyang and Lanzhou indeed showed up more often as tourist destinations in popular print media in the mid-1930s.[182]

Zhang Henshui's tourist perspective does not mean that he was ignorant of the national crusade to fully incorporate the northwestern frontier. Ending his long journey at Lanzhou, Zhang could not help but examine the unique location of this Yellow River city. A town "incorporated into the map of China as early as the Han dynasty," Lanzhou had long been a city of military importance.[183] He asked his readers to draw a cross over a map of China, and they would find Lanzhou located exactly at the center of the cross. By pointing this out, Zhang underlined the city's position as the center of China's geobody and the entryway to the far Northwest.[184]

Although Zhang Henshui never cited any slogans from the Head to the Northwest campaign in his travel essays, he was clearly aware of the moment of national crisis that had inspired similar travels. Just as Bing Xin told her readers that she had witnessed "the solemnity of the motherland" in her Northwest travels, Zhang urged readers to connect the dots of cosmopolitan coastal cities, ancient capitals in central and western China, and frontier military towns like Lanzhou in their tourism adventures.[185] From publications with national followings such as *Eastern Miscellany* and *Liangyou* to lowbrow weekly magazines like *Saturday*, or from political publications issued by the Nationalist Party organs to professional photography magazines, texts about and images of the Yungang Grottoes, the Tibetan Buddhist monasteries in Suiyuan, scenery on Mount Huashan, and the iron bridge over the Yellow River in Lanzhou abounded.[186] And many of them bore a similar title—"Head to the Northwest!"—indicating that the familiar saying was not only a political catch-phrase but also actual advice to travelers.[187]

Xiao Qian's 1934 essay about his travels along the Ping-Sui Railway included an unexpected detail. Upon leaving Suiyuan for Baotou, Xiao learned from the local newspaper that the Beijing University professor Liu Bannong—the same Liu Bannong who had orchestrated the renegotiation with Hedin in 1926–27—was also in town conducting field research into regional dialects in Inner Mongolia. Xiao went to Liu's hotel to greet him, and Liu mentioned that he was heading to Bailingmiao.[188] This brief encounter between teacher and student on their respective field trips demonstrated that travels in the Northwest had become so common for Beijing academics and university students that their paths could cross even deep in western Inner Mongolia. But Xiao's inclusion of this detail in his travelogue was not random. Liu Bannong passed away suddenly in Beijing in July 1934, a month after the chance encounter. He died from a relapsing fever because he had not received timely treatment after being bitten by lice in Bailingmiao, a cruel reminder of the travel conditions in the Northwest.[189] Well-known for his early literary accomplishments in Shanghai, his fierce advocacy for vernacular language during the New Culture movement, and his achievements in spearheading linguistics and phonetics in China, Liu Bannong was not particularly remembered or celebrated for his crucial role in China's westward movement. Nevertheless, as this chapter has demonstrated, Liu could be viewed as a common thread in different stages of the Head to the Northwest movement, stitching together the early intellectual impulses undergirding the joint explorations and the later popular idea of saving the Northwest through travel. His personal tragedy aside, Liu Bannong was not an outlier. He was one of many Republican intellectuals and writers—like Xu Xusheng, Bing Xin, Zhang Henshui, and others across the ideological spectrum—who had participated in the knowledge production of the Northwest through travel.

Through the travel and travel writing of such figures, this chapter delineates China's westward movement in the 1920s and 1930s—a movement in which eastern urbanites sought to engage more closely with the country's vast northwestern frontiers. From an intellectual undertaking to modernize national studies via scientific travels to a popular campaign to regenerate the republic by encouraging average urbanites to travel to the Northwest, this broad westward movement helped transform the Qing western frontiers into a defining part of the national space of modern China. Prominent China scholars, both in China and the West, have acknowledged that the Republic of China had inherited its territorial definition from its Qing predecessor.[190] Yet neither the warlord regimes nor the Nationalist government could maintain a multiethnic rule over these lands. This contradiction between claim and reality provides the core question underpinning this chapter. If the Republican state's real political reach was often vaguely defined and subject to challenge, how could the popular consciousness of China's national space—one largely shared by the Han majority—unwaveringly adhere to the image projected by the Qing imperial domain?

As demonstrated in this chapter, the narrative that the Chinese republic inherited its national territory from the Qing, or that the Manchu imperial domain was simply carried over to the Republican era, was carefully manufactured. Indeed, we can see that Republican-era Han Chinese travelers and tourists to the Northwest actively evoked notions of science and modernity, colonial civilizing missions, Chinese historical discourses of frontiers, and the industrial capitalist goal of constant expansion. Clearly, the national space of modern China was envisioned and produced not only by political leaders and state apparatchiks but also by university professors, railway bureaucrats, freelance writers, magazine publishers and readers, and adventurous tourists.

Chapter 4

Facilitating the Exodus

Wartime Travel and the Southwest

In November 1938, *China Traveler* published a special issue on China's Southwest.[1] A region that had seldom registered on the radar of travelers and tourists from the east coast, China's Southwest—encompassing Sichuan, Guizhou, Yunnan, and Guangxi Provinces—began to be featured prominently in popular travel publications during the Second Sino-Japanese War (1937–45). As the full-scale Japanese invasion transformed China's eastern seaboard into an expansive war zone, the Nationalist government relocated itself to Chongqing in Sichuan Province in 1938. At the same time, China witnessed a massive wartime migration, during which factories, schools, and other institutions, along with large segments of the coastal urban population, were exiled to the wartime "home front area" in the Southwest. Although China had been "half destroyed," as one author proclaimed in the *China Traveler* special issue, "it is fortunate to have the Southwest as a 'big base camp' from which it can operate to recover lost territory."[2] Transformed from a remote interior region to the heartland of "Free China," the Southwest became the center of China's wartime mobility and a newfound source of national imagination for a republic that was fighting for its survival.

Wartime proved to be a period of great mobility in China. While estimates of war refugees vary, at least millions, if not tens of millions, of Chinese relocated to the Southwest.[3] The majority were urban professionals affiliated with state apparatuses, major industries, and cultural institutions. By examining the travel experiences of these elite refugees during the war, this chapter traces the drastic reconfiguration of transportation networks in China and East Asia. For an internal exile, the relocation from the east coast to the Southwest was by no means merely a domestic journey. Taking advantage of the semicolonial nature of the regional travel network, elite refugees largely relied on the triangular system connecting Shanghai, Hong Kong, and Vietnam or Burma. Once centered on the Pacific coast, the non-Japanese-controlled travel network in East Asia during the war pivoted on China's Southwest and its neighbors in colonial Indochina. This transition not only kept pathways between the occupied coastal regions and southwestern provinces unblocked; along with the massive relocation, it also signaled a shifting relationship between China and the Euro-American presence in Asia, as well as the renewed connection between the east coast–based Nationalist regime and its southwestern populations.

As Rana Mitter has suggested, the war disrupted China's modernization process, which had undergone a successful acceleration during the Nanjing Decade.[4] Despite the tremendous obstructions on China's eastern seaboard, however, the wartime relocation allowed the Nationalist regime to assert some meaningful control over its southwestern provinces for the first time since its establishment.[5] The development of the Southwest Motorways (*xi'nan gonglu*), an interprovincial motorway system, and other wartime state-building efforts testified to the growing Nationalist presence in the area. CTS, the prewar promoter of leisure travel, also emerged as a major facilitator of wartime relocation, assisting their prewar clientele from east coast urban centers during their exile. With these developments, elite refugees were able to maneuver across different national and colonial borders to reenter China from its southwestern border to reach Kunming, Chongqing, Guilin, or other major cities in Free China.

As they detailed every step of their ordeals, elite wartime travelers contrasted what they saw on the road—in colonial Indochina and the Southwest—to the homeland they left behind. On the one hand, as various wartime travel narratives have demonstrated, the transnational nature of China's wartime travel network reminded Chinese travelers of the semicolonial condition in which the country was embroiled. While they were concerned with China's uphill fight for survival against Japanese invasion, the utter "otherness" of the Southwest evoked an acute sense of loss. On the other hand, the lengthy sojourn enabled transplants from the east coast to have access to tourist sites in the Southwest. This exodus, despite its complications, strengthened a sense of the internal integrity of China, making the ties between the national core and the southwestern periphery more tangible than ever before. Following these developments, this chapter challenges the notion that China's nation-building processes were interrupted by the Second Sino-Japanese War. Instead, wartime mobility and travel culture suggest that although the war did abruptly shut down the nation-building efforts in the core regions of China, it also opened up opportunities for the Nationalist government and China's coastal elites to engage with the Southwest, a region that had been largely beyond their reach prior to the war.

Modern Travel and Tourism in the Prewar Southwest

Because of a lack of direct modern travel routes, traveling from eastern China to the southwestern provinces was complicated in the early twentieth century. Located on the upper reaches of the Yangzi River, Sichuan was relatively accessible from Shanghai by ship. After the signing of the Yantai Treaty in 1891, Chongqing was open to foreign steamers.[6] However, accidents happened frequently along the numerous gorges, rapids, and dangerous river shoals on the Yangzi in Sichuan, and until 1908, rather than extending to Chongqing, steamship lines still ended in Hankou or Yichang in Hubei Province.[7] According to the Sichuan guidebooks published by CTS, even after steamship navigation techniques improved in Chuanjiang—the section of the Yangzi River that flows through Sichuan—direct passage between Shanghai and Chongqing was available only during the summertime when water levels were high.[8] Because of the lack of railway lines (resulting from geographical and political limitations), to reach places other than port cities in Sichuan, tourists would have to take long-distance buses or utilize more rudimentary means of transportation, such as sedan chairs and mule carts.[9] The difficulty of traveling to Sichuan was infamous in imperial China, and the situation did not change much in the early twentieth century.[10] However, air travel became available between Chongqing and Chengdu in 1933, making touring within Sichuan more efficient for elite travelers.[11]

While major cities in Guangxi Province were more accessible than other locations in the Southwest, travelers from northern and eastern China still had to transfer at Guangzhou (Canton) or Hong Kong before proceeding on to Guangxi if they wanted to use modern transportation. Connected to the Pearl River Delta by the West River, Wuzhou in Guangxi was usually within one or two days' travel from Hong Kong and Guangzhou.[12] A rudimentary motorway system connecting other major Guangxi cities allowed travelers to reach Nanning and Guilin, the political and cultural centers of the province. In a travelogue about Guilin appearing in *China Traveler* in

1936, author Qian Hua suggested two routes between Shanghai and this scenic southwestern city. The first one followed the Yangzi River upstream from Shanghai to Changsha, from which people could travel further to Guilin. However, that route required considerable time and money, and road conditions between Changsha and Guilin were subpar. The second path, which Qian called the common route, went through Hong Kong, where one could take a steamship to Wuzhou in Guangxi. Taking a motorboat from Wuzhou to Pingle, people then could travel to Guilin through the motorway via Yangshuo. The whole journey took six to seven days.[13] Aside from the waterways and motorways, air travel also took off in Guangxi in the 1930s. In 1933, supported by the Guangxi warlord Bai Chongxi and the Guangdong warlord Chen Jitang, the Southwest Aviation Corporation (*Xi'nan hangkong gongsi*) was established, the third airline in China but the first set up solely by Chinese. Beginning with flights between Guangzhou and Nanning via Wuzhou, Southwest Aviation soon added routes to other cities in Guangxi as well as to Haikou on Hainan Island and Hanoi in Vietnam in 1934.[14]

Vietnam, a French colony since 1887, occupied a crucial juncture in China's southwestern travel network. Coastal Chinese traveling to Yunnan Province often relied on the Yunnan-Vietnam Railway (*Dianyue tielu*) connecting Haiphong, a port city in French Indochina, and Kunming, the provincial capital of Yunnan. Travelers could first travel to Hong Kong via steamship or railway. Through Hong Kong, they could travel by steamship to Haiphong, where they were able to transfer to the Yunnan-Vietnam Railway, a lifeline for Chinese domestic travelers trying to reach the southwestern corner of China via modern transportation. Without this detour to Vietnam, travel between eastern or northern China and Yunnan was nearly impossible.[15] Dubbed one of the "eighteen oddities of Yunnan" (*Yunnan shiba guai*), this reality was described in a local adage as "trains that go abroad instead of inland" (*huoche butong guonei tong guowai*). As the only modern transportation route between eastern and northern China and Yunnan, this triangulated route funneled many travelers to Yunnan through Hong Kong, sometimes resulting in a long wait there for an available berth on a ship. For example, when Zhang Yunfen was accompanying her husband from Shanghai to Kunming, they waited eight days in Hong Kong. The usual wait time was five days, according to a Yunnanese friend they met on their way to Kunming. But they had asked the hotel in Hong Kong to book tickets for them, and the hotel took its time in order to charge the inexperienced travelers for more days of lodging.[16]

To traverse this cross-border transportation circuit, domestic travelers had to apply for a French visa before they set out to Yunnan.[17] Colonial officials at the border, on both the Vietnamese and Chinese side, tended to examine Chinese travelers strictly. When Ding Wenjiang, a newly minted geologist trained in Cambridge and Glasgow, returned to China through Yunnan in 1911, the colonial officers at the Sino-Vietnamese border interrogated him despite the fact that Ding was carrying a passport issued by the British Consulate.[18] In an article published in 1934, the writer complained that French colonial customs officers checked their luggage carefully when they arrived in Haiphong from Hong Kong. The customs taxed any new clothing or shoes, and even confiscated cameras from transiting passengers.[19] Shaped by the colonial geopolitical order in the region and the semicolonial setting of China, modern transportation routes from China's eastern coastal cities to the Southwest presented, once again, a humiliating journey for Chinese translocal travelers in the Republican period.

Crossing the border into Yunnan was not the most difficult part of Ding Wenjiang's 1911 journey. In order to reach his hometown in Jiangsu Province, Ding had to travel across Yunnan, Guizhou, and Hunan Provinces.[20] Guizhou, blocked off by the Yunnan-Guizhou plateau, had very limited modern transportation routes connecting it to the rest of the country. Because of this, Ding had to hire a professional guide and eight luggage bearers from the largest travel agency in the southwestern region, called *Maxiangyue*.[21] Established by a Sichuan sedan chair bearer in the mid-nineteenth century, *Maxiangyue* was the largest enterprise providing transportation service across Sichuan, Guizhou, and Yunnan, delivering remittances, letters, and goods, and providing guides and bearers to assist long-distance travelers throughout the Southwest.[22] Ding was carried in a *huagang*, a kind of open sedan chair made of bamboo, traveling in a fashion not so different from that of his imperial predecessors. In fact, to avoid unnecessary attention, Ding even grew a mustache, wore a fake queue, and replaced his Western suit with a mandarin jacket. His journey across Guizhou took more than a month. As he traversed the geographically peripheral provinces in China's national territory, a cosmopolitan traveler like Ding must have felt he was also traveling back in time, an experience many elite refugees would also face during wartime. The rudimentary transportation infrastructure did not improve significantly in Guizhou, even though Guiyang, the provincial capital, became accessible via air routes during the Nanjing Decade. For average travelers from the east coast, the most convenient way was traveling by steamship to Chongqing, from which they could move forward on sedan chairs to Guiyang. This last leg of the trek normally took fifteen days in the mid-1930s.[23]

Even as the China Travel Service expanded its business network rapidly during the Nanjing Decade, the Southwest remained a blank in its tourism empire. This was not due to a lack of interest. In fact, the first two guidebooks published by the Shanghai Bank's Travel Department, the forerunner of CTS, were travel directories to Yunnan and Sichuan. Appearing respectively in 1923 and 1924, the two handbooks not only detailed different routes and fares available to travelers to the Southwest from coastal China but also listed basic information about every city along these routes.[24] However, CTS failed to open any branches in the area even during its peak years from 1931 to 1937.

The political fragmentation of China was the key obstacle. Although they swore loyalty to the Nanjing regime, the southwestern warlords, such as Lin Sen and Liu Xiang in Sichuan and Bai Chongxi and Li Zongren in Guangxi, maintained their local autonomy through the late 1920s and early 1930s. At the same time, preoccupied with consolidating its authority in the lower Yangzi region and North China, Chiang Kai-shek's government could not spare the energy needed to garner any real control over the Southwest. As discussed in chapter 1, CTS's expansion in the 1930s was catalyzed by a conjunction of infrastructure developments, especially the Nationalists' railway and motorway projects. Establishing branches along existing and newly built railway lines, motorways, and water transportation routes, CTS was a beneficiary and promoter of the Nanjing government's infrastructure projects and other nation-building efforts. However, because of the fragmentation of political power, these projects were concentrated mostly in eastern and northern China, and CTS's development accordingly focused on these regions as well.

After the Manchurian Incident in 1931, the Nanjing government consciously tried to extend its influence into the western provinces in preparation for a potential retreat. Selecting Luoyang in Henan Province and Xi'an in Shaanxi Province as secondary capitals, the Nationalist government focused its attention on the Northwest, rather than the more prosperous Southwest, where the local militarists' allegiance to the central government remained inconstant. Taking the Red Army's passage through the region as an excuse, however, the Nationalists' military forces entered Sichuan and Guizhou in 1934. After Chiang Kai-shek took over Guizhou from the warlord Wang Jialie, the Sichuan warlord Liu Xiang opened up Sichuan to Chiang in exchange for support from the national army, which Liu used to overpower other regional warlords in Sichuan.[25] With the Japanese tightening their grip on Manchuria and Inner Mongolia in the mid-1930s, the Southwest began to replace the Northwest as a more advantageous inland base for Chiang Kai-shek's government.

At this critical prewar juncture, the central government began to fund infrastructure projects in the Southwest. Aware of the absence of direct travel routes from the east coast to the southwestern interior, the Nanjing government finally extended modern motorway routes into the region in early 1937. The Sichuan-Shaanxi motorway connected Xi'an to Chengdu. It mainly served military and freight transportation, however, and it was uncommon for ordinary travelers to use this route.[26] By threading together a system of provincial motorways connecting Jiangsu, Anhui, Jiangxi, Hunan, Guizhou, and Yunnan, the state completed the Nanjing-Kunming motorway (*Jingdian gonglu*) in April 1937, shortly before the Japanese launched their full-scale invasion of China.[27] Halving the travel time between Nanjing and Kunming, this motorway system tied the Southwest to central and coastal China, aiming to supplant the semicolonial and colonial travel networks.[28] Its completion also showed the Nanjing regime's resolution to extend infrastructure projects into the Southwest, where travel conditions lagged behind the core regions of the country.

Today, China's Southwest is regarded as a tourist destination. However, during the Nanjing Decade, only a few sites in the Southwest registered in the consciousness of coastal tourists. Skimming through *China Traveler* or *Liangyou* and other pictorials before the War of Resistance, readers would see the Southwest represented by only a handful of places, including the Three Gorges, Mount Emei, Guilin, and Kunming.

Between Hubei and Sichuan Provinces, the Three Gorges (*sanxia*) was an iconic landscape on the Upper Yangzi River composed of chains of mountains narrowing the Yangzi into angry torrent. Mount Emei, one of China's "famous mountains" (*mingshan*), was celebrated as a sacred Buddhist site associated with the bodhisattva Puxian, attracting pilgrims and sightseers through the Republican period.[29] Both of these sites in Sichuan had long been canonized in classic poems and *youji* prose.[30] Further south in Guangxi, Guilin was known for its dramatic landscape of karst hills and otherworldly spires and caverns, which had enchanted literary figures since the Tang and Song dynasties. Kunming became one of the most-visited places in Yunnan because of its railway access.

Even though the travel destinations in the Southwest did not receive as many tourists as those in coastal provinces, their symbolic landscapes and scenery were meaningful to coastal readers. This was not just because these natural and cultural landmarks were canonized in classic literature and had become fixed in people's imagination.[31] Modern inscription technologies—

namely, photographs reproduced in modern books and magazines—also helped popularize these iconic sights while having them function as shorthand for a province or even the entire Southwest. Featuring multiple photographs of the Three Gorges in its main text, the Sichuan travel guide published by Shanghai Bank's Travel Department also used one such image for its cover.[32]

Travel magazines and travel columns also made extensive use of photographs of the Three Gorges. In the first installment of his Travelogues of China column, Wang Xiaoting focused his lens on the Three Gorges (figure 8).[33] As the title of the travel column suggested, he presented the Three Gorges as the "natural barriers" (*tianxian*) separating Sichuan from Eastern China. The textual narratives highlighted the difficulty of the journey through the gorges ("treacherous"), the many historical sites scattered throughout the area ("as numerous as stars"), and the shocking sights of the towering mountains and treacherous torrents ("eye-striking and heart-startling"). Arranged into a dynamic collage, the photo spread gave credence to Wang's dramatic tone. Consisting of long shots of varied subgorges, close-ups of odd-shaped rocks and famous caves, and bird's-eye views of towns nestled behind high mountains, the collage reinscribed the timeless landscape photographically.

FIGURE 8. "The Three Gorges as a natural barrier." *Source:* Wang Xiaoting, "Sanxia tianxian," *Dazhong huabao* [The cosmopolitan], no. 1 (January 1933): 14–15.

Other similar visual homages to *sanxia* in popular print media also injected signs of modern travel by, for example, including modern highways or a steamship in the frame.[34] While magnifying the breathtaking scenery, this kind of inclusion allowed the travel magazines and travel columns to send a subliminal message that tourists could in fact find a way to be in the scene themselves.

While *Liangyou* and *China Traveler* used a similar tactic to offer timeless photos of the iconic landscapes of Guilin, a different approach was adopted when the *Liangyou* travel column *Liangyou* Readers' Travel Train featured Kunming in 1935.[35] Rather than distanced panoramas, this piece offered "a few of the most interesting [images] of this not-often-visited city," the English caption read.[36] Presented like a series of postcards, the *Liangyou* column included a range of city attractions and provided short captions. In the top right-hand image, for example, readers could see a train hurtling through a mountain tunnel and learn from the caption that the Yunnan-Vietnam Railway line included more than 170 tunnels. On the opposite site, a grandiose structure appears out of place among the Chinese-style architecture. The caption explains that it is the mausoleum of Yunnan warlord Tang Jiyao, who was commemorated by the locals for his resistance against Yuan Shikai's monarchical revival in the early Republican period. To honor "this great man from the Republic's revolutionary days," a magnificent grave resembling "an ancient Roman building" was constructed in the center of the city. Through these snapshots, the photographic travel column created a sightseeing experience of the remote Southwest for armchair travelers.[37]

However, for average coastal tourists, visiting these famous destinations in person remained cumbersome in the prewar era. The lack of convenient and affordable modern transportation was a major obstacle. As mentioned above, traveling to Kunming or Guilin from Shanghai could be a long trek even with modern transportation. In the case of the Three Gorges and Mount Emei, which all required travel through the Upper Yangzi, steamship journeys were available but restricted. The Maritime Customs, according to an author in *China Traveler*, stipulated that the length of the so-called *Chuanjiangchuan*—the steamships capable of traversing the Yangzi River in Sichuan—could not exceed 220 feet, making these ships much smaller than the ships serving in the lower Yangzi. Yet their engines must be stronger than those of larger ships, in order to navigate the rapid current. On top of this, while ships could operate day and night along the Yangzi from Shanghai to Yichang, to ensure safety, *Chuanjiangchuan* had to anchor at night. These restrictions resulted in high costs.[38] Additionally, the same author pointed out that shipping on the Chuanjiang had long been controlled by foreign companies that paid little attention to passenger business, and even less to Chinese tourists.

Tourism in the Southwest did show some signs of improvement in the 1930s. The rise of Lu Zuofu's Minsheng Shipping Company in Sichuan dramatically transformed the shipping landscape and passenger business on the Upper Yangzi River.[39] With a mission slogan of "attracting people from other provinces to travel to Sichuan; persuading Sichuanese to explore other provinces," the company offered precisely numbered berths, quality meals, and convenient luggage services.[40] Travel businesses also set up shop in Sichuan. Asides from the CTS branch in Yichang, travelers could rely on the Chuanjiang Travel Service (*Chuanjiang lüxingshe*) to make basic travel arrangements to the Three Gorges. Established in 1931 in Chongqing, the Chuangjiang Travel Service was a semiofficial enterprise under the auspices of the Upper Yangzi Navigation Bureau in the Sichuan warlord Liu Xiang's regime.[41] At Mount Emei, the Chengdu-Jiading Transport Company and Chengdu YMCA formed the Mount Emei Travel Service (*Eshan lüxingshe*). The agency offered a range of services, including bus services between Chengdu and Mount Emei, sedan chair service on the mountain range, pamphlets of mountain maps and guides, and financial and communication businesses.[42]

With these developments, travel to the Three Gorges and Mount Emei became more regimented. Gao Bochen, for example, outlined a four-day itinerary from Yichang to Chongqing, with recommended lodging for each day.[43] The Unison Travel Club also organized a month-long tour to Sichuan in September 1935. Its itinerary included both the Three Gorges and Mount Emei as well as Chongqing and Chengdu, the two largest cities in the province.[44] Even with prices set at 330 yuan for first class and 185 yuan for second class, the tour still attracted twenty-five participants.[45] With the national motorway construction extended to the Southwest in 1937, Guilin was added to the Unison Travel Club's South China tour as well.[46] By the end of the Nanjing Decade, there were indications that the tourism radius for the affluent in modern China was slowly expanding into the Southwest.

The Making and Remaking of the Wartime Travel Network

Shortly after the Marco Polo Bridge Incident in July 1937, the Tianjin-Beijing region fell into Japanese hands. Shanghai and Nanjing soon followed, as did the entire Yangzi Delta—the most economically developed region in China. With the main railway lines in eastern China—such as the Tianjin-Pukou

Railway, the Shanghai-Nanjing Railway, and the Shanghai-Hangzhou-Ningbo Railway—cut off by the Japanese, CTS concentrated its early wartime efforts on finding transportation routes by sea and by river for Chinese fleeing the war zone. It extended its service, albeit temporarily, along newly emerging routes of retreat. For example, it opened guesthouses in Ganxian and Ji'an in southern Jiangxi, a rural area that had been a Communist base in the early 1930s.[47] As the Japanese army advanced southward and westward to take over Guangzhou and Wuhan in October 1938, any hope to hold on to coastal China quickly faded.

The Japanese invasion did have the effect of abruptly putting an end to the seesawing battle between the central government and the southwestern warlords. As Chiang Kai-shek was airlifted to Chongqing, anti-Chiang forces agreed to set aside their differences and unite under Chiang to fight the Japanese. At the same time, CTS inevitably was forced to retreat as well. With the exception of Beijing, Tianjin, Qingdao, and Shanghai, all its facilities in the war zones, which accounted for nearly 70 percent of CTS business in the prewar period, were shut down.[48] The headquarters of the Shanghai Commercial and Savings Bank, under whose umbrella CTS operated, also moved from Shanghai to Hong Kong in 1938. Further, CTS, together with the Daye Trading Company, a subsidiary trading company under the Shanghai Bank, merged their administrative board with the Shanghai Bank. These three branches—the travel agency, the trading company, and the bank—joined forces to facilitate travel and transportation between occupied China and Free China, concentrating on the movement of people, goods, and money, respectively.[49]

An imperative for the retreating Nationalist state was to keep transportation routes open for Free China. By the end of 1938, steamship routes in China were also restricted from the coast and the lower stretch of the Yangzi River because of the Japanese blockade. Unless one chose to forgo modern means of transportation—as many retreating college students and other poor refugees did—those fleeing to the Southwest became reliant on passage through the colonial and semicolonial locations of Shanghai, Hong Kong, and French Vietnam. The Shanghai-Hong Kong-Vietnam route had been a common itinerary for business travelers before the war, but for average travelers from coastal China, wartime retreat was the first time they had to cross borders during a domestic journey.

Since the late nineteenth century, China's Southwest had been at the center of colonial competition between France and Britain. After colonizing Vietnam and Burma respectively, they vied for control over trade routes between mainland Southeast Asia and southwestern China and explored the possibility of expanding long-existing overland trade routes—dubbed by some historians as the Southwest Silk Road—that had connected Southeast Asia, China's Southwest, South Asia, and Tibet.[50] While many ambitious proposals were made by Anglo-French colonists, the modernization of transportation routes in the area turned out to be difficult. Launched in 1910, the Yunnan-Vietnam Railway was the only fully usable industrial transportation line connecting colonial Southeast Asia and semicolonial China overland.

A key channel for the imperialist penetration of China's Southwest, the Yunnan-Vietnam Railroad nevertheless became a lifeline for China during World War II, as the retreating state lost its access to the Pacific coast. While the triangulated route of Shanghai-Hong Kong-Vietnam turned into the most commonly used modern transportation option for retreating coastal elites, other regional travel routes through colonial areas complemented it. Guangzhouwan (today's Zhanjiang), a French concession, for example, served as the transit port between Hong Kong and Vietnam.[51] The British colonies of Rangoon in Burma and Singapore also became transit hubs for Chinese wartime migrants, serving as alternatives to Vietnam. This was also the case with Manila in the Philippines, where US influence remained strong.[52]

Shortly after the Vietnam route became a busy thoroughfare for Chinese refugees, CTS established branches in Haiphong and Saigon and a guesthouse in Hanoi, while hoping to expand to the Philippines by establishing an office in Manila to supplement its existing branch office in Singapore.[53] This travel network was also connected by air. Hong Kong replaced Shanghai to become the center of China's air traffic before the Japanese invasion in December 1941. Two of the biggest airlines in China, the China National Aviation Corporation and the Eurasia Aviation Corporation, moved their headquarters to Chongqing. The air routes between Chongqing and Hong Kong and those between Chongqing and Hanoi became the fastest transportation lines connecting Chongqing with the outside world.[54]

The war also revived the old dream of modernizing the Southwestern Silk Road to further integrate China's southwestern provinces with Southeast and South Asia. Having failed to complete a railway between Yunnan and Burma, the Chongqing government rushed to construct the *Dianmian gonglu* (the Yunnan-Burma road). Commonly known as the Burma Road, this route was one of the most difficult wartime construction projects. Fu Hua Chen, the chief engineer of the project, recounted the extreme terrain in his memoir:

> The alignment of the Burma Road was such that we had to cross the Horizontal Cut Mountains. The course of the road crossed three major

> mountain ranges and two rivers, the Mekong and the Salween. The road reached an elevation of eighty-five hundred feet, then dropped sharply to three thousand feet across the Mekong. The route rose sharply again to nearly eight thousand feet, only to drop again to the Salween at an elevation of three thousand feet. One way to visualize this was to imagine a road crossing the Grand Canyon twice, although the Grand Canyon, which is about a mile deep, is shallower.[55]

Put into use in September 1938, the Burma Road soon replaced the Yunnan-Vietnam Railway as China's lifeline to the outside world, as the triangulated system through Vietnam fell apart in the early 1940s. First, France was defeated in Europe in June 1940, which, as the historian Parks Coble emphasizes, "gave the Japanese more leeway in dealing with French colonial possessions in Asia."[56] The Japanese army soon demanded that the French colonial government close the Yunnan-Vietnam Railway. In September 1940, the Japanese occupied Vietnam and immediately cut off the travel routes between it and Hong Kong. When the French colonial authorities forbade the Chinese government from using the Haiphong-Kunming line to transport foreign aid, arms, and ammunition into China, the Burma Road, which linked Kunming to Wanding on the Sino-Burmese border and further connected with a railway line between Lashio and Rangoon in Burma, became the most important international supply lines for China's defense.[57] Keeping up with these developments, CTS also set up offices in both Lashio and Rangoon.

Just as the travel network was rerouted after the Japanese occupation of Vietnam, the outbreak of the Asia Pacific War once again shattered the fragile system. Only hours after the attack on Pearl Harbor, the Japanese advanced into Shanghai's foreign concessions, Hong Kong, and the Philippines on December 8, 1941. After British Malaya fell into Japanese hands, the Japanese military attacked British Burma and Singapore in early 1942 and effectively cut off the Burma Road in May. This completely destroyed China's wartime connection with the outside world and further isolated the Nationalist government in exile. Portuguese neutrality made Macau the only relatively safe port to travel through in South China, but with significant numbers of Japanese "advisors" in Macau, it was a port of neutrality only in name. As a result, Free China looked to British India as its last resort. Before a Sino-India motorway was completed in January 1945, China relied on airlifts to haul necessary supplies from British India to China's Southwest.[58] Flying over "the Hump" of the Himalayas to enter China, these aircraft from both the Allied forces and the China National Aviation Corporation opened an aerial lifeline for Nationalist China.[59] Chinese travelers, especially students going to study abroad, could travel to Calcutta and Bombay, where they could catch American navy vessels bound for the United States. To follow these shifting travel routes, CTS established short-term offices in Macau between 1942 and 1945, and opened a guesthouse in Calcutta, a branch in Bombay in 1944, and a liaison office in the United States toward the end of the war.[60]

The Nationalist state also prioritized the building of transportation infrastructure within the Southwest itself. They began construction to upgrade and link provincial roads as soon as war broke out in July 1937, and completed the Southwest Motorways in early 1938.[61] This motorway system linked Guiyang, the provincial capital of Guizhou, with Changsha, Chongqing, Kunming, and Liuzhou, creating a coherent system covering the entire Southwest.[62] Then in 1940, the Nationalist government built the Sichuan-Yunnan motorway (*Chuandian gonglu*), a direct route between Kunming and Luxian in southern Sichuan, which significantly shortened the distance of the supply lines to Chongqing.[63]

To highlight the immense scale of this modern motorway system, the management office of the Southwest Motorways produced a map in 1940, comparing the size of the areas reached by this road system to the European continent. Key cities along the motorways were matched with capitals and important cities of the European powers, indicating the scale and importance of this transportation system. In the map, Chongqing was matched with Berlin, Changsha with Budapest, Liuzhou with Rome, Guiyang with Geneva, and Kunming with Paris. Juxtaposing China's home front with the war-torn territory of Europe, the Chinese government demonstrated the scope of China's War of Resistance, as well as hinted at its crucial position in the global antifascist war.

The hasty pace and huge scale of the construction along with the extreme geography of the southwestern provinces limited service facilities. Aware of this challenge, the Ministry of Transportation and provincial authorities commissioned CTS to sell long-distance bus tickets and provide room and board along these new motorways.[64] Among the forty guesthouses run by CTS in 1940, only eleven were established by CTS itself. The rest were commissioned by the bureau of Yunnan-Burma Motorways, the bureau of Northwest Motorways, Chinese Transportation Company (a state-run company overseeing the motorways in the Southwest), the Shaanxi Provincial Government, and the Construction Department of Jiangxi Province. The majority of these guesthouses were located in the Southwest as well as in Vietnam and Burma.[65]

Challenging the colonial order in Asia, the Japanese invasion of China and its subsequent escalation in the Asia Pacific region disrupted the circuits of contact hinging on the Pacific coast. However, these routes were reconfigured through China's Southwest and mainland Southeast Asia, relying partially on the existing Anglo-French colonial designs of the region. As China's travel networks shifted from the Pacific coast to the Southwest, the focus of CTS also moved westward. Whereas the agency had promoted modern lodging and other services to bring tourists into China's inland provinces during the prewar period, its operations along the Southwest Motorways became an organic part of the wartime state-building projects and the government's modernizing efforts. It is not an exaggeration to say that between 1937 and 1945, CTS became an intrinsic part of the Chinese state's wartime apparatus. Its growing presence also refashioned the forms of state penetration. More than simply building infrastructure, the state endeavored to provide the basic services necessary to connect Free China with the outside world and to engage the local population in nation-building projects during wartime.

Navigating the Triangulated Travel Route

Wartime retreat often involved intense modes of travel. Even though elite travelers did not have to complete the ordeal on foot, the complicated transfers through Hong Kong and Vietnam and the meager infrastructure in interior China made the trek to the Southwest both physically demanding and psychologically taxing. Between 1938 and 1940, CTS disseminated much-needed details about this route through print media. An early guide in the February 1938 issue of *China Traveler* provided wartime travelers with information such as price, schedule, and duration of different steamship, train, bus, and air travel options, as well as hotel information and passport and visa requirements.[66] An entire chapter of a CTS guidebook to Kunming issued in 1939 was dedicated to the journey between Haiphong and Kunming.[67] She Guitang, a CTS staff member stationed in Hanoi, wrote two pieces for *China Traveler* in 1940, summarizing the basic conditions in Vietnam and the newly completed Burma Road.[68] However, even with the newly available information, passing through localities beyond the Chinese borders and deep into the southwestern hinterlands was not an easy task. As Keith Schoppa has suggested, wartime refugees experienced both spatial and temporal displacement.[69] The dual displacements sharpened the sense of loss but also gave rise to new visions of the Chinese nation-state within which the Southwest was an integral and tangible part.

When the domestic travel routes between coastal China and the Southwest diminished in 1938–39, elite Chinese fleeing occupied China turned to the new itinerary via Vietnam. Among them were cultural figures leaving Beijing, Shanghai, or other urban centers in occupied China. They either followed the institutions they served to the Southwest or left their old positions and started new projects in Free China. The writer and literature professor Zhu Ziqing and the historian Qian Mu, for example, retreated from Changsha to Yunnan with hundreds of faculty and students at *Lianda* (the National Southwestern Associated University).[70] The writer Bing Xin and her husband, Wu Wenzao, however, left for Kunming via Vietnam on their own, even as Yenching University remained functioning in Beijing before 1941. As these wartime travelers proceeded from occupied China to semicolonies, from semicolonies to colonies, and from colonies to sovereign China, they constantly negotiated their self-identities while ruminating on China's national fate. They often recalled their difficult journeys in memoirs and elsewhere later in their lives. However, not many travel accounts about the triangulated route were published during or shortly after the war. A notable exception was Sa Kongliao's *From Hong Kong to Xinjiang*, a unique narrative of retreat through Hong Kong, Vietnam, and China's Southwest. Sa's travelogue not only revealed wartime travel conditions but also demonstrated the spatial relationship between China's national center and its southwestern provinces. A close reading of his account illuminates how the wartime relocation refashioned the way coastal elites viewed the Chinese national space.

Sa Kongliao had an impressive résumé as a journalist. After working for several important newspapers and journals in Beijing and Tianjin, in 1935 he became the editor in chief of *Libao*, a new daily in Shanghai. One of the most circulated newspapers in China, *Libao* moved its operations to Hong Kong when the war broke out in 1937. Subsequently, Sa also relocated to Hong Kong in the early stages of the war. With the massive influx of wartime refugees, Hong Kong—much like the "lone island" in Shanghai (referring to Shanghai's unoccupied foreign concession area)—experienced an abnormal period of prosperity. "No matter how chaotic the interior [of China] is," one *China Traveler* author pointed out in 1938, "Hong Kong is still Hong Kong."[71] Out of touch with the war-ravaged homeland, however, *Libao* in Hong Kong was far less influential than it had been in Shanghai. At the same time, Sa Kongliao was frustrated by his isolation from China's wartime struggle and yearned to travel into China's interior.[72] When his friend Du Zhongyuan, a left-leaning war correspondent, returned to Hong Kong from Xinjiang in 1938 to recruit more likeminded people to China's western frontier, Sa decided to join Du.[73]

Sa's first journey took place before the fall of Guangzhou and Wuhan in late 1938, and was relatively uneventful. From Hong Kong, Sa and Du took a steamship to Wuzhou in Guangxi via the West River, where they arranged a ride from some acquaintances and traveled to Hengyang in Hunan by car. A train took them further to Hankou, from where they flew to Dihua (Urumqi) on a chartered plane.[74] If the articles in *China Traveler* were any indication, routes from Hong Kong to Wuhan or Chongqing via Guangxi were still unimpeded in late 1938. Nevertheless, these routes were not entirely safe. Many travelers either encountered Japanese air raids personally or witnessed the damage inflicted on their travel routes. Sa, for example, experienced air raids in Wuzhou and Hankou, witnessing the violence of the war for the first time.[75]

Eager to participate in China's wartime efforts, Sa was persuaded by the Xinjiang warlord Sheng Shicai to serve as the associate editor in chief of *Xinjiang Daily*. In order to bring his family to Xinjiang, to purchase necessary equipment, books, and stationery, and to recruit more technical personnel, Sa returned to Hong Kong and stayed there for three months to wait for his purchases and recruits from Shanghai to arrive. As the Japanese army advanced southward in early 1939, travel routes between Hong Kong and the Southwest quickly deteriorated. Thoroughfares to the Southwest via the West River, Canton Bay, and the Canton-Hankou Railway were no longer available. It was even rumored that the Hong Kong-Vietnam route, which had become the sole modern passageway between occupied China and Free China in early 1939, would be closed under Japanese pressure.[76] With steamship tickets to Vietnam becoming scarcer by the day, Sa left Hong Kong for Haiphong on March 10, 1939, together with technical personnel and other travel companions and their families. Although he anticipated a herculean journey, Sa did not foresee that it would take them six months to travel from Hong Kong to Xinjiang through the Southwest, with four months spent in Vietnam and the southwestern provinces. His travel notes, first published in 1943, trace how tortuous such wartime retreats were even in some relatively peaceful areas. Furthermore, as will be discussed below, his travel account sheds light on how an elite Chinese viewed the war-torn republic and his fellow countrymen in the Southwest as he maneuvered through colonized territories in Southeast Asia and traveled across southwestern provinces.

The colonial setting of French Indochina had a significant impact on the Chinese wartime travel experience. Maneuvering past various borders, travelers retreating via this transnational travel route to the Southwest faced ever stricter border policing by the British Hong Kong government and the

French Vietnamese authorities. Acquiring different travel documents was one of the first obstacles people faced. In fact, the first travel document Chinese travelers had to obtain was a Chinese passport, which was not difficult in large cities like Shanghai. However, once China's Foreign Ministry withdrew to Chongqing, Chinese citizens in war zones could not apply for passports issued by the Nationalist government. Without a passport, one could not acquire a French visa to enter Vietnam. To solve this conundrum, CTS procured stamped blank passports from the Foreign Ministry in Chongqing. Serving as a proxy for the Chinese government, CTS could approve and issue Chinese passports in Shanghai by filling in the applicants' names on the prestamped passports. In order to avoid suspicion by the Japanese, however, instead of issuing the passports immediately, they usually made the applicants wait for two weeks and claimed that the passports were in fact delivered from Chongqing by airmail.[77] Aside from a Chinese passport, in order to travel through Vietnam, one also needed a French visa, which had to be obtained at the French consulate in Shanghai or Hong Kong. Proof of inoculation was also required, since colonial doctors checked these documents before letting passengers disembark in Hong Kong or Haiphong.[78]

Even more cumbersome were colonial customs. French customs officers in Vietnam were notorious for their pitiless inspections and tyrannical attitude toward Chinese travelers. Working only four hours a day, they would lock passengers in dark customs warehouses and leave for their three-hour lunch breaks during the hottest time of the day. Given that he was traveling with a formidable assortment of personal luggage and merchandise, Sa worried that French customs officers would create all sorts of obstacles. Therefore, before their departure, Sa sent a friend to Haiphong to help prepare for their arrival there. His friend obtained a name card from the Chinese consulate in Haiphong to verify that they were government employees who requested preferential treatment, and also hired staff from a trustworthy local travel agency owned by a *huaqiao* (overseas Chinese) in Vietnam to assist them in passing through customs. Perhaps because of the name card and the French-speaking staff from the travel agency, Sa's group was let through customs without much trouble.

Their travails in Vietnam were far from over, however. With the huge influx of Chinese citizens and goods, the work performance of the colonial customs officials had degraded. Sa, for instance, encountered difficulties in locating all his merchandise in the four customs warehouses in Haiphong. As the local *huaqiao* explained to Sa, even though Haiphong had become "the only import and export port" for Free China after the war broke out, the French colonial government was hesitant to expand its customs office to accommodate the increased volume of people and goods passing through Vietnam, as they were aware that the chaotic situation might be transient. This observation points to two seemingly contradictory developments. On the one hand, instead of asserting influence over southwest China, especially the bordering provinces of Yunnan and Guangxi, as it had in the prewar years, colonial Vietnam began to be integrated into a transportation and travel system orbiting around China's wartime needs. On the other hand, acknowledging the effects of the reconfiguration of regional travel networks on Vietnam, French colonial authorities—either due to lack of interest in actively supporting China's wartime efforts or out of the fear of Japan's inevitable advance into Indochina—held on to the old status quo even as it was obviously no longer working.

This dysfunctional situation meant that many Chinese wartime travelers experienced long delays in their transit through Vietnam. In order to transport all the trucks, machines, and goods he had purchased for Xinjiang across the border, Sa had to pay both an export duty in Vietnam and an import duty in China. On top of that, because the French colonial government stipulated that only drivers with Vietnamese government-issued licenses could drive in Vietnam and that vehicles running in Vietnam could only use gasoline sold locally, Sa needed to hire both Vietnamese and Chinese drivers and obtain gasoline from Vietnam and China in order to complete his journey.[79] To control the Chinese personnel and goods flowing into Free China via Vietnam, the French colonial government designed such regulations to protect their colonial interests against the influx of Chinese refugees and to preemptively curb Japanese objections. For Chinese refugees traveling to their own country during a foreign invasion, these regulations felt like unnecessary inconveniences, and triggered a tremendous sense of humiliation.

Sa, like many other Chinese refugees, was forced to remain in Vietnam longer than he had expected, visiting various Chinese and French government agencies in Haiphong and Hanoi. Although Haiphong had long served as the gateway to Yunnan before the war, it was hardly a regional transit center. The war changed its status in terms of the travel networks in China and Asia. On the ship from Hong Kong to Haiphong, for example, Sa met several Chinese and foreign travelers, including the president of CTS, Pan Enlin, who was on an inspection tour to Kunming, as well as a French Catholic missionary traveling from Shanghai to Saigon, a US missionary returning to Nanning in Guangxi after escorting his family to Hong Kong, and a Soviet diplomat traveling from Japan to Moscow via Chongqing. Without the war, these travelers, with diverse backgrounds and different destinations, would likely never have taken the same route. The Japanese invasion and the Chinese retreat to the Southwest transformed this quiet port into an important crossroads of East Asian travel. As a result, the number of hotels in Haiphong doubled, and as the city became a gathering place for Chinese wartime migrants, merchants, and officials, it also experienced a boom in Chinese restaurants, dancing halls, gambling houses, and brothels.[80]

As a gateway to China's Southwest, colonial Vietnam was the first unfamiliar place many Chinese wartime travelers encountered. Vietnam was also a previous tribute state of Qing China that then became a French colony after France defeated the moribund Qing in the Sino-French War (1884–85). Its colonial condition was a reminder to Chinese of China's previous humiliation under imperialist encroachments, and an alarming example of the wretched conditions of a colonized nation. While enjoying the Parisian-style coffee shops in Hanoi, Sa noticed a Chinese-style temple nearby. Possibly a remnant of the precolonial period, the pavilion was decorated with stone tablets with Chinese characters and reminded him of the Confucius temple in Beijing. By juxtaposing Hanoi's French flair with such visible Chinese touches, Sa evoked a palpable sense of loss. He noted that Vietnam's countryside looked similar to the countryside along the West River in Guangdong Province. The peasants shared similar rituals and customs. Historical relics influenced by Chinese culture exhibited in the local museum, as well as the Cantonese-speaking local *huaqiao* population, indicated the historical and contemporary ties between China and Vietnam.[81]

Chinese elite travelers also criticized the colonial condition in their wartime travel narratives. One recurrent critique of French Vietnam by Chinese travelers was the Latinization of the Vietnamese language. Commenting on the fact that the Vietnamese were forced to use Latin script to spell their own language, the literature professor Li Changzhi lamented this painful reality of cultural imperialism by underscoring the fact that "not only are they [Vietnamese] economically exploited, but they are culturally, spiritually, and morally robbed as well."[82] Similarly, Sa highlighted Vietnam's "cultural barrenness" by pointing out that there was only one bookstore in Haiphong. The local population, both Vietnamese and overseas Chinese, he noted, had no access to modern education. Although Sa—who was from a Sinicized Mongol family in Beijing—had similar plans to publish Latinized newspapers in Uyghur, Kazakh, Mongolian, and other "small ethnic languages" in Xinjiang, which was a blatant civilizing project not that different from the French tactics in Vietnam, he clearly viewed the elimination of the Vietnamese script and the limiting of people's access to modern education as a colonial act, forcing the colonized to lose their own native tongue, national identity, and entrance to modernity.[83]

Vietnam's colonial status was more than a cautionary tale for Chinese wartime travelers. Sa met with revolutionary journalists and students in Vietnam, and highlighted the national salvation struggles shared by the Chinese and Vietnamese in his travelogue. "The [Vietnamese] are not an innately backward nation. [Vietnam's] enslavement was not its people's fault but rather its previous ruler's responsibility," Sa commented. "This nation's liberation probably won't be too far off as long as the people rally together and strive for it."[84] This statement about the Vietnamese struggle against the French could also be read as Sa's expectation for the Chinese people, who had also undergone colonial and semicolonial subjugation and were in the middle of fighting against a foreign invasion. Moreover, connecting the French colonization of Vietnam to China's descent into a semicolonial condition, Sa implicitly located the fight against colonialism in Vietnam as an integral part of China's national struggle against imperialism.

Whereas wartime Chinese travelers consistently treated their passage through Vietnam as a symbolic experience of victimization by imperialism, the depiction of the Southwest during the War of Resistance often contained conflicting messages. On the one hand, as China's wartime home front, the Southwest was portrayed as a "newly revived territory" with exceptional scenic beauty, a long history of ties to the *zhongyuan* (a term in classical Chinese referring to the "central plains," the core regions of China), and a record of speedy development during wartime.[85] On the other hand, wartime travel writing underlined the "otherness" of the Southwest vis-à-vis the coastal regions. Through a spectrum of manifestations of the contrast between the core and the periphery—the geographic distance, the economic gap, and the cultural gulf, the wartime travelers emphasized the peripheral status of the Southwest even as they acknowledged its indispensability to the survival of the republic.

As an enthusiastic promoter of China's wartime development, CTS championed the Southwest as the center of China's revival. In addition to upbeat editorials in *China Traveler*, CTS published a photographic album called *Scenic Beauty in Southwest China* (*Xi'nan lansheng*) in 1939. Consisting of masterpieces by accomplished photographers with captions in both Chinese and English, the album celebrated the Southwest's topographical features, great antiquities, and modern amenities. In fact, the album provided such a positive image of the wartime home front that the Chongqing government sent copies to foreign dignitaries as official presents during the war.[86] More than an unoccupied part of China, the Southwest represented China's only path to national salvation.

As travelers crossed over the Sino-Vietnamese border to reenter China, they often took note of the Chinese soldiers vigilantly guarding the border areas and rejoiced that they had finally returned to their own country. Nevertheless, even as authors enumerated the positive aspects of the Southwest, it was inevitable that they would notice the distance—both spatial and temporal—between these wartime home fronts and coastal urban centers. Constructed in haste, the quality of much of the new infrastructure and many of the cultural institutions was subpar. The combination of poverty and a lack of hygiene posed a significant threat to the public. Even the youthful energy of the locals could be seen as merely roughness and a lack of refinement. The most negative portrayal of the Southwest, however, came from writings about areas outside of the cities. Lacking basic infrastructure and travel services, the hinterlands in the Southwest were deemed to be lagging well behind the advances of modern civilization.

After a month in Vietnam, Sa Kongliao and his travel group entered China at Zhennanguan in Guangxi. From there, it took them two weeks to traverse Guangxi and Guizhou to reach Chongqing. During their journey in these two provinces, Sa and his traveling companions, most of whom were from coastal urban centers, witnessed the dangerous conditions along the newly completed Southwest Motorways. They also complained about the scarcity of clean and modernized lodging facilities. Some local inns were so antiquated that they still hung lanterns outside printed with the old adage "seek lodgings before dark."[87] Although the facilities in provincial establishments such as the Lequnshe in Guangxi and the Sichuan Travel Service (formerly known as Chuanjiang Travel Service) in Sichuan were considerably improved during the war, hostels owned or managed by CTS were still the first choice among elite refugees. Hygienic, equipped with modern amenities, and staffed with attentive employees, CTS guesthouses seldom had vacancies. Sometimes, unable to secure a room at a CTS facility but unwilling to endure bedbugs and other insanitary conditions in a local facility, Sa and his fellow travelers preferred to spend the night in their trucks.[88]

Despite their standing as an experienced travel business, CTS found the task of facilitating travel along the Southwest Motorways daunting. When Tang Weibin, vice president of CTS, recalled this episode of wartime operation after the war ended, he cited the backward travel conditions as one of the biggest obstacles for travelers in the Southwest. He wrote, "Before the War of Resistance, the southwestern provinces were very inaccessible because of inconvenient transportation. Inns and eateries, the most basic facilities for travel, were especially crude and foul. Full of bugs and lice, they were not hygienic, let alone comfortable. It is not strange that travelers regarded [going through the Southwest] as a dangerous route and hesitated to proceed."[89] As the government regarded transportation as its first priority in the development of the Southwest during the War of Resistance, Tang pointed out, providing travelers with clean meals and comfortable rooms was of great importance. With little time to build new cafeterias and guesthouses along the road, CTS dispatched staff to different lodging stations. They chose local inns and eateries in relatively good condition to run as special contracted guesthouses and restaurants for CTS. Loaning clean bedding and furniture to these local inns, CTS also sent its own staff, who retreated with CTS from the coastal provinces, to these local facilities to supervise their operation.[90] By improving the quality of the service, CTS created a competitive environment for other local restaurants and hotels, which "envied the success [of the contracted inns] and imitated their services." Because of the efforts of CTS, Tang concluded, the contracted hotels "were able to gradually get rid of their eighteenth-century model."[91]

Major southwestern cities were also constantly attacked by the Japanese air forces. Arriving in Chongqing in May 1939, Sa encountered one of the most massive bombings to happen to the city, during which thousands of people were killed, and over two hundred thousand people lost their homes.[92] Luckily, Sa Kongliao survived when a roof collapsed above him. He wrote about huge blazes breaking out in the streets, refugees desperately trying to escape the bombarded city, and mothers devastated by the loss of their children. Despite the danger, however, Sa Kongliao valued the experience a great deal. "Not only did we witness the disastrous suffering of the people," he pointed out in his travel journal, "but we experienced the pain they experienced as well."[93] As a journalist, Sa viewed his journey as the only way to experience wartime China and to participate in the anti-Japanese struggle. By detailing these traumatic wartime experiences and the people who refused to yield to them, Sa presented a China that refugees in the international settlements in Shanghai, Hong Kong, or Vietnam seldom experienced.

It was not uncommon for travelers to compare the origin and destination of their respective trips. Unlike leisure travelers in the prewar era, however, wartime travelers like Sa Kongliao tended to reminisce about their cities of origin by projecting their sense of loss into the comparisons they drew. Through this sense of loss, the meaning of the Southwest, together with its ties to Free China, was reinforced. With these comparisons, travelers negotiated the meaning of China by highlighting regional disparities and discerning subtle distinctions. In the process, they were also redefining the substance of the nation. If the lost territories in coastal China represented the fallen center of the republic, the Southwest symbolized a newly discovered heartland crucial to the survival and revival of the Chinese nation-state. Like many elite wartime refugees, Sa Kongliao experienced not only the geographical but also the economic and cultural distance between the two Chinas. At the same time, just as his transnational journey demonstrated the underlying connections between occupied China and Free China, through his comparisons Sa reminded himself and his readers of China's oneness above and beyond its size and diversity.

Wartime Travel Narratives and Tourism in the Southwest

In the postscript to the 1938 special issue on the Southwest, *China Traveler's* editor presented the wartime as an "opportunity" for China. Quoting essayist Sun Fuxi, the editor argued that the Japanese invasion was like a jolt, awakening Chinese to a harsh reality. But more effective than the "forty years of new education," this painful reality made Chinese more aware of "the idea of country and nation" and forced them to "develop the Southwest."[94] This notion of wartime as a prime time for national unity was not simply nationalist propaganda. As described above, the wartime crisis resulted in an unexpected wartime mobility allowing people from East, North, and South China, especially those from the upper and middle classes, to move across geographical and geopolitical barriers to the Southwest.

As the War of Resistance against Japan stretched into an eight-year struggle, people from many regional Chinas intermingled and interacted with each other, creating a new political geography. Highlighting China as one inseparable whole, Sun Fuxi, in a piece written for the same *China Traveler* special issue, called for his compatriots to shatter the long-existing regionalism. Understanding the potential differences, Sun encouraged those in the Southwest to welcome the coastal "refugees" as their "brothers and sisters." Simultaneously, he warned the people relocating from the more affluent lower Yangzi region not to become "high-class refugees" who regarded the Southwest with contempt.[95]

As the war reshaped the political geography and regional divisions in China, travel and tourism also became an important medium for citizens to embrace the new reality and to identify with the home front as the core of the nation. War and leisure travel are not entirely incompatible. During the war, for example, Japan continued to encourage domestic tourism. Not only did many Japanese settlers and locals from its colonies tour the national heritage sites within Japan proper, but Japan also advocated tourism in Taiwan, Korea, Manchuria, and Japan's other exotic overseas possessions. Leisure travel fanned the heightened sense of patriotism in Japan.[96] War did not halt leisure travel in China either. Even in the middle of the retreat, refugees took short breaks to sightsee when possible.[97] With the improvements in transportation infrastructure in the Southwest, the relatively peaceful home front and the dramatic landscapes in the Southwest also became attractive for sojourners from other parts of China.

These new developments were reflected in the wartime issues of *China Travel*, which did not stop its publication during the war. Continuing its operations in occupied Shanghai, the magazine moved to Guilin after Pearl Harbor and began to publish there in 1942 until it relocated again to Chongqing in 1945. Emulating Britain's "business-as-usual" policy, *China Traveler* encouraged its readers to continue their interest in tourism during the wartime retreat. Emphasizing the utility of tourism to national defense, one author argued that tourism revenues could help cover the cost of "wartime administration and construction," while excursions and sightseeing would strengthen the body and stimulate the "patriotic spirit."[98]

Wartime retreat also changed the *China Traveler*'s coverage of domestic tourist destinations. Among the thousands of travelogues published during the Nanjing Decade, only a few dozen had been dedicated to the tourist sites in southwestern provinces, and the attractions featured in those articles were predictable.[99] Wartime *China Traveler* steered its focus to the Southwest. This shift diversified the itineraries for armchair travelers and also catered to those readers from China's eastern seaboard who had relocated to southwestern provinces. Consumers of tourism during the prewar years, these new sojourners in the Southwest yearned to discover local tourist sites to replace the destinations in the occupied regions.

Hardly venturing beyond the lower Yangzi region in the prewar era, coastal Chinese had largely viewed the Southwest as a distant and exotic land. As one *China Traveler* author from Yunnan wrote half jokingly in 1938, for coastal visitors, the name of Yunnan conjured up images of "a barren and savage land" where the locals had "long hair all over their body or even tails."[100] As the war drew the Southwest from the margins of China's national politics toward its center, however, travel writers found it necessary to rectify these stereotypes and to paint a more positive image of the region. Kunming, the Chinese terminus of the Yunnan-Vietnam Railway and home to a number of research and higher education institutions relocated from occupied China, was one of the most written-about and photographed southwestern municipalities in the wartime *China Traveler*.[101] Authors writing about Kunming acknowledged the conventional images of the city as remote, savage, and full of "harmful miasma."[102] To counter this profile, wartime travel writers highlighted the connections between Kunming and China's core regions. Li Qiyu, for example, argued that the locals were mainly descendants of Han migrants from the provinces along the Yangzi River, and so the local culture and customs were not that different. Mandarin was the common tongue. Also, Kunming played a significant role in the history of the republic, due to the fact that the Yunnan warlord Cai E was among the first to oppose Yuan Shikai. At the same time, Li downplayed the "otherness" and "backwardness" that had often been associated with Kunming and Yunnan. Acknowledging the local practice of growing opium as an economic crop as well as for local consumption, Li pointed out that the government had implemented a prohibition policy in recent years. While he admitted that Yunnan was home to many ethnic non-Han peoples, Li emphasized that many of them had already been Sinicized.[103] Travel writers also contrasted the prewar stagnation of the city with its latest developments. Before the war, newspapers in Kunming would simply copy stories from newspapers in Hong Kong or Shanghai, which often arrived in Kunming ten days after being published. With the establishment of air travel routes as well as a radio communication system during wartime, news could be released in a timely manner, and Kunming became more connected to the rest of the country.[104]

The completion of the Southwest Motorways also eased Kunming's isolation, making it more accessible as a tourist destination.[105] Besides the two different travel guides for Kunming published by CTS during wartime, introductions to Kunming's tourist attractions also appeared often in *China Traveler*, following the format of conventional travel guides.[106] Divided between the city proper and its outskirts, Kunming's tourist sites, such as Cuihu Lake, Wuhua Mountain, Jinbi Park, Yuantong Mountain, Kunming Lake (also known as Dianchi), and so on, were listed as the prime attractions.[107] Similar to the guides to Hangzhou or Beijing, the wartime *China Traveler* travelogues about Kunming included information such as the location, history, and distinct features of various attractions, while photographs were often published alongside the texts. Some travelogues also suggested unique local experiences, such as enjoying the unique Kunming delicacy *jizong* mushrooms.[108] Ubiquitous in this tea-loving city, teahouses at Kunming's many scenic spots allowed tourists to relax with tea, cigarettes, and snacks while enjoying the beautiful scenery. Interestingly, some authors mentioned Kunming's notoriously ill-mannered teahouse waiters, a phenomenon deemed an unusual tourism feature in the city.[109]

Group tours, a popular travel practice among coastal middle-class urbanites, also gained some traction in the Southwest. Liu Zhiping was an architect involved in the famous Institute for Research in Chinese Architecture (*Zhongguo yingzao xueshe*), which was relocated to Kunming from Beijing during wartime and continued to conduct field surveys of historical architecture in the Southwest. Liu published a travelogue in *China Traveler* about a five-day hiking tour around Lake Kunming.[110] Organized by the local YMCA, the tour attracted more than forty university students and office workers. With the YMCA staff guiding the tour and taking care of the logistics, the group moved according to a fixed schedule and lodged at schools and other public institutions overnight. As the tour passed through smaller towns near Kunming, Liu, out of professional habit, paid attention to township layouts and local architecture. He made a visual component a conspicuous part of the travelogue with both photographs and sketches. Whereas the photographs gave readers direct representations of the temples and buildings they visited along the tour, Liu's skillfully sketched drawings of building details offered an unusual glimpse into an architect's view of Kunming and Yunnan (figure 9). The professional gaze cast by an architect suggested how the previously ignored peripheries of the country were increasingly put under the microscope of modern industrial disciplines.

If Kunming owed its newfound fame to its crucial location on the retreat route and the cultural institutions relocated there during the war, Chongqing's frequent appearances in wartime travel writings were the result of its

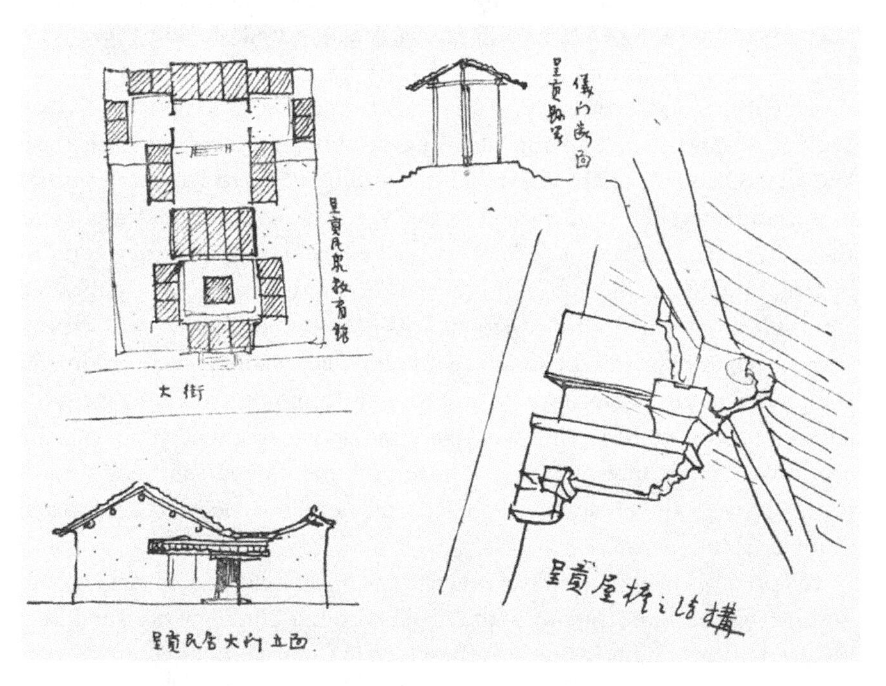

FIGURE 9. Architectural sketches by Liu Zhiping. *Source:* Liu Zhiping, "Kunming hu huan xing ji" [Around-the-lake tour in Kunming], *Lüxing zazhi* [China traveler] 14, no. 6 (June 1940): 5.

new status as China's political center. Cloistered by mountains from different sides and surrounded by rivers in three directions, Chongqing was a city built on narrow spaces. Its distinctive vertical spectacle made an impression on many wartime refugees from downriver (*xiajiang*) on the Yangzi. The famous writer Zhang Henshui was awed by how the locals built multistory structures on the sides of cliffs, which often left the entrance on the top rather than the bottom of the buildings. "In other places, people must go downstairs to get out [of buildings]," Zhang wrote, "but here, it is common to go upstairs to exit."[111] Another author, Shen Tao, reminisced about the surprising sight of city lights on the evening of his arrival. Scattered from the foothills to the mountaintops, Chongqing's city lights, in Shen's eyes, resembled the starry night sky.[112] The foggy mountain climate, the hundreds of stairs that passengers and porters needed to climb up and down at the docks, and the steep or winding city streets constituted many outsiders' first impression of this "mountainous city" (*shancheng*).[113]

Different *China Traveler* authors also recorded "strange" local customs. While often covering their head with a piece of white cloth to keep warm in the misty weather, Chongqing natives, especially middle- and lower-class men, seldom wore socks and simply put on straw shoes over their bare feet. Zhang Henshui found this mismatch in the local dress code puzzling. Other details, such as unique dishes in the local cuisine, strange expressions in the Sichuan dialect, the local wedding customs, the bathhouses on boats, and even the open sedan chairs made of bamboo, were foreign to *China Traveler*'s coastal readers. Underscoring the unfamiliarity in the daily life there, travel writers attributed the lack of mutual communication between the locals and the outsiders to poor transportation conditions. Since 1938, a huge segment of the coastal population had retreated to Chongqing, Chengdu, and even smaller towns and villages in Sichuan. As travel writers pointed out, living among and interacting with the locals during wartime, the *xiajiang* migrants gradually became accustomed to Sichuan customs. At the same time, the coastal lifestyle seeped through the barriers, having a visible impact even in the countryside of Sichuan.[114]

Travel writers took note of wartime China's state-building efforts centering on Chongqing. Alongside the influx of a huge population, more than 150 coastal factories moved inland to Chongqing in 1937–38. To meet the demand for electricity, a new high-capacity power plant was built. Among the new transportation developments, air travel made the most noticeable leap. More than ten air traffic routes linked Chongqing to regional hubs such as Chengdu and Guilin as well as major international cities such as Hong Kong and Hanoi. Energized by the many intellectuals and media workers gathering in the city, Chongqing's cultural life was dynamic as well.[115] Whereas these endeavors brought positive changes to Chongqing, wartime shortages and population increases caused spikes in the price of food, essential goods, and rent.[116] Frequent Japanese air raids also made the city vulnerable, and air shelters became a fixed feature of the wartime narrative of Chongqing.[117]

Compared to Kunming, Chongqing did not offer many resources for leisure travel. Even though CTS had organized tours to the North Hot Spring in Beipei, the South Hot Spring was closer to Chongqing and more convenient to reach, and so proved to be the more popular tourist destination for day-trippers and weekenders from Chongqing.[118] Two *China Traveler* authors wrote about their experiences at the South Hot Spring during wartime.[119] While the baths and the swimming pool were the main draw, natural scenery in the area also provided pleasant sights for urban dwellers. The must-see spots listed by these authors included mountains, caves, and waterfalls, as well as manmade structures such as pavilions and steles. Reminding seasoned travelers of the Xiaotangshan Hot Spring near Nanjing and the Huaqingchi Hot Spring near Xi'an, the South Hot Spring was also popular among holidaymakers from Chongqing because the infrastructure at the resort was adequate. Although unrefined in many coastal sojourners' view, it at least had postal and telephone services. The caves also served as natural air raid shelters.[120] With cleaner air and enjoyable scenery, it attracted people who were escaping air raids in Chongqing.[121] Leisure travel was therefore motivated both by self-preservation and self-indulgence during wartime.

As provincial capitals and the linchpins of the Southwest Motorways system, Chengdu, Guilin, and Guiyang were often featured in wartime *China Traveler* as well. Dubbed the "little Beiping [Beijing]" by the newcomers, Chengdu in Sichuan, as one author pointed out, is a city much older than Beijing.[122] Before the war, even though few had visited Chengdu, many coastal Chinese were familiar with the city through classical poems and prose written by literary giants such as Li Bai (701–62), Du Fu (712–70), and Lu You (1125–1210). Immortalizing the city, these widely known literary works also became associated with famous historical sites, making Chengdu a town rich in tourist interest. CTS even opened a touring office there in 1942 that provided travel information and tour guides for members of the Allied forces.[123] Made more accessible by the Southwest Motorways, Chengdu was the subject of a few lengthy travelogues in the wartime *China Traveler*.

No matter whether included in a thorough list enumerating Chengdu's many tourist attractions or as a concise itinerary for those who hoped to see the most sites in a limited amount of time, the three best known attractions—

Wuhou ci (the Wuhou Temple), *Du Fu caotang* (Du Fu's thatched cottage) and *Wangjiang lou* (the River-Viewing Pavilion)—remained the highlights.[124] All were linked to famous historical figures who had taken refuge in Sichuan at times of crisis.[125] They were inscribed with historical resonance, allowing wartime tourists—especially the refugees—to reflect on their own fate while identifying with the Southwest through a collective memory of the heroes, poets, and beauties from the past. Aside from its rich history, Chengdu was also famous for its lively urban life and street culture. Yi Junzuo, an author who visited Chengdu from Chongqing, recorded his seven-day visit in the city, which was filled with sightseeing and socializing events at teahouses and restaurants.[126] From drinking tea in the intellectual hub of Shaocheng Park to eating *chaoshou* dumplings at the famous Wu Chaoshou restaurant to seeing Sichuan opera with friends, these tourist activities also turned wartime Chengdu into a vibrant town.[127]

Guilin's recognizable karst hills and verdant waters appeared often in the travel writings and travel photography in popular magazines such as *China Traveler* and *Liangyou* even before the war broke out. Wartime *China Traveler* continued to feature well-known scenic spots and recommend classic tourism activities in Guilin, while including more information about local customs and wartime state-building efforts.[128] Even travelers pressed for time would squeeze sightseeing into their schedule in Guilin, and few were left unimpressed by its picturesque tranquility. The standard itinerary included famous scenic spots such as the Solitary Beauty Peak, Crescent Hill, Flower Bridge, Elephant Trunk Hill, and Seven Star Cave.[129] At the famous Seven Star Cave, one could hire one of the local guides waiting at the entrance with torches. The cave was huge, and it took twenty to thirty minutes to walk from the entrance to the exit. It was also pitch black inside. Guides with torches would point out the different rock formations that were thought to resemble certain animals or figures.[130] A boat ride to Yangshuo, a town in the southeast of Guilin, was also a must. One visitor explained in his travelogue the origin of the name Picture Mountain (*huashan*) in Yangshuo. "The so-called Picture Mountain is a cliff dyed yellow, white, and black," he wrote. "The yellow is from the iron, the white is from the limestone, and the black is from the erosion caused by the acidity from the vegetation."[131] Quoting the famous adage "Guilin's scenery is the best under heaven" in their travelogues, many wartime authors emphasized the iconicity of Guilin in the Chinese tourism imaginary.

A city perched on the Yunnan-Guizhou plateau at the center of the Southwest Motorways system, Guiyang also underwent a period of quick modernization. Like other major southwestern cities, Guiyang received waves of émigré

Chinese fleeing from war-torn regions. To serve the growing population of the city, new stores, restaurants, cinemas, and bookstores opened, many just like those in Shanghai.[132] Modern schools, such as the Xiangya Medical College from Changsha and Daxia University from Shanghai, also relocated there, attracting young men and women from all over the country. These new businesses and modern institutions changed the image of this mountainous city from a remote backwater to the key place for a national renaissance.[133]

In contrast to Guilin, Guiyang had barely registered on the tourist radar before wartime. In 1935, *Liangyou* featured a series of photographs of Guizhou Province by the Guangzhou photographer Liu Tizhi in its column *Liangyou* Readers' Travel Train. Among the images of Guizhou's famous scenic spot Huangguoshu Waterfall and landscapes of mountain ranges was a single picture of Guiyang. Setting the tiled roofs of the residential houses against a faint outline of mountains in the background, the image conveyed well what the short text in the column had summed up about Guizhou: it was a beautiful yet secluded corner of China.[134]

With its combination of impressive natural landscapes and rich cultural heritage, Guiyang turned out to be a rare gem in wartime tourism. Situated in a valley surrounded by mountains, the city of Guiyang often surprised outside visitors with its stupendous scenery. Mount Qianling and Mount Fufeng were two of the most popular tourist sites. Ascending by a series of zigzagging stairs, travelers could visit a Buddhist temple, explore a mountain cave, and enjoy an excellent view of the city from the top of the mountain.[135] Aside from the natural scenery, Mount Fufeng was famous among tourists because of the Ming Confucian scholar Wang Yangming (1472–1529), who had meditated there after being banished to Guiyang. A shrine commemorating him became the main tourist attraction where visitors could admire Wang's statue and calligraphy.[136]

Unlike the fleeting depiction of Guiyang in its prewar travel column, *Liangyou* published an entire spread titled "Kweiyang [Guiyang] in War Colour" in 1939, focusing closely on the wartime reality as well as tourism intrigues (figure 10).[137] The upper section of the spread was devoted to the images of the YMCA Rural Service Corps, which, according to the captions, were very active in involving the local population in the War of Resistance. Two smaller images framing the title in the middle section depicted a scene of Japanese air raids on the city center and Guiyang's dignified city gate draped with patriotic slogans. And in the lower half, two larger scenic photographs—one portraying terraced fields and the other depicting Jiaxiu Tower, a local landmark and tourist attraction—showcased Guiyang's unique

FIGURE 10. "Kweiyang [Guiyang] in War Colour." *Source:* "Qianjin zhong zhi Guiyang," *Liangyou* [Young companion], no. 142 (May 1939): 28–29.

landscape. Covering a miscellany of topics, these photographs portrayed the city's daily life during wartime, and made tourism an intrinsic part of it.

Travel narratives of the Southwest were the mainstay of *China Traveler*, *Liangyou*, and other print media during wartime. In the early years of the war, essays recounting lengthy retreats from coastal China to the southwestern interior provided readers with transportation and logistical information crucial for the planning and execution of their own relocation. For evacuees from war-torn regions, the Southwest represented both hope and the unknown. As many of the wartime refugees became more settled in 1938–39, leisure travel resumed in the southwestern cities, bringing back a sense of normalcy. Conventional travelogues of the Southwest began to appear more often in newspapers and magazines. In travelogues about southwestern cities, old tropes were replaced by new narratives. Guilin, once depicted as a faraway idyll and a dream destination that was hard to reach, was celebrated in wartime travelogues as a must-visit spot on the home front. Previously seen as too exotic to set foot in for east coasters, Kunming became a paradisiacal place, and even cities like Guiyang that were too obscure to register on a tourism map in prewar years became desirable destinations.

Wartime popular travel publications also presented the Southwest as a collective whole. In the *China Traveler* special issue on the Southwest, for example, a four-page photographic segment titled "the scenic line in the Southwest" (*xi'nan fengjing xian*) featured twenty photographs of major tourist attractions in Changsha, Guiyang, Guilin, and Kunming. The segment, together with the rest of the special issue, suggested a scenic thread linking travel sites such as the Flower Bridge in Guilin, the Shuangming Cave in Guiyang, and the Kunming Lake (figure 11).[138] Some of these images included tourists: a family strolling down a bamboo grove, a touring group taking a group picture at a mountain cave, and two friends posing for a camera at a pavilion. These wartime tourists in the Southwest looked not that different from peacetime travelers in eastern China, depicting a sense of normalcy during wartime disruption.

Largely written by and for coastal refugees, wartime travel narratives also emphasized the connection between the southwestern home front and an occupied China. Aside from the nicknames of "little Shanghai" and "little Beiping" for Chongqing and Chengdu, refugee travel writers often mentioned how the local commercial streets in Kunming or Chongqing reminded them of business hubs in Shanghai and Nanjing. Writer Sha Ou

FIGURE 11. "The scenic line in the Southwest." *Source:* "Xi'nan fengjing xian," *Lüxing zazhi* [China traveler] 12, no. 10 (October 1938): n.p.

emphasized the scenic beauty of Kunming, arguing that it combined the landscapes of Qingdao and Mount Lushan, while the Yuantong Temple and Cuihu Lake in Kunming reminded him of Beihai and the Biyun Temple in the Western Hills of Beijing.[139] Downplaying the cultural distance between the different areas, wartime *China Traveler* writers emphasized the homogeneity of the Southwest by leaving non-Han ethnic populations largely out of their narrative, even though at the time, an increasing number of relocated Han Chinese intellectuals began to treat the Southwest as an ethno-laboratory, and popular media like *Liangyou* continued to engage their Han Chinese audience with texts and images of the ethnic minorities.[140] Emphasizing the oneness of China, the wartime travel narrative of the Southwest enhanced the perceived relationship between the state and its components.

This chapter has traced the transition of the Southwest from a hard-to-reach backwater region to the home front of wartime China. After the Second Sino-Japanese War broke out in 1937, the China-centered travel circuits were reconfigured. Losing safe access to the Pacific coast, Chinese wartime refugees had to rely on a triangulated system connecting Shanghai, Hong Kong, and Vietnam (later also Burma and India) in order to retreat into China's Southwest. Republican wartime refugees from war zones shared similar travel experiences along this newly established retreat route. Even as the staple tourism activities of CTS were threatened and disrupted by war, as the leading operator and promoter of tourism in modern China, it forged on to provide travel services for a nation on the move. In this process, it became a crucial component of China's wartime mobilization.

The transnational route for a domestic retreat also allowed China to preserve a supply line from the outside world. The irony of this chaotic circumstance is that during a time of foreign imperialist invasion, the semicolonial setting in which China was embedded helped maintain the country's partial sovereignty. Furthermore, as a war with roots in Japan's imperial ambition to expunge Euro-American control over China and other parts of Asia, the Second Sino-Japanese War forced China to rely even more on these competitors to Japan and reinforced the preexisting semicolonial and colonial geopolitical order. The leftist journalist Sa Kongliao's wartime journey from Hong Kong to Xinjiang via Vietnam and the Southwest demonstrated the delicacy of this itinerary. Recounting their experiences in French colonial Vietnam and the remote Southwest, Sa and other coastal elites' wartime travel writing also underscored how this unique route exposed the colonial context in which China was embedded and triggered the longing for internal cohesion and a nation strong enough to resist imperialist invasion.

Another paradox of China's wartime mobility was that wartime simultaneously disrupted and stimulated travel and tourism. For CTS, while its main operations in China's eastern seaboard ceased, its expertise in travel logistics allowed it to participate directly in wartime operations. It also expanded into the Southwest, a tourism market it had yearned to enter since the prewar era. Similarly, for average tourists from eastern China, famous travel attractions in the Southwest, such as the Three Gorges and Guilin, were often difficult to reach and largely consumed via travelogues and travel photographs in popular magazines before the war broke out in 1937. Their wartime retreat and sojourn in the home front area allowed them to visit these cultural landscapes in their own backyard.

Even though there is no doubt that war had a major destructive impact on tourism in China, it did not end the practice. Rather, it changed the direction and nature of the travel. This is similar to the situation in other countries during World War II. As Bertram Gordon's study of tourism in wartime France suggests, tourism and vacations "provided a sense of normalcy" and helped people "survive and comprehend their wartime privations."[141] Moreover, wartime developments in Chinese mobility helped forge a sense of geographical oneness in China, even as the country was divided by foreign intruders.

Chapter 5

Between Empire and Nation-State

Tourism and Travel in Manchuria and Taiwan, 1912–1949

As publishing giants and travel businesses in Shanghai were busy expanding their guidebook franchises in the 1920s and 1930s, only a few made headway in publishing travel guides about Manchuria and Taiwan. The reason was simple. These two regions were part of the Japanese colonial empire, or its sphere of influence, for most of the first half of the twentieth century. While the rise of Japanese imperial tourism to Manchuria and Taiwan in the 1920s and 1930s is well-documented, Chinese tourists' experiences visiting these regions remained less visible even as modern travel routes between China proper and Manchuria and Taiwan were largely uninterrupted before 1931.[1] One of the main reasons for this disparity is that Chinese tourism to these regions was either less organized (in the case of Manchuria) or restricted (in the case of Taiwan). As will be discussed below, this was dictated by geopolitical factors.

Unlike the Northwest and the Southwest, Manchuria and Taiwan were not hard to reach or navigate via modern transportation. On the contrary, nearly 40 percent of the railway tracks built in China in the late Qing and Republican periods were concentrated in Manchuria.[2] Seaports in Manchuria and Taiwan were also amply connected to various Chinese coastal cities and major ports in Japan and colonial Korea. However, Manchuria and Taiwan not only occupied peripheral places in Republican China's nation building but also had come within the sphere of Japanese imperial outreach since the late nineteenth century. If Manchuria's fragmented semicolonial railway zones rendered it an internal outsider in China's national space, Taiwan's colonial position made this former province of the Qing a loosely attached yet separated domain in China. Manchuria and Taiwan nonetheless exerted a unique attraction for Republican Chinese travelers.

Examining the place of Manchuria and Taiwan in Chinese travel culture in the first half of the twentieth century, this chapter explores the impact of the Japanese colonial empire on the imagination of China's national space. Even though myriad obstacles imposed by the Japanese colonial order prevented the expansion of Chinese tourism in these two regions during the majority of the Republican period, there was still a steady stream of travel writing about Manchuria and Taiwan that circulated in the popular print media.

There were two short windows of time when Republican Chinese were enthusiastic about visiting Manchuria and Taiwan. For Manchuria, the upsurge in travel occurred in the late 1920s, when the launch of the Northern Expedition by the Nationalists triggered a fervent anticolonial nationalistic movement. This picked up steam after 1928, when the local Chinese warlord declared his allegiance to the Nanjing government, ending Manchuria's de facto independence from the central government. Using their travel experiences in Manchuria to raise an alarm about Japanese and Russian colonial influences, popular Chinese travel writers emphasized Manchuria's place in China's newfound anticolonial national identity.

In the case of Taiwan, while commercial publishers had produced popular geography books on the island and published a few travel writings by Chinese intellectuals in the 1920s and 1930s, the uptick in mainland Chinese travelers to the island did not happen until the Nationalist government's twilight years between 1945 and 1949, when the island was returned to China after Japan's defeat in World War II. Tourism played a central role in the decolonization process. The return of Taiwan to China soon turned this long-lost national territory into the newest tourist attraction for elite mainland travelers, and accounts of postcolonial Taiwan began to occupy prominent spaces in travel magazines and other popular print media in mainland China. However, mainland travelers' celebratory tone about the island's return was often mixed with their suspicions of the deep colonial imprint left on the land and the people of Taiwan, echoing the unpredictable future of China's fate after World War II.

Manchuria and Taiwan, two geographic and political peripheries of China's national space, were caught between the Chinese nation-state and the Japanese colonial empire. Through popular travel narratives, Chinese travelers and tourists attempted to incorporate these areas into their imagined national territory at different critical moments during the Republican period.

Manchuria as a Thoroughfare and a Contact Zone

During China's transition from empire to nation-state, Manchuria occupied a peculiar place in the Chinese national imagination. Known as the homeland of the Manchus, it had been off-limits to the Han Chinese after the Qing conquest of China proper in the mid-seventeenth century. But by the end of the nineteenth century, the once carefully maintained boundary between the Manchu homeland and Han Chinese provinces became less rigid. With the lifting of the ban on Han migration in 1895, a large number of Han Chinese from North China migrated to Manchuria's agricultural heartland, effectively transforming the region into predominantly Chinese provinces by the time the Qing empire crumbled in 1911.[3]

The increase in Han migration to the region in the last years of the Qing dynasty was also triggered by the intensification of imperial encroachment on the region, as Manchuria became the central stage for competition between Japan and Russia at the turn of the twentieth century. With the signing of the Treaty of Shimonoseki in 1895, China ceded the Liaodong Peninsula in southern Manchuria to Japan, an area that included the strategically important port of Lüshun. However, the Triple Intervention by Russia, France, and Germany forced Japan to return the Liaodong Peninsula to China. A year later, Russia approached Beijing and brokered an agreement to build a joint railway across Manchuria, the Chinese Eastern Railway, linking the Trans-Siberian Railway to Vladivostok. Shortly after the agreement, Russia seized the Liaodong Peninsula ports of Lüshun (Port Arthur) and Dalian (Dalny), which formally became Russian concessions in 1898. However, when Japan emerged triumphant from the Russo-Japanese War of 1904–5, not only did Russia lose the Liaodong Peninsula to Japan, but it also ceded the rail line of the Chinese Eastern Railway from Changchun to Lüshun to Japanese control. From that moment on, three empires—the Qing, the Russian, and the Japanese—became entangled in their struggle for supremacy over Manchuria. This entanglement persisted through the Japanese annexation of the Korean peninsula in 1910, the establishment of the Chinese republic in 1912, and the Russian revolution in 1917. These tumultuous events only further cemented the fluid identity of Manchuria as a region caught between China proper, the Russian Far East, and Korea under Japanese rule.

For Western travelers and tourists in the early twentieth century, however, Manchuria symbolized adventures and opportunities. The concentration of imperial railway zones in the region also turned Manchuria into one of the most convenient and intriguing travel destinations in East Asia. Everard Cotes, a British journalist traveling through China proper, Manchuria, Korea, and Japan shortly after the Russo-Japanese War, recognized the great potential for Port Arthur to become "a rendezvous for profitable globe-trotting traffic," while "its commercial brother, Dalny" could turn into another door to the "treasure-house" that was Manchuria.[4] Cotes's 1906 travel route—which took him from Beijing and Tianjin to Dalian via steamers, and to northern and southern Manchuria and the Korean peninsula by way of Japanese-controlled rail lines—soon emerged as a standard itinerary for Western tourists.

This heightened mobility gave Manchuria a new spatial identity. Japanese colonial operators and Chinese nation builders collaborated to maintain the region's newfound status as an international hub, particularly by facilitating through rail and steamship traffic. In the 1910s, Japanese and Chinese transportation authorities inaugurated joint conferences held in turn in China proper, Manchuria, Korea, and mainland Japan. These conferences were attended by agents from Japanese state railways, Korean colonial railways, the South Manchuria Railway, various independent railways under the Chinese government railways, and steamship companies operating in China and Japan.[5] They resulted in the emergence of through traffic and circular tour services from China proper and Manchuria to Korea and mainland Japan.

According to two pamphlets issued by the Chinese government railways in the 1910s, travelers departing from China proper could purchase "through tickets" to various destinations in Manchuria, Korea, and Japan at train stations in major cities like Shanghai, Beijing, Hankou, and Tianjin.[6] Four different itineraries were available for "China-Japan circular tours," which not only offered reduced rates on passenger tickets but also allowed through checking for luggage. While Mukden (known as Fengtian and later as Shenyang in Chinese) was the only stop in Manchuria on these "circular tours," customers were encouraged to extend their tours from Mukden to Dalian with significant fare discounts.[7] Globetrotters could also book overland trips from major Chinese cities to Europe, traveling via the South Manchuria Railway and Trans-Siberian Railway.[8]

Although Japanese colonial operators had begun to build up Japanese hegemony in Manchuria ever since 1905, Western tourist accounts and travel guides which appeared in many popular venues in the 1910s and 1920s highlighted Manchuria as a place with a fluid spatial identity, not only

Chinese and Manchurian but also Japanese and Russian.[9] It was a thrilling place to visit precisely because of this hybrid nature. In a 1910 travelogue appearing in *National Geographic*, the iconic travel writer Eliza Scidmore marveled at how Manchuria and Korea were "overrun by tourists," who were "in search of excitement, the picturesque and the unexpected." Calling it "the grand detour from the grand tour of the world," she identified a standard tourist route from Japan to Manchuria via Korea, following the steps of "General Kuroki"—the Japanese navy commander who led Japan to its triumph during the Russo-Japanese War.[10] Emily G. Kemp, a British adventurer and writer who had already published a travelogue on China in 1909, went on a four-month tour of Manchuria, Korea, and "Russian Turkestan" in 1910.[11] Two European-language commercial guidebook series, the Cook's Handbook series issued by Thomas Cook and Son and the Madrolle's Handbook series resulting from a collaboration between French explorer Claudius Madrolle and the French publisher Hachette, also produced guides to North China and northeast Asia around the same time. These commercial travel guides detailed varied travel plans to destinations in North China, Manchuria, Korea, and the Russian Far East, all within the same volume.[12]

In these Western guidebooks and travel writings, Manchuria's hybrid spatial identity was presented through the varied textures of its major cities and their attractions. Mukden, variously depicted as "a bit of 'real China'" or as the location of a "pure Manchurian" architectural style by *Cook's Handbook*, was often portrayed as a typical walled city with a distinct history tied to the rise of the Manchus.[13] Similar principal sights related to the Manchu ruling house—such as the Imperial Palace and Governor-General's Office within the city walls and the East and North Mausoleums on the outskirts—were featured in both the Cook's and the Madrolle's handbooks.[14] Calling Mukden "the permanent camp of the all-conquering Manchus," Eliza Scidmore made sure her readers in the United States would clearly identify Mukden as the epitome of a Manchu city by comparing the city's layout, its palace, buildings such as the drum tower and bell tower, and the street crowds to those of Peking (Beijing)—the other "Tartar city." She further emphasized this connection between Mukden and Peking by suggesting that the Manchu ruling class had always treated Mukden as their safety plan for "a rainy day," that is, when they were expelled from China proper. The legend that the city was full of buried treasures left by the Manchus was repeated by Scidmore to add more mysterious color, and she called the story "so plausible, so characteristic, so Manchu."[15] Similarly, in her travel writing about Mukden, Emily Kemp paid particular attention to the tombs and the Imperial Palace.

She compared the tombs outside Mukden to the Ming tombs outside Beijing. With a special permit and in the presence of officials, Kemp and her travel companions were shown the valuable collections of bronzes and porcelain, which were usually locked away in glass cases in the palace. However, complaining about how different items were crammed in too small a space to be viewed properly, Kemp suggested that the Chinese should learn from "the arrangement of our museums."[16]

If Mukden epitomized the Manchu-ness of the region through its historical sites, the port cities of Lüshun and Dalian displayed an ineluctable Japanese and Russian presence just north of the Great Wall. When Cotes visited Manchuria in 1906, he pointed out that Russian influences in Port Arthur (Lüshun) remained visible everywhere: from the remnants of the old Russian fort on the hill to the sunken Russian war vessels remaining in the harbor, and from the omnipresent droshkies (a Russian-style open carriage) to "Russian tea, knives and forks stamped in Moscow" at the main hotel in town. Simultaneously, the Japanese takeover was also obvious, evidenced by the newly erected Japanese fortress, the Japanese shopkeepers and hotel managers, and "a comfortable kimono beside one's bed, to wear on the way to a copious hot Japanese bath" in the former Russian hotel.[17] The two-day itinerary for Port Arthur from the 1910 Cook's guide was peppered with battlegrounds, commemorative monuments, and abandoned fortresses and forts, demonstrating the imprint left by the Russo-Japanese War.

Compared to the military presence at Port Arthur, "Dalny" (or Dairen in Japanese) represented the speedy modernization of Manchuria under the auspices of Japan's South Manchuria Railway. As Cook's guidebook boasted, the city was equipped with state-of-the-art infrastructure—"miles of good macadamized roads with wide pavements," "a perfect drainage system," "public gardens," "tramways," "electric light, telephones—local and long distance," "a hospital," "an exceedingly well appointed club-house," and "spacious and comfortable hotels." The principal sights included the Electric Park, a Western-style amusement park, and "Ro-ko-tum" Beach and the "Ko-ku-se-ki-sho" seaside resort.[18]

The Western guidebooks and travelogues conveyed the nature of Manchuria as a contact zone by describing the "Chinese quarter," "Japanese town," and "Russian settlement" that composed different Manchurian cities and towns. However, the segregation and antagonism between different communities were also palpable. "No Chinese vehicle is allowed in the Russian quarter," Kemp noted about "Kharbin" (Harbin), the northern Manchurian city with a significant Russian influence.[19] This separation between jurisdictions is illustrated best by the required switching of trains as people traveled along railways. Kemp, for example, summed up the transition from the Russian-controlled Chinese Eastern Railway to the Japanese-owned South Manchuria Railway in a brief paragraph:

> The first section of the railway line running southwards is still in the hands of Russia, and one's attention is continually arrested by the large numbers of soldiers who are kept all along the line to guard it. Kwan-cheng-tze is the terminus of the Russian line: it is not quite half-way from Kharbin to Moukden. The Japanese call their station at Kwan-cheng-tze Changchun, which is rather puzzling to the traveler who is unaware that the place boasts two names. All passengers have to change trains here.[20]

Along with the changeover of trains, the personnel serving in the carriages, the hospitality and tourist facilities along the lines, and the currency used to buy tickets all changed. For Western travelers, however, the incongruous characteristics of different parts of Manchuria resulting from the geopolitical jigsaw of Chinese, Japanese, and Russian spheres of authority were perhaps what made Manchuria mysterious and attractive. Recounting the rapid development of Manchuria in an article for *National Geographic*, the American journalist Frederick Simpich compared Manchuria in the early twentieth century to "the early boom days of the American West."[21] This narrative epitomized how Manchuria registered in the imaginations of Western travelers—a land of adventurous imperialist settlers, competitive energies, and inexhaustible opportunities.

Western travelers and tourists were not the only adventurers in Manchuria. Since 1906, when Japan sent its first tour to Manchuria and Korea, Japanese imperial tourism there continued on into the late 1930s. At the same time, Japanese communities in China were also eager to visit Manchuria. For example, Tōa Dōbun Shoin, or the East Asia Common Culture Academy, a Japanese educational institution in Shanghai, organized a series of field trips called "Grand Travels." Starting in 1907, the summer senior students at the Shoin were divided into groups of three to five people, and each traveled to different parts of China. Along the way, students conducted investigations and kept travel journals.[22] In the summer of 1915, seven students went on "Grand Travels" to North China and Manchuria before reaching Vladivostok in the "Russian Far East." Traveling at a time when anti-Japanese boycotts and demonstrations were springing up in major Chinese cities, the Japanese students noticed that the closer they got to Manchuria, the more hostility they sensed from the Chinese.[23] In Lianshan and Jinzhou, the first two towns they reached after entering Manchuria, local Chinese authorities refused to let them pass despite their travel visas. Only when they reached Mukden did the unfriendliness fade away. In Mukden, like their Western counterparts, the students visited the Qing Imperial Palace and the battlefield of the Russo-Japanese War. But unlike the Westerners, the students also went to a Japanese-run school and hospital in the city. Noticing the pidgin Japanese used by the Chinese peddlers on the street, these young Japanese travelers felt that they were finally in their "colony."[24] They then moved on to visit Fushun and Tieling, both of which exemplified the strong influence of the Japanese military. Businesses named after generals who had fought in the Russo-Japanese War, such as Ōyama and Tōjō, could be found everywhere.

Although their travelogue glorified the advances Japan had made, the students expressed their concerns about the Japanese in Manchuria. Taking the Chinese Eastern Railway for Harbin, the students noticed that many of their fellow Japanese passengers on the train were prostitutes and drug dealers, not real merchants. Disturbed by the fact that these illicit dealings had become part of Japanese commercial activities in Manchuria, the students were afraid that these examples would harm Japanese interests there. Their second anxiety concerned the ever-present Western powers in the region. Whereas in North China, European powers, although distracted by World War I, were still the major players in various treaty ports, in Manchuria, cities such as Harbin had a huge Russian population. Right across the Sino-Russian border lay Vladivostok, which was a heavily garrisoned Russian naval port. In addition to the Russians, German merchants had a strong presence in both Harbin and Vladivostok.[25]

Moreover, Japanese travelers also identified Manchuria as a place representing the potential risks Japan faced in China. Taking advantage of the outbreak of World War I in Europe, Japan had deepened its penetration into Manchuria and North China. However, by joining the league of the Great Powers, as the Shoin students recognized in their travel writing, Japan also became a rival and therefore a target of Western powers, as well as arousing more acute resentment among Chinese nationals. Trying to verify Pan-Asianist ideals through their journeys to Manchuria, Japanese student travelers also discovered the gap between aspirations and reality. They found that rather than symbolizing a Sino-Japanese alliance, Manchuria also embodied the antagonistic reactions to Japan's rise.

The Travel Boom in Manchuria, 1926–1931

When Chen Guangfu established the Travel Department of the Shanghai Bank in 1923, he viewed Manchuria as a key market. Yet, as discussed in chapter 1, Russian and Japanese railway imperialism curtailed his ambition to expand his travel business to the region. Chen managed to open a branch in Mukden only with the support of Chinese warlord Zhang Zuolin. With the mounting success of the Nationalist Party's Northern Expedition (1926–28), however, the possibility of a more unified national market reignited the Travel Department's interest in Manchuria.

In 1926, the Travel Department published *Dongsansheng lüxing zhinan* (*Travelers' Guide to Manchuria*), one of the earliest guidebooks produced by the travel agency. In its preface, the editor stressed that because "the general course of the Far East is tied to Manchuria," more Chinese should visit the region.[26] The core of the guidebook was a forty-day itinerary. Targeting travelers from Shanghai, the itinerary suggested taking steamships from Shanghai to Dalian via Qingdao and returning via the South Manchuria Railway and the Chinese Eastern Railway. In addition to the chief cities along these trunk lines, such as Dalian, Lüshun, Mukden, and Harbin, the itinerary also included smaller towns reachable through other branch rail lines and river routes. The majority of these recommended tourist sites overlapped with those in Western and Japanese travel guides, which included ones embodying the Japanese and Russian imperial presence in Manchuria. Dalian's main wharf area built and managed by the South Manchuria Railway, for example, was described as "one of the only in East Asia" for its "grand scale" and "meticulous design." The guidebook also suggested that travelers visit the battlefield sites and the war memorial museum of the Russo-Japanese War in Lüshun.[27] Photographs of transportation and travel facilities and notable tourist attractions were scattered throughout the guide, many of which bore both Japanese and English captions. This detail suggests that the Travel Department had likely copied these images from foreign tourist guidebooks about Manchuria and indicates that the Travel Department considered many sites featured in foreign guidebooks to be potentially attractive to Chinese tourists as well.

In 1928, Japanese Kantō army officers assassinated the Manchurian warlord Zhang Zuolin. In retaliation, the "Young Marshal" Zhang Xueliang, the son of Zhang Zuolin, signed an agreement with the Nationalist government in December 1928, bringing Manchuria, at least nominally, under the control of the Nanjing regime. This political move allowed the Nationalists to draw Manchuria closer to the orbit of the Chinese national core. With a renewed enthusiasm for national unification, travelers from the core regions of China proper developed an interest in traveling to Manchuria shortly after this decision.

Educational institutions, governmental agencies, and the press were at the forefront of this movement, spearheading group tours to the Northeast.

In 1929, for example, the Central Military Academy in Nanjing, the top academy that trained the elite officer corps for the Nationalist Party, organized an investigation tour to Manchuria.[28] While an industry investigation tour, cosponsored by six universities in Beijing, Tianjin, Nanjing, Canton, and Manchuria, government and nongovernmental agencies, and news media, set out for Manchuria in 1929, the Geological Institute in Beijing dispatched a geological expedition to the Northeast in April 1930.[29] At the same time, political and business elites and other individuals with means also went to visit and investigate Manchuria on their own. One of the prominent examples was Lu Zuofu, the founder of the Minsheng Shipping Company, who toured Manchuria's main industrial sites in 1930.[30]

Responding to this upsurge in interest, in 1930, the Nationalist Party argued that travel to Manchuria was imperative to "facilitating expansion and colonization" and "understanding the frontier situation" in the area. It urged the Ministries of Railways, Communication, and Education to promote travel to Manchuria in more meaningful ways.[31] The Ministry of Railways issued an eight-point measure to offer group travel discounts to investigative tours to the "eastern three provinces" (*dongsansheng*). Based on this measure, after obtaining a letter from their university or the provincial Chamber of Commerce, any student or business group with a minimum of three members could apply for preferential travel passes for their journey to Manchuria, which would guarantee a half-off discount on their train fares on any government railway.[32] New tourism development initiated by Chinese institutions also surfaced, especially in Mukden. The Beijing-Mukden Railway, the only Chinese government railway partially in Manchuria, built a branch line from Mukden to the North Mausoleum on the city's outskirts and ran special tourist trains during holidays.[33] In 1929, the North Mausoleum Park Administrative Office issued a pamphlet, introducing different structures, tombstones, and ten famous views in the park via texts and photographs.[34] The Shen-Hai (Shenyang to Hailong) Railway, built by the warlord regime in the 1920s, operated a similar special tourist train to the East Mausoleum during the peak summer season. Sensing a potential travel boom in the Northeast, CTS sent a manager to northern Manchuria to research the possibility of opening a new branch in Harbin.[35] At the same time, *China Traveler* frequently featured travel writings about the key cities in Manchuria, such as Dalian, Mukden, and Harbin.[36]

The Manchurian Incident, however, abruptly ended this travel boom in late 1931. In 1932, the Japanese established Manchukuo (*Manzhouguo*), "the land of the Manchus," and installed Pu Yi, the last and deposed emperor of the Manchu Qing dynasty, on the throne of their puppet regime. A de facto border at the Great Wall between Manchuria and China proper emerged, preventing Chinese travelers from entering the region as conveniently as before. The Chinese customs houses in Niuzhuang (known then as Newchwang in English, today's Yingkou) and Harbin were forced to close and relocated to the Great Wall. The Manchurian segment of the Beijing-Mukden Railway was taken over by the Japanese.[37] Travelers between Beijing and Mukden had to disembark at Shanhaiguan and switch to a Japanese train. Jointly managed by the Chinese and the Russians, the Chinese Eastern Railway in northern Manchuria was also taken over by Japan after it was sold to the South Manchuria Railway by the Soviet Union.

In this new political climate, CTS's role in Manchuria shifted again. After the cease-fire agreement between China and Japan was reached in May 1933, the Nationalist government, while continuing to protest the Japanese occupation of Manchuria, began to negotiate the terms under which the through rail traffic between Beijing and Mukden could be resumed. In order to avoid any presumption of China's recognition of Manchukuo, the Ministry of Railways hoped to commission a private commercial company to oversee the operation. CTS became the obvious choice. Even though the company was aware of the political sensitivity of the issue and the potential fallout from its collaboration with the Japanese, CTS management found it impossible to reject the government request while hoping to retain and expand its network in North China.[38] So it agreed to take up the responsibility on a six-month trial basis.[39]

Consisting of staff members from CTS and the Japan Tourist Bureau, a joint travel agency named *Dongfang lüxingshe* (Oriental Travel Bureau) was founded at Shanhaiguan to handle through traffic. Zhang Shuiqi, a manager from CTS, served as its chief manager, while the Japanese representative Hirayama Sadanari assumed the role of the associate director of the agency. Rather than a representative from the Japan Tourist Bureau as publicized in the agreement, however, Hirayama was in fact sent by the South Manchuria Railway. To prevent the misapprehension that China recognized the imperial railway and the puppet Manchukuo regime, the final contract was signed by Zhang and Hirayama as individuals rather than on behalf of their parent companies.[40] The trains on the through-traffic service would switch their locomotives and their operation staff at the Shanhaiguan station. While Chinese staff served on the trains between Beijing and Shanhaiguan and their Japanese counterparts would take over between Shanhaiguan and Mukden, all serving staff onboard wore the uniform of the Oriental Travel Bureau. Railway passengers would be inspected at Shanhaiguan too, following a simple inspection procedure drawn up by the Beijing-Mukden Railway, the

South Manchuria Railway, the Chinese Customs, the Customs of Manchukuo, and the Oriental Travel Bureau.[41]

The outcry against this arrangement as appeasement of the Japanese was immediate. The first Mukden-bound train from Beijing in three years was bombed on July 1, 1934, killing two Chinese passengers.[42] A call to boycott CTS surfaced in Shanghai and elsewhere, and protesters threw a rock through the large window of the CTS headquarters one night. Newspapers also called Zhang Shuiqi a "traitor to China" (*hanjian*), a term reserved for collaborators with the Japanese.[43] In early 1935, CTS decided to withdraw from the joint venture.[44]

Although a blemish, this unique episode in CTS history still conformed to the agency's consistent role in serving the Nationalist state's interest. At the same time, CTS involvement in resuming the through traffic between China proper and Manchuria also suggested that movement between the two was not completely prevented by the Manchurian Incident and the subsequent demarcation at the Great Wall. Just as the Chinese state endeavored to maintain the connections between China proper and Manchuria, Chinese travel businesses continued to consider the Northeast to be a part of the national tourism market, even as the political environment underwent drastic changes.

Travel Narratives about Manchuria during the Travel Boom

Although Manchuria was not marketed as a leisure travel destination like Huangshan or the beach resorts in Qingdao and Beidaihe, it gained a foothold in Chinese national tourism in the late 1920s and the early 1930s for its crucial place in the imagination of China's national space. *China Traveler*, as mentioned above, featured a string of pieces about Manchurian cities, including a series of essays penned by its chief editor, Zhao Junhao, who took part in an Observation Tour to the Northeast organized by the newspaper union in 1929. Between 1929 and 1932, nearly every issue of *Liangyou* included news of the Northeast, along with travel essays and photographs.[45] After accompanying the Central Military Academy investigation group to Manchuria in the summer of 1929, Ma Hetian, a Nationalist political agent and an expert on Northwest affairs, published his travelogue on the Northeast first in *New Asia* magazine and subsequently as a book in 1934.[46] Lu Zuofu, the owner of the Minsheng Shipping Company, conducted a tour of Manchurian industrial sites in 1930. His travel journal appeared in multiple publications.[47]

Like their Euro-American counterparts, many Chinese travelers were drawn to the exotic elements of Manchuria as well. Such details even popped up in the most mundane parts of their journeys. As the majority of the transportation apparatuses in Manchuria were either owned by the Japanese or managed by the Russians, their amenities prioritized the needs of travelers from non-Chinese backgrounds. On the Japanese steamship SS *Sakaki Maru* from Shanghai to Dalian, for example, Zhao Junhao noticed that the majority of the books and magazines available in the reading room were in Japanese or Western languages. In contrast, the only Chinese publications there were a few old volumes of *Eastern Miscellany*.[48] The washrooms on the South Manchuria Railway trains were done up either in Western or traditional Japanese style. At the same time, the meals served on the Japanese-owned steamship and trains were often either the Western-style *xican* or Japanese-style bento boxes and *donburi* (Japanese rice bowls).[49] The Japanese or Russian train conductors and service staff spoke little Chinese.[50] At the spacious South Manchuria Railway station in Dalian, as Lu Zuofu observed, Japanese policemen and other Japanese personnel were omnipresent, which "makes you wonder where on earth you are."[51] Similarly, Xu Hongtao, a traveler from Zhejiang, commented on the prevalence of Japanese police, conductors, and inspectors on the South Manchuria Railway train he took from Changchun to Dalian.[52]

In Chinese popular travel narratives, the hybrid influences from Japan and Russia were presented as part of the intrigue of tourism. At Dalian, Zhao Junhao called readers' attention to different facilities built by the Japanese. At the wharves, he was drawn to the vast waiting room and the huge semicircular entrance pavilion of the passenger wharf. Visiting the famous beach resort Xinggepu (known as Hoshigaura in Japanese), Zhao saw the aquarium and praised the well-built asphalt road leading to the beach site, which were both lined with trees. The nearby mountain was called the "Fuji of Dalian" by the Japanese, Zhao noted, because its shape resembled that of Mount Fuji, a sacred mountain and national symbol of Japan.[53] Ma Hetian was also impressed by Dalian's beach, where one could find "Chinese, Japanese, English, and American" tourists among the many beach-goers there. The Japanese-run aquarium in Dalian, Ma wrote, was even better than the aquarium in the Philippines opened by the Americans.[54]

Whereas Dalian was depicted as a quintessential modern Japanese town, Chinese travelers considered Harbin as a city with international flair. Recounting the history of its remarkable transformation from "a small hamlet along the Songhua River" to the "little Paris of the Orient," Wang Yuting pointed out that foreign residents in Harbin consisted of seventy thousand to eighty thousand Russians, roughly three thousand Japanese, and a handful of other Euro-Americans.[55] "Just standing in the streets for a short while makes you feel like you are in a foreign country," Zhao Junhao wrote in his travelogue.[56] Some travel writers even claimed that Harbin was more cosmopolitan than Shanghai, with parks that were better designed than the public gardens there.[57] With the easy access to the Trans-Siberian Railway, travel from Harbin to Moscow took seven days, and Harbin to Paris a mere twelve days. Accordingly, all kinds of consumer goods from Europe and the United States could be found in Harbin, some of them "even more trendy than what appears in Shanghai."[58] The prevalence of dance clubs, coffee shops, and movie theaters further confirmed Harbin's high standard of living and cosmopolitanism, at least in the foreigners' sector. The city's European influence was so dominant that the Chinese businessmen in Harbin could speak Russian and would take off their hats to shake hands when meeting friends, following the European manner of greeting.[59]

The Russian population decreased in northern Manchuria as it came under Chinese rule in the mid-1920s, when the Eastern Provinces Special District Administration was established to oversee the railway concessions recovered from the Soviet Union. Nonetheless, the apparent Russification of Harbin was still striking to Chinese travelers. Not only was Russian the more common language in Harbin, but travel service providers, such as hotel managers and taxi drivers, were also predominantly Russian.[60] Woken up by a chorus of church bells ringing on a Sunday morning, Xu Hongtao found that Russians—men and women, old and young—were rushing to the Orthodox churches scattered around the town.[61] Visiting the waterfront of the Songhua River in summer, Lu Zuofu witnessed how the Russians in Harbin enjoyed swimming and sunbathing along the shore and on Sun Island (*Taiyang dao*) in the middle of the river, as well as dancing and spending their leisure time at coffee shops and ice cream stands along the riverbank. With the exception of a few Chinese men and even fewer Chinese women, Lu observed that the majority of the beach-goers were Russian men and women, especially couples, who did not feel too self-conscious in their minimal swimwear to publicly display their affection.[62]

Like Western tourists, Chinese travelers in Manchuria in the late 1920s found themselves experiencing the peculiar sense of being in a cosmopolitan alien land with hybrid foreign influences. This could also be alarming: to them, Manchuria was an integral part of the Chinese nation-state especially after 1928, and nothing caused more anxiety among Chinese travelers than the obvious Japanese and Russian presence in Manchuria. As travelers arrived in Dalian, for example, Japanese quarantine doctors would come onboard to examine passengers before they could disembark the steamship, while the passengers' luggage would be inspected and individuals questioned by Japanese customs officers. These measures implied a border carefully guarded by the Japanese imperial authorities.[63]

Dalian was where the South Manchuria Railway Company's headquarters were located. Whereas "all important institutions in Dalian were under the auspices of the Mantetsu," Xu Hongtao commented, there was neither Chinese political presence nor sizable Chinese businesses in town. Bemoaning the clutch Mantetsu had on Manchuria, Chinese travel authors also compared it to the British and the Dutch East India Companies of an earlier day, alluding to the imminent danger of Japanese colonization of Manchuria.[64] At the Museum of Products from Manchuria and Mongolia (*Man-Meng ziyuan guan* in Chinese, or *Man-Mō shigen kan* in Japanese) in Dalian, the industrialist Lu Zuofu was shocked to find specimens of plants, animals, and minerals from all over Manchuria and Mongolia. Not only were there specific objects on display, but the Japanese also made statistical tables to indicate the annual outputs of various agricultural and mineral products. Exclaiming that the Japanese had gathered all the treasures of the "Three Eastern Provinces" in a few rooms "as if they belonged to them," Lu also underlined that these Japanese exhibition facilities had targeted Japanese visitors in order to stimulate business-related interest in Manchuria. "What should we Chinese do?" Lu wrote with apprehension.[65] Visiting the same museum in 1929, Xu Hongtao noticed an exhibition hall showcasing an archetypal family in Manchuria: "Women with bound feet, old men with queues, wearing peacock feathers and hat buttons, and next to some water pipes [for tobacco] and opium pipes." These details were clearly intended to taunt and vilify Chinese people.[66] Chinese visitors also noticed that the special collections at the South Manchuria Railway library were labeled "colonization," "Manchurian-Mongolian management" (*Man-Meng jingying*), and "transportation" respectively, displaying Japan's treacherous intent.[67]

The imprint of foreign imperialism on Manchuria could be found well beyond Dalian. The Japanese war memorial in Shenyang (Mukden)—called *chūreitō* in Japanese (memorial monument to loyal spirits)—was a site that could trigger bitterness among Chinese travelers. Commemorating Japan's triumph in the Russo-Japanese War, the memorial tower was a bomb-shaped monument alongside a huge cannon. Literally representing Japanese militarism, its existence on Chinese soil, as one author commented, was an affront to China and Chinese people.[68] After a tour of the Cultural Relics Institute in Harbin, Zhao Junhao bemoaned the fact that this establishment once attached to the Russian-controlled Chinese Eastern Railway demonstrated the "sinister intentions" of the Russians. A seemingly cultural institution, it was a Russian enterprise in charge of investigating, surveying, and then collecting, editing, and assembling data about Manchuria and Mongolia. "Half of it was unknown to us Han Chinese," Zhao emphasized. While he criticized the imperialist ambitions embedded in these Russian data-collecting activities, Zhao also lamented the incautious attitudes of the Chinese, which had allowed these foreigners to "map and measure our great rivers and mountains" in the first place.[69]

While Chinese travelers were alarmed by the colonial institutions in Manchuria and the expansionist intent they represented, they were also sometimes impressed by the "scientific" methodology and the propagandist function embodied in the imperial tourist sites, and considered them a valuable lesson. This ambivalence was common in the popular travel narratives of Manchuria. Zhao Junhao, for example, was very vocal in criticizing the Japanese presence in Manchuria. He called the South Manchuria Railway a "sharp weapon of Japanese imperialism" and "maggots attached to our bones."[70] Yet Zhao was in awe of the Japanese constructions in Manchuria, observing that the high-quality infrastructure indicated an "intention for them to last a long time."[71] This passing comment conveyed a simple message: the more attractive and grandiose the Japanese projects in Manchuria, the more dangerous a threat Japanese imperialism posed for China. When Lu Zuofu visited the Japanese-operated coal mine in Fushun, he discovered that the management anticipated that the site could be utilized for at least three hundred years, which, Lu alerted his readers, was how long the Japanese planned to exploit Manchurian resources.[72]

Chinese travel writers also focused on new Japanese place-making tactics. Zhao Junhao, for example, lambasted Japan for coining the term *nanman* (South Manchuria) and spreading the false claim that the northeastern provinces had not belonged to China originally, which obscured China's sovereignty and related rights in Manchuria.[73] Ma Hetian, too, insisted that Chinese should stop use terms like *Manzhou* (Manchuria) and *Manmeng* (Manchuria and Mongolia). Rather, Chinese should call it *Dongbei* (the Northeast) instead, since the Northeast had been part of China "since ancient times."[74] Traveling from Dalian to Lüshun, Lu Zuofu emphasized that the area was very much "domesticated" by the Japanese. Called *Kantō-shū*, this Japanese-leased territory had adopted an administrative system from Japan. He further warned readers that the area was already painted the same color as the rest of Japan on the maps produced by the Japanese. Lu's travel companions joked that since Chinese cartographers still painted this part of Manchuria as "belonging to China," the Chinese and Japanese seemed to be able to avoid conflict by each enacting their own claims separately.[75] Although a passing witticism, this comment summed up Manchuria's double spatial identity at this critical historical moment.

While condemning the Japanese institutions and actions in Manchuria, Zhao Junhao pointed out that there was a gap between Japanese attention to Manchurian tourism and that of his fellow Chinese citizens. Zhao cited the fact that Japanese institutions organized more than 250 tour groups to Manchuria in 1928, which brought in more than 10,600 Japanese tourists. This number did not even include individual travelers. The South Manchuria Railway Company also conducted surveys, on which many of its publications were based.[76] "As soon as the Japanese came here," Zhao noted, "they learned the local dialects and investigated the social conditions."[77] These colonial tourism programs and institutions demonstrated the role tourism and travel played in deepening imperialist penetration in China. Zhao exhorted his fellow Chinese citizens to head to Manchuria to counter imperialism.

Tourist sites emphasizing Chinese influences or state presence, therefore, became an indispensable part of the itineraries of Chinese tourists. In addition to the mausoleums and the Manchu Imperial Palace in Shenyang, Chinese tourists toured military facilities, arsenals, radio stations, and Chinese-run universities and military academies, while attending banquets and meetings hosted by different provincial leaders, including Zhang Xueliang, the warlord turned regional leader. In Harbin, Zhao Junhao and his fellow travelers toured the telephone company, the railway bureau, the forestry company, and the University of Industry.[78] Even though many of these facilities had been initiated and funded by Chinese warlords before they swore allegiance to the Nanjing government, they nevertheless presented a visible Chinese state presence, which was particularly desirable for those yearning for a unified China. Just as the Japanese student travelers from Tōa Dōbun Shoin had visited Japanese hospitals and schools in Dalian during their "Grand Travels," these Chinese travelers went to see Chinese communications infrastructure, military facilities, and educational institutions. At the same time, with Japan encouraging its colonial subjects in Korea to migrate to Manchuria, Zhao Junhao argued that "in order to save the Northeast [from the Japanese], there is no other solution than immigration."[79]

Although the Manchuria travel boom only lasted for a short period of time, popular travel narratives about Manchuria played a crucial role in reconciling its semicolonial status and its emerging spatial identity as part of Nationalist China. In these travel narratives, while different authors from the national core denounced the imperialist institutions and colonial programs, their writings also demonstrated a subtle identification with what defined the modernity of Manchuria. For them, Manchuria's rightful place within the Chinese nation-state needed to be restored by resisting the Japanese advances. However, they also suggested that the vibrant modern identity of Manchuria should be preserved by learning from the Japanese and other colonial powers. From mimicking Japanese scientific methods in establishing different economic, social, and cultural institutions, to initiating similar activities such as tourism and migration, they considered these colonial strategies to be compatible with and even indispensable to China's national modernization efforts in Manchuria.

Even though Chinese tourism to the region was reduced by the Japanese occupation of Manchuria between 1932 and 1945, popular travel narratives about Manchuria continued to function as a discursive strategy to include Manchuria in the imagination of China's national space. As the fragile political ties between China proper and Manchuria that had emerged in the late 1920s were drastically severed, these narratives continued to circulate in popular print media.[80] Japan's occupation of Manchuria remained a great source of anti-Japanese nationalism in Republican China, which further strengthened the Chinese assertion of sovereignty—in travel narratives, at least—over the Northeast.

Japan's "Island-Colony" in Chinese and English Travel Narratives

Taiwan, located across from Fujian Province, was formally incorporated into the Qing empire in 1683 and made a province in 1885. Taiwan attracted attention from Qing official travelers, who produced a variety of texts about and images of the island's history, land, and peoples. As Emma Teng has argued, these texts shaped the Chinese attitude toward Taiwan and heightened the Qing colonial control of the island.[81] By the end of the nineteenth century, however, the situation changed. After China's defeat in the first Sino-Japanese War in 1895, Taiwan was ceded to Japan and remained a Japanese colony until 1945.[82]

During the Republican period, Taiwan's position in China's travel network and its place in China's national space were ambiguous. As a Japanese colony, Taiwan's ties to China had been officially broken off before the establishment of the republic. Because the island did not have a contiguous land border with mainland China or any other country, it was relatively easy to cordon it off from the Chinese tourist flow, while allowing controlled access to other travelers. In fact, during the Japanese colonial rule, non-Japanese tourists, especially Chinese, needed an official invitation from the Governor-General's Office or other approval from Japanese institutions in order to enter Taiwan. Whereas Manchuria's semicolonial status from 1905 to 1931 and its location abutting northern China made it unambiguously part of China's national space, Taiwan's colonial status placed it out of bounds to Chinese travelers.

This does not mean that Taiwan was left out of the imagination of China's national space. After Taiwan became Japan's colony, however, the loss of the island conjured up a shift in the Chinese popular imagination. Immediately after 1895, the defeat by Japan and the cession of Taiwan were often cited in Chinese journals and political circles as the symbol of China's impending *wangguo* crisis and as the source of a great humiliation. A term literally meaning "lost country," *wangguo* was used by late Qing thinkers to denote colonization.[83] As Teng has pointed out, in contrast to the early Qing travel writings about Taiwan's "entering the map" when the island was conquered by the Qing empire, "Chinese nationalists . . . did not speak of Taiwan's 'exiting the map'" after 1895. Instead, they considered Taiwan a limb "severed" (*ge*) from China's body. Teng further emphasizes that this notion of Taiwan being severed from China indicated that the island was indeed perceived as an integral part of China.[84] Triggering a sense of crisis in the late Qing period, the loss of Taiwan continued to be a symbol of national humiliation into the Republican period. This paradox of Taiwan—never having been part of the Chinese republic, yet imagined to be integral to China's national space—shaped Chinese travel narratives about Taiwan in the Republican period.

Taiwan did not attract as many Western and Chinese tourists as Manchuria in the early twentieth century. As one British travel writer claimed in the early 1920s, this "island-colony of the Japanese" remained "far from the beaten track of the tourist" and had "yet been penetrated by the myrmidons of Thomas Cook."[85] The majority of the Euro-American visitors who had the opportunity to tour the island were either social and political elites invited by the governor-general of Taiwan or were on diplomatic or missionary duties in Japan or on the island.[86] Western elite travelers also ventured to Taiwan after visiting Japan proper, while nontourist travelers, including anthropologists and ethnologists, sojourned in Taiwan to study its "savage headhunters."[87] Despite or perhaps because of its obscurity compared with Manchuria in global tourism circuits, English travel writings painted colonial Taiwan as one of the most exotic and intriguing destinations of the time.

Similarly, between 1895 and 1931, only a handful of Chinese travelers had the opportunity to visit Taiwan, and most were from a distinguished background. While still in exile in Japan, Liang Qichao, for example, visited Taiwan in 1911 with an introduction letter from a Japanese politician.[88] In 1916–17, when the Taiwan Industrial Exhibition (*Taiwan kangyō kyōshinkai*) was held in Taipei (known as Taihoku in Japanese) to celebrate the twentieth anniversary of Japanese colonial rule, official delegations from Jiangsu and Fujian Provinces attended the event and toured the island.[89] After the establishment of the Nationalist government, even fewer Chinese travelers ventured to Taiwan. Disguising himself as a Singaporean merchant, Huang Qiang, a military officer in the Nationalist Revolutionary Army, visited Taiwan in late 1927 and published his travelogue in 1928.[90] Jiang Yong, who penned the only article about Taiwan found in *China Traveler* before 1945, made a fundraising trip to Taiwan in 1929 as the representative of the US-sponsored China International Famine Relief Commission.[91] A professor and the chair of the Department of Chinese Studies at McGill University, Jiang Kanghu (better known as Kiang Kang-hu in the West) toured Taiwan in the summer of 1934.[92] This handful of eyewitness accounts of Taiwan under Japanese rule sustained a public interest in Taiwan among Chinese urban readers.

By the late 1910s, outside travelers to Taiwan, Japanese or non-Japanese, tended to follow similar travel routes. Entering at either Keelung (Kiryū in Japanese) or Kaohsiung (Takao in Japanese)—the chief ports of northern and southern Taiwan—tourists moved along the western coast via the Taiwan Government Railway and made stops at Taipei, Taizhong (Taichū in Japanese), Chiayi (Kagi in Japanese), Tainan, and Kaohsiung, which were the main stations along the railway line. A comparison between the itinerary suggested by the Japan Tourist Bureau, the recommended sights in British and US guidebooks, and the travel sites documented by Chinese visitors reveals many commonalities. In all of them, cultural institutions and public places such as museums, botanical gardens, and parks were often recommended, along with Japanese Shintō shrines, such as the Taiwan Shrine in Taipei and the Kaizan Shrine in Tainan. The most celebrated and commonly visited natural sites were the Sun-Moon Lake (*Riyue tan* in Chinese, and *Jitsugetsu tan* in Japanese) and Mount Ari (*Ali shan* in Chinese and *Ari san* in Japanese). Japanese and non-Japanese tourists also frequented some of the largest sugar refineries, sawmills, and other industrial production sites in Taiwan.[93]

Chinese, Western, and Japanese travel narratives also shared common themes. Like their Japanese counterparts, English guidebooks and travelogues tended to draw readers' attention to the island's "rare beauty and charm," especially in its natural landscape and diverse vegetation.[94] They attributed the Western name for Taiwan, Formosa, to the sixteenth-century Portuguese voyagers who had spotted the island as they skirted its west coast and called out, "Ilha Formosa" (Beautiful Island). Some English guidebooks produced by Japanese institutions juxtaposed the Western name "Formosa" with the Japanese name "Takasago," which had been given to the present-day Taiwan by Japanese visitors "in olden times." As one book claimed, Takasago also "mean[s] 'Beautiful Island.'"[95] Associating Taiwan's natural beauty with its fateful "discovery" by Western and Japanese explorers, these naming stories emphasized the perceptiveness of the explorers, foreshadowing the island's potential for colonization.

Indeed, besides Taiwan's natural beauty, the economic value of its "bright green ricefields," "plantations of waving sugar cane," and "mysterious jungles of mighty treeferns" was not lost to travel authors writing in English.[96] In a travelogue appearing in *National Geographic*, Alice Ballantine Kirjassoff used more than seven pages of text and photographs to introduce the production of camphor in Taiwan, which "holds a practical monopoly in the world's market of this valuable drug."[97] Just as the contemporaneous Japanese guidebooks emphasized Japan's colonial vision of modernity by showing tourists how Taiwan's natural resources were made into profitable commodities, English-language travel narratives linked Taiwan's exotic natural beauty to the potential for direct economic benefits, rendering the colonial intent of the Japanese exploitation into a self-evident response of modernity toward nature.[98]

Chinese travelers' reactions to Taiwan's rich natural resources and their transformation into remunerative enterprises were not as unambivalent. Chinese visitors did praise the business ventures in Taiwan. They viewed Japan's success in Taiwan as a road map for China's own modernization. When Liang Qichao visited the island in 1911, he listed ten specific topics he planned to investigate, which included industry and commerce. As the majority of Taiwan's population was ethnic Chinese, Liang noted, such investigations would provide valuable lessons for reform agendas in China, especially on China's northeastern and northwestern frontiers.[99] In the 1910s and 1920s, other official tourists were eager "to investigate its industry and commerce."[100] Chinese professional schools sent students to tour Taiwan with a focus on its agricultural and mining facilities.[101] While outright praise of the Japanese colonial rule was rare in Chinese officials' travel reports, details of the staple products and commodities of each prefecture of Taiwan were carefully recorded. One visitor to the Industrial Exhibition in 1916 compared the outputs of sugarcane, gold, and coal in Taiwan between 1899 and 1913, which indicated increases of more than tenfold.[102] After touring the Monopoly Bureau in Taipei, Huang Qiang argued that China should follow the model of Taiwan and take control of key commodities. At Mount Ari, Huang painstakingly recorded the forestry management, logging businesses, and the operation of sawmills in the area. The model of Taiwan's forestry management could be transplanted to Hainan Island, Huang argued.[103]

Meanwhile, some Chinese travelers pointed to troubling issues raised by Taiwan's economic development. Liang Qichao viewed the monopoly and other economic policies as discriminatory toward the native Taiwanese population. While people's incomes seemed to have increased under colonial rule, he commented, the cost of living rose precipitously, because the Japanese authorities imposed high tariffs on Chinese goods and nearly all the everyday consumer goods on the island were imported from Japan.[104] Jiang Yong, a jurist and the former vice-minister of justice in the warlord government, singled out the risks of excess production. He pointed out that with the German invention of synthetic camphor, the future sale of Taiwanese camphor became unpredictable. When touring the Taiwan Sugar Company in Pingtung (Heitō in Japanese), the largest sugar refinery in Taiwan, he mentioned that because of the excess production in recent years, the company's stock price had plunged, causing negative ripple effects in Japanese and Taiwanese financial markets.[105]

The degree of Taiwan's economic success also became a measurement of how big a loss China had suffered from ceding Taiwan to Japan. The more successfully the Japanese exploited Taiwan's riches, the more poignantly Chinese travelers felt the loss. As Jiang Yong stressed in his travel essay in *China Traveler*, because the production of rice, sugar, and camphor in Taiwan was "unparalleled throughout the world," Taiwan's annual income had reached 120 million yen. China had lost all this "just because of one battle," he added, "and what a pity it is!"[106]

Taiwan's aboriginal people were also often featured in both English and Chinese travel pieces. Many general guidebooks and travelogues used scientific vocabulary, especially about race, to describe the origin of Taiwan's aborigines. The original inhabitants of the island were of "Malayan or possibly Polynesian origin," as one travel author claimed, with blood ties to "the original Filipinos" and "the Dyaks of Borneo."[107] Chinese travelers, too, identified the race of the *fanzu* (the aborigines) as different from the Chinese settlers and related instead to the indigenous people in *Nanyang* (a Chinese term for today's Southeast Asia), who belonged to the same race as the Malays.[108] Witnessing two aboriginal women at the "Pavilion of Savages" at the Industrial Exhibition in 1916, a Chinese official traveler noticed their indigo facial tattoos and exclaimed how valuable they were as research materials for ethnologists.[109]

Overlapping with this scientific terminology were stories about and images of headhunters. Travel narratives' highlighting headhunting in Taiwan as a trope denoting the savagery and lawlessness of the indigenous peoples was not new. Whereas the Qing travel writers and cartographers marked the divide between the civilized and the barbaric by presenting the bloodthirsty nature—the "rawness"—of the headhunting savages, early twentieth-century travel writers were curious to witness this within the safe environment ensured by the Japanese colonial institution.[110] When the colonial government set up programs to tame the indigenous people, it also encouraged tourists to visit a model savage village at Mount Jiaoban (*Jiaobanshan* in Chinese and *Kappanzan* in Japanese) in north Taiwan after the late 1910s. Accompanied by police escorts, visitors would see the villagers' houses and family life, while visiting the Japanese school for savage children and buying local woven garments made for tourists.[111] Some Western travel writers tracked the earlier problems the aborigines had caused to the Japanese and how the Japanese launched punitive expeditions and built barbed-wire fences charged with electric currents to fend off the raw savages. These tactics reportedly finally led the savages to yield to the Japanese.[112] For Western travelers, the thrill was obvious. One English guidebook summed it up this way: "the possibility of safely hobnobbing with savage head-hunters who secretly covet the visitor's head but are prevented by Japanese law and watchfulness from taking it unless the traveler is willing!"[113]

Even though most Western visitors were excited to observe the villagers, some were a little disappointed. In her account of the trip to "Kampanzan," Alice Ballantine Kirjassoff wrote about the calm village life she had witnessed. Upon their arrival toward dusk, the village women were making dinner and children were running around. "[T]heir sires, one-time braves, but now mere blear-eyed phantoms of savagery" was how she described village men. Afterward, at the school for savage children, they saw the children singing the Japanese national anthem and making speeches. Impressed by the children's ability to perform, she was amused by the fact that the speeches "were so obviously the product of the teacher's pen," which followed the same script and praised the "kind Japanese."[114] Even though the Japanese educational program for the aboriginal children impressed her, its performative side did not pass without notice. Yearning to witness savage warriors and authentic indigenes, she was disappointed by the ordinariness of the indigenous village.

If, for Western travelers, visiting the model savage village was an opportunity to view the Taiwanese aborigines up close, Chinese tourists visiting Kappanzan regarded it as the perfect location to observe how the Japanese "control the barbarians" (*lifan*).[115] For example, when Huang Qiang visited Kappanzan in 1927, he was particularly interested in the Japanese institutions and their function in the pacification of the savages. He explained their significance this way: "The *Li* people of Hainan island in our Guangdong Province are a huge problem for the governance of Hainan. [The government] has long established the 'Bureau of *Li* Pacification.' But because it has yet to grasp the keys [to pacification], [such work] has yet to achieve much progress. I believe that the Japanese method in pacifying the savages [in Taiwan] shares similar threads with our efforts to pacify the *Li* people."[116] He detailed the shift in Japanese attitudes toward the savages by tracing their early antagonistic approaches and the more recent policies of pacification. Whereas the former stage involved methods such as military campaigns, the fencing program, and the confiscation of weaponry, the latter referred to programs such as teaching farming, husbandry, and silkworm rearing; providing modern education and medicine; promoting handicraft trade; making films about them; and facilitating tourism.[117] Just as Chinese travelers considered their journey through Taiwan's industrial and agricultural production sites to be a learning experience for those interested in Chinese economic modernization, they understood that visiting the savage village could help them improve similar programs in China aimed at taming the savage ethnic minorities.

Despite their many overlaps, one thing separating Chinese and Western travel writing about colonial Taiwan was their portrayal of Taiwan's pre-Japanese history and the depictions of the ethnic Chinese population, the majority ethnic group, on the island. Like their Japanese counterparts, Western tourists seldom interacted with the Han Chinese colonial subjects during their journey in Taiwan. The few comments in English guidebooks and travel writing regarding the history of the Chinese rule and the Chinese population often cast them in a negative light, in contrast to their generally positive view on the Japanese colonists. In their historical sketches or passing comments about Taiwan's history, the period when the Qing had control of the island was often categorized as "a period of gross misrule."[118] Even as they expressed their admiration for the legendary Koxinga (Zheng Chenggong), whose shrine was located in Tainan, this history was largely considered an aberration from the island's predestined fate to become a colony of a European-style colonial power.[119] In their view, the local Chinese or Formosans—a term referring to the descendants of the early Han settlers from China—were deemed "ignorant," "greedy," and evincing "all the intense conservatism of their race."[120]

Chinese travelers, on the other hand, were particularly interested in finding the residual "Chineseness" in Taiwan. Aside from Taiwan's industrial sites and aboriginal peoples, the Shrine for Koxinga often earned the most elaborated descriptions from Chinese travelers. They seldom mentioned that the shrine had been converted into a Shintō shrine by the colonial authorities, and emphasized instead that the Taiwanese had a long tradition of worshiping Koxinga. By recounting Zheng Chenggong's campaign against the Dutch and his status as a Ming loyalist resisting the Qing, Chinese travel writers implied that political loyalty to China had long existed on the island. Taking pictures beside Zheng Chenggong's calligraphy on display in the shrine, Huang Qiang argued that Zheng had driven out the Dutch in order to use Taiwan as a base to resist the Qing Manchu regime. In this sense, "Zheng was the real progenitor of the revolution, which aimed at overthrowing the Manchus."[121] Dedicating a poem to the site, Jiang Yong pointed out that Zheng's "brave soul" would be gratified to learn that the mainland belonged to Han Chinese again. By evoking a notion of anti-Manchu nationalism, these travel narratives suggested that Zheng's unfinished mission had been finally realized in the 1911 revolution, which overthrew the non-Han regime that had lost Taiwan to Japan. And because Zheng Chenggong—according to this narrative—was the forbearer of the Han nationalist revolution and used Taiwan as his base, the island, they insinuated, not only had a place in China's national space but also symbolized the origin of Han nationalist struggle.

Alongside the Chinese historical figures, Chinese travelers were more interested in the local Taiwanese population than were the Western or Japanese tourists. Labeled by the Japanese as "islanders" (*hontōjin* in Japanese, *bendao ren* in Chinese), the majority of the Taiwanese were descendants of the Chinese who had migrated from Fujian or Guangdong over the centuries. Chinese travelers were eager to point out that Chinese and Taiwanese were "born of the same roots." "The residents of Taiwan are all of our race," Liang Qichao wrote in his travelogue, "and their character and customs are the same as those in China proper."[122] A Fujian native himself, Jiang Yong pointed out in 1929 that many Taiwanese still spoke the same dialect and retained the same daily habits as their forefathers.[123] This emphasis on the "sameness" was encapsulated in the term *tongbao*, which Chinese travelers employed to refer to the Taiwanese. Literally meaning "same womb," the term was used to indicate siblings birthed by the same mother. In the age of the nation-state, it came to denote members of the same nation. In a speech he made in front of a Taiwanese audience in 1911, the official traveler Shi Jingchen drove home this analogy between family and nation: "Gentlemen, you and I are genuine *tongbao* brothers. But because our parents were poor and weak, as a last resort, they entrusted you, my brothers, to our uncle, who has fostered you on their behalf. We as brothers should understand the difficulties of our parents. As for the arrangements our uncle made for you, my brothers, some are quite good, while others are not."[124] Rather than a heartless abandonment, Shi depicted China's loss of Taiwan as an unfortunate tragedy. At the same time, using the family analogy to kindle a sense of shared national identity, he expressed an underlying fear that the Taiwanese might harbor resentment against their "parents," China.[125]

Although we do not know how Shi's Taiwanese audience responded to his speech, many elite travelers recorded that visitors from mainland China were warmly received by the local Taiwanese. Jiang Kanghu, an advocate of the revival of classical Chinese, was invited by Taiwanese literary societies to make speeches, attend informal discussions, and enjoy tea parties and banquets.[126] On a fundraising trip for famine relief in mainland China, Jiang Yong fraternized with the most influential Taiwanese business families during the Japanese colonial period, including the Banqiao Lins and the Lugang Gus.[127] For average Taiwanese, especially those outside the capital, the arrival of mainland travelers often stirred up great excitement. In Chiayi, for example, when the Fujian official Wang Yang toured the famous Mount Ari in 1916, the locals welcomed him warmly, not having seen any mainland officials since 1895. Wang noted in his travel journal that one sexagenarian who had settled in Taiwan for forty years could still speak Mandarin (*guoyu*), a rare thing in Taiwan, as the majority of Taiwanese spoke Hokkien.[128] Just as Japanese imperial tourists had viewed the colonized subjects' ability to speak Japanese as a measure of their place in the Japanese empire, Chinese travelers considered their Taiwan *tongbao*'s capacity to speak Mandarin or other Chinese dialects as a sign of the preservation of the Chinese identity on the island.[129]

As Shi Jingchen hinted in the aforementioned speech, Chinese travelers also believed that some Japanese colonial policies were not beneficial to the Taiwanese. Many took aim at the colonial education system. Wang Yang observed in 1916 that separate educational tracks existed in Taiwan's colonial education policy. Whereas the Japanese children in Taiwan could attend elementary and middle schools with the same curriculum as in Japan proper, ethnic Chinese islanders and aborigines had to go to the "common schools" or "common schools for aborigines," in which Japanese language and Japanese nationalism were prioritized. When Wang Yang paid a visit to the First Common School in Tainan, there was a Japanese-language lesson in session, and he noticed that several phrases in *kanji*—Japanese vocabulary in logographic Chinese characters—were written on the blackboard. The two most striking phrases were *Dai nippon teikoku* (the Great Japanese Empire) and *kokutai* (national essence). Here, the emphasis on Japanese language went hand in hand with Japanese nationalism. "Among all the policies in Japanese colonial rule of Taiwan, nothing is more alarming than this fundamental policy," Wang commented, "because no one will know the historical past [of Taiwan] within twenty years."[130]

By the 1920s, although the quality of common schools was improving steadily and some children of Taiwanese elites attended schools for Japanese colonists, the gaps in educational opportunities persisted.[131] Jiang Yong reported that compulsory education—a policy in Japan proper—was not implemented in Taiwan, and many Taiwanese children were not sent to school. Visiting the two common schools for boys and girls in Taipei in 1929, Jiang told his readers that all the textbooks for these schools were compiled by the governor-general of Taiwan. Different from the textbook used in Japan proper, the *Moral Conduct* textbook for common schools in Taiwan, for example, was filled with repetitive messages about "loyalty to the emperor and patriotism to the nation."[132] Although it was likely that the textbook used in Japan proper contained similar phrases, Chinese travelers were sensitive to this language, as it reflected not only the Japanization policy but also the de-Sinicization effects of the Japanese colonial education.

There were few opportunities for Taiwanese to pursue higher education as well. There was no teachers' college for women in Taiwan, so Taiwanese girls who had completed their education at higher girls' schools had no way to pursue further education, unless they went to study in Japan proper.[133] A university professor himself, Jiang Kanghu wrote extensively about Taiwan's education system in his 1934 travelogue. He noted that the majority of the school administrators and faculty in Taiwan were Japanese. Among all the schools in Taiwan, he remarked, the higher the level of the school, the more Japanese and fewer Taiwanese students it had. For example, more than four-fifths of the students at Taipei Imperial University, the highest institution of learning in colonial Taiwan, were Japanese.[134] "As the only university in Taiwan, it admits fewer than forty Taiwanese students," Jiang wrote, "how strange that is!"[135] Given that 4.7 million of the 5 million people in Taiwan were ethnically Chinese—a statistic cited by Jiang—they were disproportionately discriminated against by the education system in colonial Taiwan.[136]

With the travel restrictions imposed on Chinese nationals, the number of Chinese travelers to Taiwan during the colonial period was small compared to those of Japanese tourists. However, those who had the opportunity to visit wrote travel narratives that helped formulate an image of Taiwan that emphasized the island's inseverable ties to its old motherland. For

Republican Chinese, the place of colonial Taiwan was hard to define. Just as a sibling could hardly reenter its natal family after being adopted (using Shi Jingchen's analogy), the island's colonial status was largely undisputed by the central governments of the republic. For Chinese travelers, colonial Taiwan's economic achievements and ethnic integration represented a model for both China's modernization and national unification. Once its membership in the Chinese nation-state was reinstated in 1945 after Japan's defeat in World War II, however, this ambiguity gave way to a celebratory tone, making Taiwan a symbol of China's national unification and sovereignty. This transition shaped Chinese travel culture in postcolonial Taiwan.

Cross-Strait Tourism and Postwar Travel Culture in Taiwan

A week after the Japanese surrendered on August 15, 1945, the editorial office of *China Traveler* began to prepare a special issue commemorating the twentieth anniversary of the magazine's launch. Still operating from Chongqing, the office sent out letters to famous writers nationwide, soliciting contributions. Among the six topics they named, the top two were about Manchuria and Taiwan. Newly recovered from Japanese control, Manchuria and Taiwan understandably captured the media's attention. "As our compatriots seemed to lack a deep understanding of them," the editor of *China Traveler* hoped the special issue would "offer a panoramic introduction to them."[137]

Immediately after the recovery of Manchuria, there was a renewed interest in travel to the Northeast. Commercial Press published a general guidebook to Manchuria in 1946, and CTS opened two guest houses in Shenyang and Changchun, which were affiliated with the Nationalist government.[138] The outbreak of the civil war between the Communists and Nationalists, however, paralyzed CTS businesses in Manchuria and North China and disrupted domestic tourism. The economic and social disorder caused by the civil war also spread into wider regions, rendering postwar leisure travelers less enthusiastic than their prewar counterparts.[139]

If there was one notable accomplishment of CTS during the tumultuous years between 1945 and 1949, it was the great efforts it made to cater to the growing interest in tourism to Taiwan. Newly returned to Chinese control, the island was also kept out of the escalating civil war on the mainland. A place with "beautiful scenery and mild climate," Taiwan had "abundant natural resources and convenient transportation," making it not only "a famous place to visit" but also "an advantageous region in which to develop industry," according to Tang Weibin, the general manager of CTS.[140]

In an article discussing CTS's grand postwar plans, Tang explained why CTS focused tremendous energy on Taiwan:

> By bringing our agency's business [to Taiwan], we can directly encourage tourism and facilitate industry and commerce; indirectly, by bringing more officials and intellectuals to visit Taiwan, we seek to increase cultural exchange and spiritual connection. As this province had fallen into enemy hands for more than fifty years, its cultural products and social institutions have undergone many changes. Today, the territory has just been restored, but it is unavoidable that there will be some incompatibilities between the two sides. If we can increase our opportunities to visit and communicate with each other, it will benefit our mutual understanding and promote the unity of the nation.[141]

Writing at a time after the "February 28 Incident" in 1947, Tang Weibin was aware that the original celebratory tone around Taiwan's return in 1945–46 had been replaced by the mutual distrust between Taiwanese and mainlanders. The conflict began in Taipei after a government tax policeman beat up a native tobacco seller peddling contraband cigarettes. The initial protest against police brutality in February 1947 soon spread throughout the island and became an escalating confrontation with the Nationalist government. The government responded with a massive suppression of the Taiwanese anti-Nationalist movement. An estimated twenty thousand Taiwanese, primarily from the elite class, were executed or disappeared in the following months.[142] However, for Tang, the promotion of tourism in postwar Taiwan was crucial to remedying the "incompatibilities" exposed by the incident. As a tool for imperial expansion and nation-building agendas, tourism also had a central place in decolonization and constructing new identities for former colonies. The February 28 Incident did not discourage mainland promoters of Taiwan tourism; instead, it confirmed the urgency of increasing mutual understanding between Taiwanese and mainlanders.

CTS once again took the lead in this mission. By 1947, CTS had opened branch offices in Taipei, Keelung, and Kaohsiung, and by 1949, it operated two guest houses in Taipei and Keelung for the China National Aviation Corporation. While only one travelogue about Taiwan appeared in the magazine from 1926 to 1944, between May 1945 and May 1949, *China Traveler* published nearly forty travel essays about the island, covering both the well-developed tourist attractions in the western coastal cities such as Taipei, Tainan, and Kaohsiung, and the less developed regions on the east coast of Taiwan.[143] Just as it issued *Scenic Beauty in Southwest China* shortly after the center of tourist mobility shifted to China's Southwest during the second Sino-Japanese War,

CTS published *Taiwan lansheng* (*Scenic Beauty in Taiwan*) in 1948, combining photographs of landscapes and scenery in Taiwan with texts about the island's history, industries, and other travel information.[144] Although CTS was not the only travel service provider in postwar Taiwan, nor was it the only enterprise promoting Taiwan tourism, its dominant status in China's tourism industry made comprehensive cross-strait tourism possible.[145]

Professors, journalists, politicians, and other professionals in urban centers took advantage of organized tours to Taiwan. They contributed articles to *China Traveler*, resulting in an eruption of travel writing on Taiwan in this immediate post–World War II period. In 1945–49, the Nationalists had adopted a cultural policy some scholars have termed "de-Japanifying and re-Sinifying Taiwan."[146] Taiwan tourism was a crucial part of this cultural policy. Chinese travel narratives of postwar Taiwan demonstrate how mainland elites followed these two major threads to reposition Taiwan within the fold of the Chinese republic.

To reconnect Taiwan's severed ties to mainland China, travelogue authors emphasized that the island's history, geography, demography, and other aspects all suggested its true identity as part of China. A special correspondent sent by *China Traveler*, the journalist Chen Qiying, for example, suggested that China had discovered Taiwan "as early as the Yin dynasty," the second dynasty of ancient China, which lasted roughly from the sixteenth century to the eleventh century BCE.[147] Another author dated China's first discovery of Taiwan to the Spring and Autumn Warring States period (770–221 BCE).[148] Moreover, rather than considering Taiwan a Qing colonial possession as of 1684, some travel writers pinpointed the original moment Taiwan "entered the map" (*ru bantu*) in the late Mongol Yuan dynasty, when a government office was established in Penghu (the Pescadores Islands, which is part of modern-day Taiwan).[149] In Chen's narrative, since the late Ming dynasty, first Spanish and then Dutch colonial forces invaded Taiwan. But they were fiercely resisted and driven out by Zheng Chenggong. Indeed, just like their prewar counterparts, postwar Chinese travelers continued to treat Zheng Chenggong as a national and nationalist hero. After Taiwan's recovery, his shrine was visited by prominent Nationalist generals like Bai Chongxi. At Chikan Tower, a Chinese structure built on the site of Fort Provintia, where Zheng Chenggong had accepted the surrender of the Dutch, tourists could view oil paintings depicting Zheng's successful campaign. "How I wish I could see Westerners bowing and bending their knees [to surrender]," one author wrote.[150]

This refashioned nationalist history of Taiwan did not end with Zheng Chenggong's success against the Dutch. Chen Qiying, for example, argued that although the Qing rule of Taiwan lasted more than two hundred years, it was a failure because of its defeat by Japan and the cession of the island.[151] But even with this misrule, Taiwan *tongbao* had "always opposed" the Japanese, he emphasized. They even established a "democratic country" in Taiwan to resist Japan, and became "the first democratic country in East Asia."[152] Albeit short-lived, the importance of this so-called Republic of Formosa was not lost on Chinese travelers. When visiting the museum in Taipei, mainland tourists often took notice of the flag of this anti-Japanese regime and a "call to arms" written by one of its leaders to resist Japan.[153] However, the idea that the anti-Japanese uprising was also an independence movement did not register in the consciousness of Chinese tourists.

While it was not unusual for travel writers to ignore the achievements made during the Japanese colonial period, some postwar writers did touch on it, claiming that Taiwan's colonial period demonstrated the malicious ambition and greediness of Japan. They argued that the deliberate efforts made by the Japanese in building commercial and naval ports, developing a modern transportation network, and facilitating industries were aimed at transforming Taiwan into a base (*genjudi*) to serve Japan's imperialist southern policy (*nanjin*) and its advances toward China's coastal provinces.[154] Taiwanese, they suggested, were oppressed by the colonial assimilation policies and deprived of economic opportunities.[155]

Many travel essays in *China Traveler* published in 1946–47 celebrated the moment of Taiwan's *guangfu* (recovery). Although Taiwan had been ceded long before the Nationalist Party garnered national power in mainland China, Chen Qiying's travelogue—excerpts of which would appear in *Scenic Beauty in Taiwan*—emphasized that the surrender of Japan and the recovery of Taiwan brought "the blue sky and bright sun" (the emblem of the Nationalist Party) back to the islanders and relieved the past humiliation faced by the Taiwan *tongbao*, who rejoiced at returning to the "free and peaceful" China. "Every family celebrated by worshipping their ancestors," he added, "joy mixing with grief while happy tears flowed into their drinking cups."[156] Unsurprisingly, the jubilation of Taiwan's *guangfu* captured in travel writings was also compared to the return of a lost child to his mother. As one author noted, after the Japanese surrender, Taiwanese people smiled just like "infants returning to their loving mother's bosom."[157] In these popular narratives, its colonial past was cast as a relatively brief aberration. Constructing a Nationalist history of Taiwan helped resituate the island within China's national historical trajectory.

Mainland travel writers also incorporated the island into the geography of China's national space. Taiwan's close proximity to Fujian and China's southeastern coast was considered a critical asset. The Taiwan strait, which separated the island from the mainland, was within the continental shallow sea and therefore simply an inland sea (*neihai*) of China. Chen Qiying called Taiwan "China's barrier in the Southeast."[158] Geographer Ju Xiaoming also argued that a recovered Taiwan should be treated as "an outpost of national defense."[159] Natural geography aside, Taiwan's demographic composition also represented its deep-rooted "Chineseness." Similar to the points prewar Chinese travelers had made, postwar travel authors also emphasized that the majority of *Taibao* (shortened from *Taiwan tongbao*) were descendants of Chinese migrants from several counties in Fujian and Guangdong Provinces and therefore shared the customs, habits, and dialects of those areas.[160] Guo Zhusong, a historian from Sichuan, argued that except for the "mountain peoples" (*gaoshan zu*) who were "our distant relatives," *Taibao* were "Hoklos and Hakkas," so "we are the same family." He added that the physique of the Taiwanese was similar to the Cantonese. Even though some of the Taiwanese customs were different from those on the mainland, the Taiwanese were likely continuing practices of Chinese ancestors that mainlanders had abandoned.[161] Although the majority of the *Taibao* under thirty could not read Chinese or speak Mandarin because of the colonial education, Hokkien was still the lingua franca among the Chinese Taiwanese. "Only 60 percent" of Taiwanese spoke Japanese, and it was used reluctantly when Taiwanese interacted with Japanese.[162] "After tremendous effort," Chen Qiying stressed, "the Japanese still could not assimilate them."[163]

However, not every travel author took such a sanguine view of the unmistakable Chineseness of the Taiwanese. Exactly what language the Taiwanese spoke was in fact the central issue. Postwar mainland Chinese tourists—who were largely from the lower Yangzi region and therefore could not decipher Hokkien or Hakka dialects spoken by Taiwanese—often took issue with the fact that the majority of Taiwanese understood Japanese but not standard Mandarin. Yufeng, a special correspondent for *China Traveler*, cited a language barrier as the main problem in postcolonial Taiwan. He suggested that everything in Japanese should be banned.[164] Guo Zhusong commented, not without irony, that the universal education during the colonial period actually helped them overcome the language barrier issue in Taiwan. Whenever they encountered a problem with verbal communication, they could rely on "pen conversation" (*bitan*). Because of the widespread Japanese language education, even "rickshaw drivers, workers, and waitresses" were literate and could write Chinese characters. However, for Taiwanese, the phrase "national language"—*guoyu* in Chinese or *kokugo* in Japanese, although both use the same Chinese characters—did not mean Mandarin but rather

Japanese. Guo found this high literacy in Japanese an unnerving aspect of Taiwanese society. He wrote:

> If *Taibao* were illiterate, we would only need to start from the beginning—teaching them characters and how to read, and then inculcating them with national consciousness and criticizing Japanese tyranny and their venomous plans. However, *Taibao* are not illiterate. Their education levels are on average higher than people from other provinces in China. The only problem is that a large part of their education is harmful to the current environment. Therefore, we must remove the poisonous fang and get rid of the venom, and then give them good medicine. Only after the idiosyncratic and dangerous thoughts inculcated by the Japanese are purged from them can we make them accept new opinions that are contrary to what they have previously learned.[165]

Traveling to Taiwan shortly after the outbreak of the February 28 Incident, Guo was preoccupied with concerns for Taiwan's current chaos, and his comments encapsulated the discomfort shared by other mainland travelers, who had attributed the uprising to the Japanese colonial education. Guo remarked that through the colonial education system, the Taiwanese only learned "skills of making a living and ideologies of betraying their motherland."[166] And "[t]hese people were the main force behind the mob involved in the 'February 28 Incident'—the student-army who wore Japanese army uniforms, spoke Japanese, and used Japanese weapons to kill Chinese."[167]

Guo's travel writing also exposed the complicated factors that triggered Taiwanese grievances. The Taiwanese soldiers who had fought for the Japanese empire in the war returned home only to face unemployment, contributing to the tension between mainlanders and Taiwanese. Resenting the monetary inflation spread from the mainland, Taiwanese, especially those in the south, called for "true democracy," a movement originating from the independence movement that had emerged when Japan's colonial rule crumbled at the end of World War II.[168] After direct conflicts surfaced in the February 28 Incident, some even suggested that a Taiwanese president should lead a New China in Taiwan, rather than the Nationalist government from mainland.[169] However, Guo's opinion was clear. The fifty-year colonial rule had trained Taiwanese to be the tools of the Japanese. Rather than blaming *neidi ren* (mainlanders) or *Taiwan ren* (Taiwanese), Guo implied that the culprit was Japanese colonialism.[170]

With all these concerns in the background, the question of how to categorize Taiwan's tourist attractions was also fraught in the postcolonial period. When Taiwan first began to attract mainland tourists in 1946, introductory information was often copied from Japanese tourist materials, including the famous list of "Eight Views and Twelve Famous Sites" (*bajing shi'ersheng*).[171] Based on the result of a popular contest run by the *Taiwan Daily* (*Taiwan nichi nichi shinpō*) in 1927, this list of tourist attractions—which were selected through readers' votes and decisions made by a mostly Japanese committee—soon became the standard tourist attractions of Taiwan endorsed by the colonial government. Scattered along Taiwan's main railway line, this collection of sites had framed the way people viewed colonial Taiwan. In fact, the original result from 1927 included two more special (*bekkaku*) sites—Taiwan Shrine (*Taiwan jinja* in Japanese) and New High Mountain (*Niitaka-yama* in Japanese). The former was the highest ranking Shintō shrine in Taiwan, and the latter was the tallest mountain in Taiwan and indeed the entire Japanese empire.[172] Given their symbolic status in Taiwan's colonial landscape, these two sites were left out of the *China Traveler* articles, while the rest of the list was replicated as representative scenic spots. After 1946, however, this list from the colonial period disappeared from *China Traveler* and the travel media, even as Chinese tourists followed similar itineraries from the colonial period. For example, Cai Yumen, a doctor from Shanghai, toured Taiwan in November 1946 and adhered to a route almost identical to the 1923 Japan Tourist Bureau itinerary prepared for Japanese imperial tourists. Not only did he visit natural attractions such as the hot springs at Beitou and Sun-Moon Lake and Mount Ari, but he also toured sugar factories and other industrial sites, as well as the ports at Keelung and Kaohsiung.[173]

Among all the attractions featured in *China Traveler* in the immediate postwar years, the most highlighted tourist site was the Sun-Moon Lake in central Taiwan. Mentioned in most prewar English and Japanese guidebooks of Taiwan and included in the "Eight Views" of Taiwan in 1927, the Sun-Moon Lake was not very accessible until the mid-1930s, when the completion of a hydraulic power plant helped improve the transportation infrastructure.[174] To celebrate the first anniversary of Taiwan's recovery, the Nationalist leader Chiang Kai-shek visited Taiwan for an inspection tour in October 1946, during which he made a stop at the hydraulic power plant and enjoyed a boat ride on the Sun-Moon Lake. Other politicians soon followed suit.[175] After that, many elite mainland tourists to Taiwan would make their own pilgrimage. In *China Traveler*, at least twelve travelogues from 1946–49 featured the Sun-Moon Lake.[176] Although this "paradise of the island" was located about 2,400 feet above the sea level and in the middle of the mountains, mainland tourists compared the Sun-Moon Lake to West Lake and the water villages in the lower Yangzi region such as Suzhou, which were tourist sites in urban settings.[177] By juxtaposing the Sun-Moon Lake to some of the most recognizable tourist attractions in mainland China, travel writers not only located Taiwan within China's national tourism market but also elevated its stature to a tourism-rich province.

While some travelers complained about its remoteness and others were disappointed because the scenery did not meet their high expectations, the Sun-Moon Lake nonetheless captured the imagination of mainland tourists for several reasons.[178] First was its natural wonder. The largest natural lake in Taiwan, it consisted of two conjoined lakes—one round and the other crescent-shaped—which were surrounded by stunning ridges of mountains covered by ancient forests. He Mingxian, a journalist from Fujian, asserted that having traveled to eleven provinces in China, he had yet to encounter anything as "marvelous and intriguing" as the Sun-Moon Lake.[179] Secondly, the manmade beauty of the hydroelectric power plant also impressed modern tourists. Wu Chenyi, a civil engineer from Shanghai, emphasized that, compared to the natural scenery, he treasured the "manmade beauty" more. Referring to the dams and other hydraulic structures built on the lake, Wu highlighted that the power plant there was the largest hydraulic power plant in China. Drawing the water from Wushe (*Musha* in Japanese), the famous site where the "mountain peoples" had led an anti-Japanese uprising in 1930, the power plant included two tunnels cutting through mountains.[180] Recognized as an engineering marvel, the hydraulic power plant was put on the list of must-see sites in Taiwan, even though it was built by the Japanese.[181]

The last intriguing aspect of the Sun-Moon Lake to mainland tourists was the so-called mountain peoples. Just as the Japanese colonial government forbade the derogatory term *seiban* (raw savages) after the Musha incident and replaced it with the neologism Takasago-zoku (the Takasago people), the Nationalist government promoted the term *gaoshan zu* to refer to the aborigines in postwar Taiwan in order to discourage the use of vulgar racist epithets like *shengfan* (raw savages), *shufan* (cooked or tamed savages), and *tufan* (native savages).[182] Similar to the umbrella term of Takasago-zoku, which collapsed the difference between indigenous ethnic groups, the term *gaoshan zu* also grouped all the tribal peoples in Taiwan together as a single object of the state's ethnic policy on the island.[183] The imitation of the Japanese colonial strategy was deliberate. Writing about the mountainous areas in Taiwan, including the Sun-Moon Lake, Chen Qiying pointed out that although there was a consensus that the aborigines of Taiwan were the island's first people, their racial origin was still open to debate. The Japanese, for example, argued that the Takasago-zoku were affiliated with the Yamato race, the native people of Japan. Chen himself then put forward a theory that the mountain peoples of Taiwan might be the descendants of the Jurchens fleeing the Mongols in the thirteenth century, and therefore could be a branch of the *Zhonghua minzu* (Chinese nation).[184] Whether or not the theory was scientific was irrelevant to Chen's point. Just as the Japanese could suggest a link between the Yamato people and the aborigines in Taiwan by renaming them, Chen seemed to insinuate that offering a new group name in Chinese could also create an association.

Yet the more ubiquitous association between Chinese and the aborigines in Taiwan was through ethnic tourism. When visiting the Sun-Moon Lake in the late 1940s, mainland tourists often called on the aboriginal village of *Shiyin fanshe* on the east coast of the lake. Taking either an indigenous canoe or a motorboat to cross the water from the Hanfenlou hotel—the common lodging choice for tourists—travelers would be met by aboriginal children selling postcards and souvenirs.[185] The tribal leader would come out to greet them and arrange a performance of their famous pestle song and dance. The performance entailed a group of aborigines singing and dancing as they beat rhythmically on a stone slab with long wooden pestles used for pounding grain (figure 12).[186] The performance was not free. In fact, according to several travelogues in *China Traveler*, the pricing was standardized. Zhang Qiqu noted that there was a price list posted on the wall and tourists were charged based on the number of the viewers. "It costs 600 yuan for more than ten people in the audience," he recorded, and "1,000 yuan if there are more than fifteen people."[187] When Xu Renhan, a high-school principal from Shanghai, visited the Sun-Moon Lake in November 1948, he recorded that tourists could also pay 1,000 yuan for a picture with the two daughters of the tribal leader. One of the "princesses" would sign and date the photo for the tourists as well.[188] Considering their dance to be inferior to the dances of the ethnic minorities in China's Southwest, Guo Zhusong was surprised that these mountain people were so business savvy, commodifying their own image and culture.[189]

Unlike the Taiwanese whose ancestors were migrants from China's coastal provinces, in Guo Zhusong's view, these aboriginal people were distant relatives rather than siblings.[190] Calling them cultured barbarians who no longer practiced cannibalism, he pointed out that their society was "centered around women, and men submit themselves to women," implying that this reversed gender hierarchy indicated their Otherness vis-à-vis the Han Chinese.[191] While some writers emphasized their primitiveness, other mainland travelers visited the old battlefield between the aborigines and Japanese and praised "the barbarian heroes" who had resisted with unyielding spirit against the Japanese colonizers.[192] Treating these aboriginals as yet another ethnic minority on China's frontier regions, the mainland travelers viewed

第二十三卷 第一期　　　　台游觀感 (67)

時返旅館，早餐畢，先至包車安放行李，然後輕裝先後參觀高山博物館、三代木、及神木。在兩木前復各攝一影，移時登車，四時半返抵嘉義。

日月潭攬勝

十一月四日，在台中鐵路飯店西餐室進早餐後，七時半許，分乘小包車三輛，向日月潭進發。日月潭爲台省唯一名湖，一名水社湖，又名龍湖，距台中約七十九公里，汽車三小時可達，海拔約八百餘公尺。（涵碧樓前八百卅公尺）周圍及深度，各書所載互異，只得從略。目力估計，潭較杭州西湖爲小，水則較深。潭旁水社大山、大尖山等，高峯圍繞，加之遍山密林，葱蘢蒼翠，潭水深碧，一片油綠，白雲松風，幽邃無比，而湖光山色，朝暉夕月，更是氣象萬千，使人萬念俱消，有絕塵之感。較之西湖太湖，幽靜多矣。恐與瑞士湖山勝景，亦難分軒輊。潭中光華島，介乎日月兩潭之間，登臨其上，可覽全潭景色。潭旁涵碧樓，瀕臨湖濱，正對光華島，或遊目騁懷，攬景怡情，或樓頭小酌，閒坐品茗，或駕遊艇，或泛木舟，放乎中流，偃仰高歌，沈醉於大自然中，流連忘返，其樂無窮，眞是世外罕有桃源，國中難得樂土。其他可遊之處，尚有文武廟、石印番社等處。文武廟背山臨潭，高踞山巔，可眺全潭，景色。廟祀孔聖關岳，故名。文武。石印蕃社有高山化蕃衆居，可聽杵歌，並觀蕃女歌舞。所謂杵歌，乃由蕃女十餘人，分執高矮粗細輕重不同之木杵一根，團繞於埋在土中之大塊堅石四周，先後疾徐，輪流搗聲，發出木石鏗鏘之音，隨搗隨歌，頗合節奏，清婉悅耳。作者好奇，獨立石中，倩沈兄耀宗代攝一影，藉留鴻爪。個中翹楚，酋長大女公主名毛玉娟，二女公主名

作者在日月潭高山族杵歌石上

FIGURE 12. A mainland tourist with Taiwan indigenous performers. *Source:* Xu Renhan, "Taiyou guan'gan" [My impression of Taiwan], *Lüxing zazhi* [China traveler] 23, no. 1 (January 1949): 67.

them not only as the Other vis-à-vis mainlanders but also as the Other vis-à-vis the Han Chinese in Taiwan. As a group of people who had tenaciously resisted the Japanese colonizers, they seemed to possess certain characteristics that the Han Taiwanese lacked. Just like their underdeveloped mountainous environment, which contained the natural resources for modernizing Taiwan, their ethnic identity also became a natural resource that could be exploited for the tourism industry.

The reclamation of Taiwan in 1945 signified the Nationalist state's legitimacy and success in leading China's anti-imperialist struggle. By facilitating cross-strait tourism, CTS also joined the Nationalist state's efforts in reclaiming a former Japanese colony and initiating the process of decolonization. Locating Taiwan at the center of their imagination of a Chinese nation-state at this postwar moment, mainland tourists helped appropriate Taiwan's colonial past while projecting their vision of Chinese nationalism on the island. While praising Taiwan's remarkable natural scenery and distinctive culture, mainland travelers also expressed their concerns about the uncertainties they encountered, especially the deep imprint left by Japanese colonialism. To mainland Chinese, Taiwan continued to represent a series of paradoxes. It was exotic and familiar, backward and modernized, a former Japanese colony and a new Chinese frontier.

Unlike in the Northwest and Southwest, Manchuria and Taiwan's peripheral status in China's tourism radius was not due to a lack of modern transportation or travel facilities. Instead, these areas' entanglement with the Japanese colonial empire both hindered and stimulated Chinese tourism to Manchuria and Taiwan. Whereas the Japanese presence in Manchuria complicated travel logistics for Chinese travelers, Japanese colonial authorities in Taiwan imposed restrictions on Chinese visitors. However, even when organized tourism to Manchuria and Taiwan was impossible for Chinese nationals, their desire to keep these areas within China's national space stimulated public interest in travel narratives about them. These travel writings sustained and satisfied the reading public's curiosity, yet also precipitated anxiety over Japan's firm grip over these purported Chinese territories. When the anti-imperialist Nationalist state successfully drew Manchuria into its political orbit in 1928 and when Taiwan was returned to China after Japan's defeat in 1945, Chinese tourists, especially the elite travelers from the nation's core, were eager to tour these areas. What triggered tourist enthusiasm was precisely the colonial condition, and tourism became a tool for Chinese nationals to express their anti-imperialist sentiments.

Chinese travel narratives about Manchuria and Taiwan indicated the parallels between colonial tactics and nation-building agendas, parallels that also existed in Chinese travel narratives about other peripheries like the Northwest. In travel writings in the late 1920s and early 1930s, Chinese travelers bemoaned the reality of the multinational influence on Manchuria. They criticized Japanese imperial tourism and Japan-sponsored immigration to Manchuria, and attacked the scientific and cultural institutions established by the Japanese and the Russians. However, as Chinese travelers envisioned the Chinese nation-building agendas in their travel narratives, similar projects—such as tourism and immigration—were proposed as approaches to restoring Manchuria from a fragmented semicolonial space to coherent Chinese provinces.

Along the same lines, the vogue of cross-strait tourism from 1946 to 1949 prompted Chinese travelers to insist on a remediation and reversal of the effects of Japanese colonial rule in Taiwan. However, when confronted with the native Taiwanese population's demand for self-rule, the nationalist tourists from mainland China resorted to colonial narrative strategies to characterize Taiwan and its people as less Chinese, and therefore subject to subjugation by a strong Chinese state. Blaming the Japanization policies imposed by the Japanese during the colonial period, Chinese tourists to Taiwan identified with the Nationalist government's standpoint and stressed the urgency to re-Sinicize the native population in their travel writings. Some of the decolonization efforts of the Nationalist state mimicked the Japanese colonial strategy as it tried to forge a national identity among the Taiwanese population.

This history of Chinese tourism and travel in Manchuria and Taiwan demonstrates that asserting the inclusion of a semicolony like Manchuria or repositioning a former colony like Taiwan within China's national space was a challenging mission. Using tourism as a tool in shaping national identity, government agencies, tourism promoters, and individual travelers stressed the inseverable links that connected Manchuria and Taiwan to China proper by enumerating different forms of evidence from history to geography to society to culture. However, they also needed to account for the influence of the colonial regimes they observed. From the scientific explorers in the Northwest to postcolonial tourism in Taiwan, we can find evidence that Chinese travelers identified with colonial modernity as the paragon of what could be achieved when a nation-state adopted a scientific and expansionist mode of development in its peripheries. In this sense, colonialism and its legacies continued to shape Chinese nationalism and the imagination of China's national space even in the postcolonial era.

Conclusion

Legacies of Republican-Era Tourism and Travel Culture

Since this book begins with a map, perhaps our story should end with a map as well. In February 1948, CTS participated in a "Resort and Holidays Exhibition" in London as the sole agent representing the Pan American World Airways in China. To broadcast the appeal of Chinese tourism, CTS staff displayed a huge poster—eight feet wide by six feet tall—of a tourist map of postwar China. Unlike the map cited at the beginning of this book, this map did not fill every province with the image of its most representative tourist attraction. Instead, it showcased an extensive yet uneven network, more developed in eastern coastal regions yet extending into the Northwest and the Southwest, as well as Taiwan and Manchuria. At the time, China was in the middle of a civil war and tourism was hardly a priority, but CTS's proud display demonstrated the accumulative effects of tourism development in the first half of the twentieth century. By superimposing the tourist network on an outline contour of China's geobody, this map also demonstrated the close link between tourism and the conceptualization of China's national space.[1]

In the first half of the twentieth century, the growing accessibility of modern transportation and the increasing prominence of tourism made it possible for middle-class Chinese to experience more of their own country. Developing along the modern transport circuits, China's tourism industry and radius underwent impressive expansions during the Nanjing Decade. A national tourist network began to take shape, expanding to encompass the natural and historic landmarks of previously less-traveled areas. At the same time, in response to China's political fragmentation and increasing Japanese imperialist aggression, peripheral regions beyond China proper became a matter of great interest to state agents, scientific explorers, journalists, and adventure tourists from the core regions of the nation. In 1937, Japan's full-scale invasion of China interrupted the momentum of tourism development. However, as many intellectuals and members of the prewar urban bourgeoisie fled to the Nationalist-controlled "Free China" in the Southwest, they also brought their way of life, including tourism consumption. Wartime travel made the Southwest more concrete to coastal citizens and tied the southwestern provinces more firmly to the national core. And while the surrender of Japan in 1945 did not bring peace to China, it brought Taiwan into the orbit of China's domestic tourism.

In the same time period, narratives about leisure tourism and other forms of travel—explorations, academic fieldwork, and wartime flight—were also an important part of modern Chinese travel culture, shaping a consciousness of China as a coherent unity, despite the political fragmentation and the central state's weakened hold over the periphery. Simply put, modern travel made modern China.

Tourism, Travel Culture, and China's National Space

Historians of tourism in other national contexts have noted that empire and tourism are intimately linked.[2] As I have described in this book, the impact of empire on tourism development in modern China was multilayered and multidirectional. The advancement of industrial transportation and the permeation of Western tourist culture in treaty ports and coastal areas helped elite and middle-class Chinese to develop a new mindset: rather than being a hardship, travel for adventure and leisure became a symbol of modernity. Yet the racial hierarchy embedded in the semicolonial transport sector and the lack of comprehensive tourist services catering to Chinese nationals hindered the development of tourism. Although the treaty port system persisted after the establishment of the republic, voices clamoring against the imperial presence in China swept through urban centers by the late 1910s. They not only energized major political parties to bring anti-imperialist nationalism into their platform but also stimulated an urban consumer culture that boycotted foreign goods and promoted national products. This paved the way for the emergence of a native travel industry that defined their travel services and package tours as national products, appealing to consumers imbued with anti-imperialist sentiment.

Compared with European treaty powers whose grip on the semicolonial transport sector directed tourism development in China's core regions, the Japanese empire had a detrimental effect on tourism in the peripheries. Japan's bellicose approach curtailed the expansion of tourism into Manchuria throughout the republican period and disrupted modern transport circuits during World War II. However, between the two Sino-Japanese Wars, Japanese transportation authorities did maintain a working relationship with the Chinese transport sector and travel agencies like CTS to allow for through traffic between China proper, Manchuria, colonial Korea, and the main Japanese islands. This collaboration continued even after the establishment of Manchukuo.

Conversely, the continuing threat from the Japanese empire also stimulated Chinese tourism to the peripheries. Whereas the Japanese dominance in south Manchuria from the early Republican years contributed to the uptick of business, journalist, and intellectual travelers there, the Japanese invasion and occupation of the entire Northeast in the early 1930s triggered a deep sense of crisis and gave rise to the vogue of adventure tourism to China's Northwest. When Japan's full-scale invasion shut down travel and tourism along the eastern seaboard, wartime tourism in the Southwest helped facilitate popular support for resistance in the Nationalist-controlled Free China. Moreover, in the immediate postwar era, the newly returned Taiwan drew popular attention from mainland travelers precisely because of its colonial past. Their travel narratives raised an alarm over the continuing effects of the Japanese colonialism in Taiwan and emphasized the urgency of decolonization.

The impact of European and Japanese empires interacted complexly with the lingering afterimage of the Qing empire. Unlike the Ottoman, Habsburg, and other empires that collapsed in the early twentieth century, the Qing empire at its demise did not crumble into several new nation-states. Since its founding, the republic asserted that its sovereign territory "continues to be the same as that of the former Empire."[3] However, this definite territorial claim was hazy in reality, as colonial encroachments, warlord separatism, ethnic nationalism, and Communist opposition all made it difficult for the changing Chinese state to exercise meaningful control over many parts of its putative national territory.

Nevertheless, more than a principle proclaimed by the state, the vision of China as a unified and bounded nation-state superimposed on the Qing imperial realm was a popular imagination widely shared among the core constituency of the republic—the urban bourgeoisie. This shared imagination of China's national space was inseparable from the increase in tourism mobility and the rise of a nationalist travel culture. As the radius of tourism and travel was extended from China proper to the peripheries, Chinese capitalists and urbanites were pulled into the nation-building effort as China was increasingly seen as a unified national market and one national space encompassing local and regional diversity.

Ruptures and Legacies of Republican-Era Tourism and Travel Culture

From the end of its wartime operations in 1945 to the shuttering of its doors on the mainland in 1954, the last ten years of CTS were punctuated with great expectations and unfulfilled ambitions. The end of war had been anticipated as a new beginning for China's tourism industry, but the reality was sobering. The outbreak of the civil war meant that aside from organized tours to Taiwan, leisure travel in China became nearly impossible. The defeat of the Nationalist Party and the birth of a Communist regime in 1949 sharply changed tourism and travel culture.

Broadly speaking, tourism was an awkward fit in the economy of socialist states. This does not mean that the tourism industry did not exist under socialism. It did.[4] However, "[t]he requirements of the tourism industry, which entailed the availability of a flexible entrepreneurial and self-financing sector responsive to changing demands and fashions," as some historians have observed, "were the antithesis of the centralized socialist economy based upon heavy industry."[5] As China was gradually transitioning to a socialist society, CTS's business model underwent significant transformations. In the 1950s, the socialist state viewed travel agencies as bastions of capitalist practices, and considered the commission system a form of exploitation. When commissions were eliminated, CTS lost one of its main income sources. At the same time, there was a shrinking market for tourism. Redefining the relationship between leisure and labor, the socialist state regarded tourism as a means to rejuvenate the labor force; it was therefore intended to serve the working class, rather than the urban bourgeoisie.

Endeavoring to keep abreast of the new ideology, CTS adjusted the discourses of tourism in *China Traveler*. In an article appearing in the June 1950 issue, for example, author She Guitang emphasized that after socialist revolution, organized tours "cannot stay the same." Instead of the conventional tourist sites identified by literati in ancient times, She pointed out, tour organizers should include places "suitable for relaxation, recreation, and sports" on their itineraries. Instead of reminiscing about famous historical figures, they should reflect on the different social conditions that the historical relics represented.[6] Even though these principles did not deviate significantly from the Republican-era guidelines, CTS carefully incorporated the rhetoric of historical materialism into this renewed theorization of tourism practice. Yet, despite its adjustments in the new political and economic environment, CTS failed to remain afloat without the revenues from railway commissions and package tours by the mid-1950s.[7] Its closing marked the end of Republican-era tourism and travel culture.

Although the transformation of tourism and travel culture under socialism and the postsocialist market economy is beyond the scope of this book, the legacies of the Republican era are important to the development of tourism and travel culture in post-1949 China. First of all, many of the scenic spots and tourist sites canonized during the Republican era constituted the must-see places in the socialist and postsocialist periods. In the context of "tourist diplomacy" during the Maoist era,[8] even though new construction completed by the socialist state was highlighted, inbound travelers, including invited foreign guests and permitted self-financed tourists (often overseas Chinese), continued to frequent tourist attractions celebrated in the Republican period.[9] The proletarian tourism of the socialist era did not start from scratch either. Operated by state agencies and labor unions, health resorts and sanitoriums for workers were often set up in the famous resorts and well-established travel sites from the late Qing and Republican periods. While proclaiming that famous tourist sites in China were no longer "reserved for the wealthy," the socialist state continued to uphold places like Moganshan, Lushan, and West Lake as the defining landmarks in Chinese leisure culture.[10] Interestingly, many pre-1949 tourism strategies also persisted in Nationalist Taiwan after 1949. These included highlighting Chineseness through tourist sites in Taiwan and promoting ethnic tourism in the communities of the aborigines.

The familiar sites established in the Republican period also continued to serve as the building blocks to domestic bourgeois tourism when it reemerged in the 1980s and 1990s in mainland China.[11] Previously included in package tours by travel agencies and tourist clubs in the 1920s and 1930s, Mount Huangshan, Mount Taishan, and West Lake once again attracted tourists in the market economy era. Today, China has one of the largest domestic tourism markets in the world and has established sophisticated categories and rating standards for tourist attractions.[12] However, rather than being a new phenomenon of the reform era, the development of popular tourist attractions in China has its roots in the first half of the twentieth century.

The practice of constructing China's national space through tourism and travel culture carried on in post-1949 China, as well. In 1949–50, a series of events determined the scope of national territory of the People's Republic of China. While a treaty signed between China and the Soviet Union finalized Outer Mongolia's independence from China, the People's Liberation Army's victories in suppressing separatist movements in Inner Mongolia and Xinjiang and occupying Tibet firmly incorporated these elusive ethnic frontiers into the socialist state. Shortly after these important alterations to China's national territory were settled, CTS issued a series of five travelogue pamphlets between 1952–53. Written by People's Liberation Army cadres and military correspondents, these pamphlets covered Outer Mongolia, Xinjiang, Inner Mongolia, and Tibet. Reflecting the new political reality, these travel narratives also served as a tool for constructing China's national space in a new context.[13] Shortly after that, in socialist-era promotional materials for inbound tourism, destinations like Urumqi and Lhasa appeared alongside major cities like Beijing, Shanghai, and Guangzhou, signaling a much closer tie between the core and the peripheries under the socialist state.[14]

In the post-Mao reform era, this practice of constructing national space through tourism and travel culture was adapted into new formats. In the Splendid China Folk Village in Shenzhen, one of the earliest theme parks in China, miniaturized iconic landscapes of China are the main attraction, including the Great Wall and the Potala Palace. The Folk Culture Villages in the park made ethnicity the main theme, highlighting performances by members of ethnic minorities.[15] Collectively, these miniature landmarks and performed ethnic culture represent the multifarious tourism resources in China on a broad scale. Moreover, in the case of Hong Kong, Macau, and Taiwan—which remain semiautonomous or independent from mainland China—China's outbound tourism to these areas has become a tool for the mainland to extend its influence and incorporate these areas into its orbit of political control.[16] Through tourism to these areas, travelers can better envision a new construction of China's national space inspired by the "One Country, Two Systems" concept.

The key elements of the Republican-era travel culture survived and continued to influence the ways in which Chinese consumed domestic tourism and perceived China's national space. The revival of bourgeois tourism in post-Mao China would also be unthinkable without the maturation of modern travel in the first half of the twentieth century. As indicated by the Hong Kong protests in 2014 and 2019–20 and the strained cross-strait relationship between the mainland and Taiwan, the contentions over the idealized national space of China have also persisted. Tourism and travel culture continue to help us track the convolutions within popular constructions of China's national space.

Notes

In citing works in the notes, short titles have generally been used. Works frequently cited have been identified by the following abbreviations:

LZ *Lüxing zazhi* [China traveler]

SSCYD Shanghai shangye chuxu yinhang dang'an [The archives of Shanghai Commercial and Saving Bank], 1915–50. Shanghai Municipal Archives.

ZLZD Zhongguo lüxingshe zongshe dang'an [The archives of the Head Office of China Travel Service], 1923–54. Shanghai Municipal Archives.

Foreword

1. "Outbound Tourism—Travel (Million US Dollars)," Knoema, World Data Atlas, accessed December 29, 2020, https://knoema.com/atlas/topics/Tourism/Outbound-Tourism-Indicators/Outbound-tourism-travel.

2. For a concise summary, as well as an extensive bibliography designed to help with further reading, see Eric G. E. Zuelow, *A History of Modern Tourism* (London: Palgrave, 2016).

3. For example, Mary Louise Pratt, *Imperial Eyes: Travel Writing and Transculturation*, 2nd ed. (Abingdon, UK: Routledge, 2008); Kathleen R. Epelde, "Travel Guidebooks to India: A Century and a Half of Orientalism" (PhD diss., University of Wollongong, 2004).

4. Zuelow, *History of Modern Tourism*.

5. F. E. Stanford, "Mokanshan," in *With Our Missionaries in China*, ed. Emma Anderson et al. (Mountain View, CA: Pacific Press, 1920), 329.

6. Shelley Baranowski, *Strength through Joy: Consumerism and Mass Tourism in the Third Reich* (Cambridge: Cambridge University Press, 2004).

7. Ellen Furlough, "Making Mass Vacations: Tourism and Consumer Culture in France, 1930s to 1970s," *Comparative Studies in Society and History* 40, no. 2 (April 1998): 247–86.

8. Diane P. Koenker, *Club Red: Vacation Travel and the Soviet Dream* (Ithaca, NY: Cornell University Press, 2013); Christian Noack, "Building Tourism in One Country? The Sovietization of Vacationing, 1917–41," in *Touring beyond the Nation: A Transnational Approach to European Tourism History*, ed. Eric G. E. Zuelow (Farnham, UK: Ashgate, 2011), 171–93.

9. Marquerite S. Shaffer, *See America First: Tourism and National Identity, 1880–1940* (Washington, DC: Smithsonian Institution Press, 2001).

10. For discussion of the "See Ireland First" program, see Eric G. E. Zuelow, *Making Ireland Irish: Tourism and National Identity since the Irish Civil War* (Syracuse, NY: Syracuse University Press, 2009), 23–24.

11. The two most frequently cited titles are Benedict Anderson, *Imagined Communities: Reflections on the Origin and Spread of Nationalism* (New York: Verso, 1991), and Ernest Gellner, *Nations and Nationalism* (Ithaca, NY: Cornell University Press, 1983).

12. Partha Chatterjee, *The Nation and Its Fragments: Colonial and Postcolonial Histories* (Princeton, NJ: Princeton University Press, 1993), 6. It is worth noting that Chatterjee's first chapter, "Whose Imagined Community?," represents a direct challenge to Anderson's treatment of the subject.

13. Eric G. E. Zuelow, "Negotiating National Identity through Tourism in Colonial South Asia and Beyond," in *Cambridge History of Nationhood and Nationalism*, ed. Matthew D'Auria, Cathie Carmichael, and Aviel Roshwald (Cambridge: Cambridge University Press, forthcoming).

Introduction

1. Wang Xiaoting, "Meili de Zhonghua" [Beautiful China], *Dazhong huabao* [The cosmopolitan] no. 12 (October 1934): 30–31. Unless otherwise noted, all translations are my own.

2. Wang Xiaoting, 31.

3. On the paradox of the Republican period, see John King Fairbank and Merle Goldman, *China: A New History* (Cambridge, MA: Harvard University Press, 2006), 259–60; Frank Dikötter, *The Age of Openness* (Hong Kong: Hong Kong University Press, 2008), 2.

4. Zhang Lili, *Jindai Zhongguo lüyou fazhan de jingji toushi* [An economic study of tourism development in modern China] (Tianjin: Tianjin daxue chubanshe, 1998), 112–20, 170–74.

5. Xiang Wenhui, *Hangzhou lüyou ji qi jindai mingyun* [Hangzhou tourism and its modern fate] (Hangzhou: Zhejiang daxue chubanshe, 2018), 30.

6. Two books on the steamship and railway industries demonstrate the link between imperialism and modern transportation institutions in modern China. Anne Reinhardt, *Navigating Semi-Colonialism: Shipping, Sovereignty, and Nation-Building in China, 1860–1937* (Cambridge, MA: Harvard University Asia Center, 2018); Elisabeth Köll, *Railroads and the Transformation of China* (Cambridge, MA: Harvard University Press, 2019).

7. The modern travel agency can be traced back to the 1840s, when Thomas Cook began organizing tours in Britain. Jill Hamilton, *Thomas Cook: The Holiday Maker* (Stroud, UK: Sutton, 2005).

8. Mount Zhong, Huangpu River, and Nanping are famous tourist sites in Nanjing, Shanghai, and Hangzhou respectively. Preface to *Huning huhangyong tielu lüxing zhinan* [Travel guide for the Shanghai-Nanjing and Shanghai-Hangzhou-Ningbo Railroads], ed. Huning huhangyong lianglu biancha ke [The compilation department of

the Shanghai-Nanjing and Shanghai-Hangzhou-Ningbo Railroads] (Shanghai: Huning huhangyong tielu guanli ju, 1918), 1.

9. For a discussion of China's territorial national image in Nationalist-era maps in geography textbooks, see Robert Culp, *Articulating Citizenship: Civic Education and Student Politics in Southeastern China, 1912–1940* (Cambridge, MA: Harvard University Asia Center, 2007).

10. "Liangyou sheying tuan," *Liangyou* [Young companion], no. 70 (October 1932): 22.

11. For a succinct survey of the field of tourism history and an example of this critique, see John K. Walton, "Tourism History: People in Motion and at Rest," *Mobility in History* 5 (2014): 74–85.

12. For studies of Western travelers in China, see, for example, Susan Schoenbauer Thurin, *Victorian Travelers and the Opening of China, 1842–1907* (Athens: Ohio University Press, 1999); Douglas Kerr and Julia Kuehn, eds., *A Century of Travels in China: Critical Essays on Travel Writing from the 1840s to the 1940s* (Hong Kong: Hong Kong University Press, 2007). For an example of the narrative of Chinese tourism having originated in the post-Mao reform period, see Chris Ryan and Gu Huimin, eds., *Tourism in China: Destination, Cultures and Communities* (New York: Routledge, 2009).

13. On traditional travel literature in the imperial period, see Richard E. Strassberg, ed. and trans., *Inscribed Landscapes: Travel Writing from Imperial China* (Berkeley: University of California Press, 1994); James Hargett, *Jade Mountains and Cinnabar Pools: The History of Travel Literature in Imperial China* (Seattle: University of Washington Press, 2018). On the history of pilgrimages in imperial China, see Susan Naquin and Chün-fang Yü, eds., *Pilgrims and Sacred Sites in China* (Berkeley: University of California Press, 1992); Brian Dott, *Identity Reflections: Pilgrimages to Mount Tai in Late Imperial China* (Cambridge, MA: Harvard University Asia Center, 2005). In addition, for the production of city guidebooks and urban touring culture in the late imperial period, see Siyen Fei, "Ways of Looking: The Creation and Social Use of Urban Guidebooks in Sixteenth- and Seventeenth-Century China," *Urban History* 37, no. 2 (2010): 226–48; Wu Jen-shu and Imma Di Biase, *Youdao: Ming Qing lüyou wenhua* [The Tao of travel: Travel culture in the Ming and Qing era] (Taipei: Sanmin shuju, 2010). For the existing literature in English on CTS and its travel magazine, see Miriam Gross, "Flights of Fancy from a Sedan Chair: Marketing Tourism in Republican China, 1927–1937," *Twentieth-Century China* 36, no. 2 (2011): 119–47; Madeleine Yue Dong, "Shanghai's *China Traveler*," in *Everyday Modernity in China*, ed. Madeleine Yue Dong and Joshua L. Goldstein (Seattle: University of Washington Press, 2005), 195–226. On the Republican-era tourism and travel culture in the lower Yangzi region, see Liping Wang, "Tourism and Spatial Change in Hangzhou, 1911–1927," in *Remaking the Chinese City: Modernity and National Identity, 1900–1950*, ed. Joseph Esherick (Honolulu: University of Hawai'i Press, 2000), 116–19; Pedith Chan, "In Search of the Southeast: Tourism, Nationalism, and Scenic Landscape in Republican China," *Twentieth-Century China* 43, no. 3 (2018): 207–31.

14. On the impact of rail travel, see Wolfgang Schivelbusch, *The Railway Journey: The Industrialization of Time and Space in the Nineteenth Century* (Berkeley: University of California Press, 2014). For the rise of commercialized travel for both the leisure

class and the masses in the context of industrial capitalism, see Eric Hobsbawm, *The Age of Capital, 1848–1875* (London: Abacus, 1995), 239–44. On tourism as an epitome of consumerism, see Louis Turner and John Ash, *The Golden Hordes: International Tourism and the Pleasure Periphery* (London: Constable, 1975).

15. On tourism's role in nation building and constructing national identity, see Baranowski, *Strength through Joy*; Shaffer, *See America First*; Zuelow, *Making Ireland Irish*.

16. Herman Towne, "Summer Resorts in China: Six of China's Major Scenic & Health Spots," *China Press*, July 8, 1936; "Group Tours Gaining Favor among Travel-Minded Chinese," *China Press*, July 8, 1936.

17. On the seaside and hill resorts created by foreigners in China, see António Barrento, "Going Modern: The Tourist Experience at the Seaside and Hill Resorts in Late Qing and Republican China," *Modern Asian Studies* 52, no. 4 (2018): 1089–133.

18. The arrival of Chinese tourists also led to the decline of Western vacationers at resorts like Moganshan and Guling. Dong, "Shanghai's *China Traveler*," 209.

19. On the development of the Yellow Mountains as a modern tourist site in the mid-1930s, see Gross, "Flights of Fancy," 135–42.

20. Shaffer, *See America First*; Baranowski, *Strength through Joy*; Koenker, *Club Red*.

21. Li Baorong, "Zhongguo wuda bishu qu" [Five largest summer resorts in China], *LZ* 7, no. 7 (July 1933): 109–13; Jiang Weiqiao, "Wuyue yu sida mingshan" [Five sacred and four famous mountains], *LZ* 10, no. 1 (January 1936): 49.

22. Wu Liande [L. T. Wu], ed., *Zhonghua jingxiang: quanguo sheying zong ji* [China as she is: A comprehensive album] (Shanghai: Liangyou tushu yinshua youxian gongsi, 1934).

23. Chen Guangfu, "Fakan ci" [Inaugural preface], *LZ* 1, no. 1 (Spring 1927): 1.

24. Turner and Ash, *Golden Hordes*.

25. Anderson, *Imagined Communities*, xiii–xiv.

26. Thongchai Winichakul, *Siam Mapped: A History of the Geo-Body of a Nation* (Honolulu: University of Hawai'i Press, 1994).

27. For the use of cartography and geography textbooks in creating China's geobody, see William A. Callahan, "The Cartography of National Humiliation and the Emergence of China's Geobody," *Public Culture* 21, no. 1 (2009): 141–73; Culp, *Articulating Citizenship*.

28. For a discussion of the horizontal dialogue between elites and the masses in tourism development and the building of a national identity, see Zuelow, *Making Ireland Irish*, xxix–xxx.

1. Travel as a Business

1. "Shanghai shangye chuxu yinhang she lüxingbu" [Shanghai Commercial and Savings Bank adds a Travel Department], *Shenbao* [Shanghai news], August 12, 1923.

2. Wen-hsin Yeh, *Shanghai Splendor: Economic Sentiments and the Making of Modern China, 1843–1949* (Berkeley: University of California Press, 2007), 13–17.

3. Yousheng lüxing tuan [The Unison Travel Club], *Yousheng lüxing tuan jianshi* [A brief history of the Unison Travel Club] (Shanghai: Yousheng lüxing tuan, 1947), 19.

4. For a sample of the literature on the significance of the concept of semicolonialism in modern China, see Shu-mei Shih, *The Lure of the Modern: Writing Modernism in Semicolonial China, 1917–1937* (Berkeley: University of California Press, 2001), 35–37; Ruth Rogaski, *Hygienic Modernity: Meanings of Health and Disease in Treaty-Port China* (Berkeley: University of California Press, 2004), 11–12; Bryna Goodman and David Goodman, eds., *Twentieth-Century Colonialism and China: Localities, the Everyday, and the World* (Abingdon, UK: Routledge, 2012), 3–7; Reinhardt, *Navigating Semi-Colonialism*, 3–7. For a summary of the locales with nonmissionary foreign presence in late Qing and Republican China, see Robert Nield, *China's Foreign Places: The Foreign Presence in China in the Treaty Port Era, 1840–1943* (Hong Kong: Hong Kong University Press, 2015).

5. Anne Reinhardt, "Treaty Ports as Shipping Infrastructure," in *Treaty Ports in Modern China: Law, Land and Power*, ed. Robert Bickers and Isabella Jackson (London: Routledge, 2016), 101.

6. Anne Reinhardt has highlighted the notion of "collaboration" to illustrate how Chinese agents preserved the country's compromised sovereignty via steam navigation. Elisabeth Köll has traced railway development in China and characterized Chinese railway companies as institutions following a hybrid enterprise model. Reinhardt, *Navigating Semi-Colonialism*, 8–14; Köll, *Railroads and the Transformation of China*, 6–12.

7. Reinhardt, *Navigating Semi-Colonialism*, 149.

8. Köll, *Railroads and the Transformation of China*, 132.

9. The China Merchants Steam Navigation Company, a commercial shipping business overseen by Qing officials, for example, made efforts to create steamship lines connecting "places without rail service" to railways in the 1900s. Nie Baozhang and Zhu Yingui, eds., *Zhongguo jindai hangyunshi ziliao, 1895–1927* [Historical materials on modern Chinese shipping, 1895–1927] (Beijing: Zhongguo shehui kexue chubanshe, 2002), 2:477.

10. Thomas Cook, *Letters from the Sea and from Foreign Lands: Descriptive of a Tour round the World* (London: Thomas Cook & Son, 1873).

11. Cook, "Fourth Letter," in *Letters from the Sea*, 31–32.

12. Thomas Cook Ltd., *Cook's Handbook for Tourists to Peking, Tientsin, Shan-Hai-Kwan, Mukden, Dalny, Port Arthur, and Seoul* (London: Thomas Cook & Son, 1910). For the locations of the Travel Department of the American Express in China, see "When You Go Home," an advertisement for the Travel Department of the American Express, *Chinese Students' Monthly* 16, no. 8 (1900): 5. Chinese Eastern Railway, *North Manchuria and the Chinese Eastern Railway* (Harbin: C.E.R. Printing Office, 1924), 431; Kate McDonald, *Placing Empire: Travel and the Social Imagination in Imperial Japan* (Berkeley: University of California Press, 2017), 50–57.

13. J. D. Clark, *Sketches in and around Shanghai* (Shanghai: Shanghai Mercury, 1894).

14. Nield, *China's Foreign Places*, 150–51, 159–60, 184–86; Barrento, "Going Modern," 1090–91.

15. Nie and Zhu, *Zhongguo jindai hangyunshi ziliao, 1895–1927*, 1:1411.

16. Jiang Weiqiao, "Taishan jiyou" [Travelogue of Mount Tai], *Xiaoshuo yuebao* [Fiction monthly] 6, no. 10 (1915): 1–3.

17. Wang Guoyuan, "Jingzhang tiedao zhi lüxing tan" [On travel via the Beijing-Kalgan Railway], *Dongfang zazhi* [Eastern miscellany] 6, no. 11 (1909): 31–34.

18. Hargett, *Jade Mountains*, 142–45.

19. Jozsef Borocz, "Travel-Capitalism: The Structure of Europe and the Advent of the Tourist," *Comparative Studies in Society and History* 34, no. 4 (1992): 708–41.

20. Reinhardt, *Navigating Semi-Colonialism*, 148–69.

21. Köll, *Railroads and the Transformation of China.*

22. For an example, see Lüsheng, "Chiren shuomeng ji: Di shi'er hui" [Chapter 12: A fool's tale of his dream], *Xiuxiang xiaoshuo* [Fiction illustrated], no. 30 (1904): 2–3.

23. Stanford, "Mokanshan," 329.

24. Zhiheng, "Bishu zhong zhi Moganshan" [Mokanshan during summer], *Jiaoyu zhoubao* [Education weekly], no. 132 (1916): 34–35.

25. For an analysis of Lü's travel poems and travelogue of Mount Lu, see Grace Fong, "Reconfiguring Time, Space, and Subjectivity: Lü Bicheng's Travel Writings on Mount Lu," in *Different Worlds of Discourse: Transformations of Gender and Genre in Late Qing and Early Republican China*, ed. Nanxiu Qian, Grace Fong, and Richard J. Smith (Leiden: Brill, 2008), 87–114.

26. Barrento, "Going Modern," 1104–5.

27. On dancing clubs and movie theaters in Republican Shanghai, see Andrew Field, *Shanghai's Dancing World: Cabaret Culture and Urban Politics, 1919–1954* (Hong Kong: Chinese University Press, 2010); Leo Ou-fan Lee, *Shanghai Modern: The Flowering of a New Urban Culture in China, 1930–1945* (Cambridge, MA: Harvard University Press, 1999), 82–119.

28. Xu Ke, *Beidaihe zhinan* [Guide to Peitaiho] (Shanghai: Shangwu yinshu guan, 1922), 2.

29. "Shouhui jigong shandi lingyi zuwu bishu zhangcheng" [Reclaiming Jigongshan and discussions about leasing out the summer resort], *Dongfang zazhi* 5, no. 11 (1908): 20–23.

30. "Moganshan bishu zhi lianyun guizhang" [Regulations of the through transportation to Moganshan], *Zhonghua gongchengshi xuehui huibao* [Journal of associations of Chinese engineers] 7, no. 4 (1920): 1–2.

31. Jiang Weiqiao, "Moganshan jiyou" [Travelogue of Moganshan], *Xiaoshuo yuebao* 11, no. 9 (1920): 1–3.

32. The German owner sold the hotel to the state railway because British and Americans tourists had boycotted it after the outbreak of World War I. Jiang Weiqiao, "Moganshan jiyou"; "Moganshan tielu lüguan" [Mokanshan Railway Hotel], *Tielu gongbao: Huning Huhangyong xian* [Railway gazetteer: Shanghai-Nanjing and Shanghai-Hangzhou-Ningbo lines], no. 41 (1921): n.p.

33. Chen Guangfu, *Chen Guangfu xiansheng yanlun ji* [A collection of Mr. Chen Guangfu's speeches] (Taipei: Shanghai shangye chuxu yinhang, 1970), 12.

34. Yuan Xijian, "Chen Guangfu de yisheng yu Shanghai yinhang" [The life of Chen Guangfu and Shanghai Bank], in *Chen Guangfu yu Shanghai yinhang* [Chen Guangfu and Shanghai Bank], ed. Wu Jingyan (Beijing: Zhongguo wenshi chubanshe, 1991), 103–4; Yao Songling, *Chen Guangfu de yisheng* [Chen Guangfu's life] (Taipei: Zhuanji wenxue, 1984), 1–9.

35. Susan Fernsebner, "When the Local is the Global: Case Studies in Early Twentieth-Century Chinese Exposition Projects," in *Expanding Nationalisms at World's Fairs*, ed. David Raizan and Ethan Robby (London: Rutledge, 2018), 176–77.

36. Weili Ye, *Seeking Modernity in China's Name: Chinese Students in the United States, 1900–1927* (Stanford, CA: Stanford University Press, 2001), 88–91.

37. Chen Guangfu, *Chen Guangfu xiansheng yanlun ji*, 12.

38. For example, the gender of the clerk was inconsistent in different versions of this story. See Zhao Junhao, "Chen Guangfu xiansheng fangwen ji" [Interview with Mr. Chen Guangfu], *LZ* 10, no. 9 (September 1936): 83–86.

39. On the "national product movement" in the Republican period, see Karl Gerth, *China Made: Consumer Culture and the Creation of the Nation* (Cambridge, MA: Harvard University Press, 2003).

40. Zhaojin Ji, *A History of Modern Shanghai Banking: The Rise and Decline of China's Finance Capitalism* (Armonk, NY: M. E. Sharpe, 2003), 120; Chen Guangfu, *Chen Guangfu xiansheng yanlun ji*, 129.

41. Brett Sheehan, "Webs and Hierarchies: Banks and Bankers in Motion, 1900–1950," in *Cities in Motion: Interior, Coast, and Diaspora in Transnational China*, ed. Sherman Cochran and David Strand (Berkeley: Institute of East Asian Studies, University of California, 2007), 81–105.

42. For the nationalization process of railways, see Köll, *Railroads and the Transformation of China*, 53–87.

43. ZLZD, Q275-1-128.

44. Chinese Eastern Railway, *North Manchuria*, 431.

45. Wu Jingyan, *Chen Guangfu yu Shanghai yinhang*, 190.

46. For a description of this practice, see Reinhardt, *Navigating Semi-Colonialism*, 138–41.

47. ZLZD, Q368-1-555.

48. "Shanghai yinhang lüxing bu zhi fada" [The development of the Travel Department at Shanghai Bank], *Yinhang yuekan* [Banker's magazine] 4, no. 4 (1924): 12–13.

49. SSCYD, Q275-1-1830.1.

50. Wu Jingyan, *Chen Guangfu yu Shanghai yinhang*, 191.

51. Zhu Chengzhang, "Zhongguo lüxingshe jianshi" [A brief history of the China Travel Service], *Haiguang* [Internal journal of Shanghai Bank] 1, no. 11 (November 1929): 3–4.

52. Quoted in Yuan Xijian, "Chen Guangfu de yisheng yu Shanghai yinhang," 113.

53. Zhu Chengzhang, "Zhongguo lüxingshe jianshi," 2.

54. Chen Guangfu, *Chen Guangfu xiansheng yanlun ji*, 8.

55. Pan Taifeng, "Ji Zhongguo lüxingshe" [On the China Travel Service], in Wu Jingyan, *Chen Guangfu yu Shanghai yinhang*, 190.

56. On the damage to the railroads by warlords, see Köll, *Railroads and the Transformation of China*, 69–71.

57. Chen Guangfu, *Chen Guangfu xiansheng yanlun ji*, 6.

58. Henrietta Harrison, *The Making of the Republican Citizen: Political Ceremonies and Symbols in China, 1911–1929* (Oxford: Oxford University Press, 2000), 69.

59. António Barrento, "On the Move: Tourist Culture in China, 1895–1949" (DPhil diss., SOAS, University of London, 2012), 51–52, 74–82.

60. "Lüxingbu zhengyou yundong jianzhang" [The general rule of membership at the travel club], *Jingwu congbao* [Jingwu journal] 2, no. 7 (1936): 5.

61. Yousheng lüxing tuan, *Yousheng lüxing tuan jianshi*, 11; Park Kyung Seok, "Minguo shiqi Shanghai de yousheng lüxing tuan he 'xiuxian lüxing'" [The Unison Travel Club in Republican Shanghai and "leisure travel"], *Minguo yanjiu* [Studies of Republican China], no. 18 (2010): 246–61.

62. "Tuanyuan xuzhi" [Notice to members], *Yousheng* [Unison] (July 1923): n.p.; Yao Yuangan, "Yousheng lüxing tuan zhi guoqu yu weilai tuanyuan zhi zeren" [The mission of the Unison Travel Club's past and future members], *Yousheng* (December 1926): n.p.; Yousheng lüxing tuan, *Yousheng lüxing tuan jianshi*, 11–12.

63. "Ben jie zhengqiu hui zhangcheng" [The rules of this recruiting conference], *Yousheng* (September 1925): n.p.

64. "Jiaru Huabei lüxing tuanyuan xingshi yilan biao" [The name list of the tour members to North China], in "Huabei lüxing zhuanhao" [Special issue for North China tour], *Yousheng* (1931): 5–8.

65. "Yousheng lüxing tuan Shanghai zongbu weiyuan yilan" [A list of committee members of the Unison Travel Club in Shanghai], *Yousheng* (November 1936): 16.

66. On Unison members' income level within the social-economic composition of Republican Shanghai, see Park, "Minguo shiqi Shanghai de yousheng lüxing tuan he 'xiuxian lüxing,'" 249–50.

67. Köll, *Railroads and the Transformation of China*, 150–55.

68. "Paiding zhaoliao guanchao zhuanche zhi renyuan" [Personnel for the special trains for viewing tidal waves], *Shenbao*, October 8, 1919; "Three Hangchow Bore Excursions to Be Held through Week-End," *China Press*, September 9, 1927.

69. Huang Weiqing, "Chongdao guilai" [Returning from Chongming Island], *Jingwu huabao* [Jingwu pictorial] 2, no. 10 (1929): 3.

70. Wu Peisong, "Canjia Moganshan denggao jingzou ji" [On my participation in the hiking competition at Moganshan], *Yousheng* (May 1935): n.p.; "Nanxiang qima jingsai" [The horse race in Nanxiang], *Yousheng* (May 1935): n.p.

71. Pan Taifeng, "Ji Zhongguo lüxingshe," 191–92.

72. Wu Jingyan, *Chen Guangfu yu Shanghai yinhang*, 11.

73. Zhaojin Ji, *History of Modern Shanghai Banking*, 167–70.

74. Yuan Xijian, "Chen Guangfu de yisheng yu Shanghai yinhang," 113–14.

75. Guomin zhengfu jiaotongbu [Ministry of Transportation and Communication], "Lüxingye zhuce zanxing tiaoli" [The provisional regulations regarding registration in the tourism industry], *Jiaotong gongbao* [Bulletin for transportation regulations] 1, no. 2 (1927): 69–71.

76. Pan Taifeng, "Ji Zhongguo lüxingshe," 191.

77. "Mingsheng guji guwu baocun tiaoli" [The regulation of protecting scenic spots, historical sites, and historical relics], *Neizheng gongbao* [Bulletins of the Ministry of the Interior] 1, no. 6 (1928): 13–16.

78. Albert Feuerwerker, "Economic Trends, 1912–1949," in *The Cambridge History of China*, vol. 12, *Republican China, 1912–1949*, ed. John K. Fairbank and Denis C. Twitchett (Cambridge: Cambridge University Press, 1983), 94.

79. Arthur Nichols Young, *China's Nation-Building Effort, 1927–1937: The Financial and Economic Record* (Stanford, CA: Hoover Institution Press, 1971), 320.

80. Pan Taifeng, "Ji Zhongguo lüxingshe," 192.

81. "China Travel Service to Link Hankow and Canton by Motors," *China Press*, June 3, 1933.

82. ZLZD, Q368-1-36.

83. ZLZD, Q368-1-477, Q368-1-36.

84. *LZ* 1, no. 1 (Spring 1927): n. p.; Wu Jingyan, *Chen Guangfu yu Shanghai yinhang*, 193.

85. ZLZD, Q368-1-599.

86. Pan Taifeng, "Ji Zhongguo lüxingshe," 196; ZLZD, Q368-1-37.

87. Wu Jingyan, *Chen Guangfu yu Shanghai yinhang*, 224.

88. "Xuzhou zhaodaisuo" [Xuzhou guesthouse], *LZ* 9, no. 9 (September 1935): 47–51.

89. C. P. Chin, "China Travel Service Takes Lead in Promoting Comfort," *China Press*, May 28, 1933.

90. "Nanjing shoudu fandian" [The Metropolitan Hotel in Nanjing], *LZ* 9, no. 9 (September 1935): 7–13.

91. "Guangrong zhi ye" [Customer letters], *LZ* 11, no. 3 (March 1937): n.p.

92. Zhang was first detained at the Xuedoushan Guesthouse in Chiang Kai-shek's hometown. When a fire destroyed the facility, Zhang was relocated to another CTS guesthouse in Huangshan. Pan Taifeng, "Kangri zhanzheng shiqi de Zhongguo lüxingshe" [The China Travel Service during the Anti-Japanese War], in *Wenshi ziliao xuanji* [Selected historical materials], edited by Wenshi ziliao yanjiu weiyuanhui [Research committee of historical materials], vol. 117 (Beijing: Zhongguo wenshi chubanshe, 1989), 170; ZLZD, Q368-1-188, Q368-1-819.

93. Zhao Junhao, "Zhongguo lüxingshe fazhan jianshi (shang)" [A brief history of the development of China Travel Service], *Lüguang* [Light of travel] 2, no. 1 (January 1941): 2. *Lüguang* was the internal publication of CTS.

94. ZLZD, Q368-1-36.

95. Zhuang Zhujiu and Xu Zhaofeng, "Zengbie you mei xuesheng" [Parting words for students going to America], *LZ* 2, no. 2 (Summer 1928): 60–69.

96. "Xinjiapo fenshe kaimu jisheng" [The grand opening of the China Travel Service Singapore branch], *LZ* 8, no. 12 (December 1934): 2–3.

97. "Overseas Chinese Tourists Make 1st Tour of Shanghai," *China Press*, October 16, 1934.

98. "Xinjiapo fenshe kaimu jisheng," 3.

99. Zhongguo lüxingshe [China Travel Service], *You Shanghai zhi Maijia* [From Shanghai to Mecca] (Shanghai: Zhongguo lüxingshe, 1933).

100. ZLZD, Q368-1-556.

101. Ming, "Zhongguo de chaojin zhe" [Chinese pilgrims], *Yuehua* [*Crescent China*] 4, no. 4 (1932): 22.

102. "China Travel Service Plans Olympic Tour," *China Press*, May 28, 1936.

103. "Group Tours Gaining Favor."

104. Jiang Huaiqing, "Canguan shijie yundonghui qu" [Let's go tour the Olympic Games], *LZ* 10, no. 3 (March 1936): 119–23.

105. "Nippon Tour Plans Are Set," *China Press*, March 27, 1936.

106. Y. T. Liang, "Open Letter to Travel Service," *China Press*, January 10, 1937.

107. C. Chen, "China's Appeal for the Tourist," *China Press*, October 10, 1936.

108. "Zhongguo lüxingshe fashou guanchao piao" [China Travel Service sells excursion tickets for tidal bore tour], *Shenbao*, September 25, 1928; Fan Dimin,

"Jinnian de jige guanchao tuan" [This year's tidal bore tours], *Xihu bolanhui rikan tekan* [Special daily for West Lake Exposition], no. 11 (1929): 3.

109. "Youhang zhuanche zhiqu" [The special tourism train to Hangzhou], *LZ* 2, no. 1 (Spring 1928): 2.

110. "Yu zhongguo lüxingshe heban gudu xiaohan youlantuan" [Our railway collaborates with China Travel Service to bring you the "winter holiday tour to the old capital"], *Tielu zazhi* [Railway magazine] 2, no. 8 (1937): 116; "Banli xiangshan youlantuan" [The Beijing-Mukden Railway organizes tour to Fragrant Mountain], *Tiedao banyue kan* [Railway bimonthly] 1, no. 13 (1936): 63.

111. "Jinpu Jiaoji lianglu huitong juban qingdao youlantuan" [The Jinpu and Jiaoji Railways organize Qingdao tour], *Tiedao gongbao* [Railway bulletins], no. 1457 (1936): 13; "Juban Beiping Qingdao liangdi lüxing youlantuan" [Tours between Beijing and Qingdao], *Tielu banyue kan* [Railway bimonthly] 2, no. 9 (1937): 150–51.

112. "Fuchun jiang lüxingtuan ji" [About the Fuchun River tour], *LZ* 4, no. 5 (May 1930): 53–58; Lu Weipin, "Fuchun jiang youji" [Travelogue of Fuchun River], *LZ* 5, no. 6 (June 1931): 55–60.

113. Xu Pudong, "Xihu bolanhui choubei de jingguo" [The process of preparations for the West Lake Exposition], *Dongfang zazhi* 26, no. 10 (October 1929): 27–28.

114. "Zhuiji xihu youlantuan" [On tour to West Lake], *LZ* 4, no. 2 (February 1930): 67–68.

115. Wu Pishi, "Xianggang qingnian hui huadong youlantuan youji" [Travelogue of Hong Kong YMCA's East China Tour], *Qingnian jinbu* [Youth progress], no. 140 (1931): 91–114. On the invention of *cuyu* as a Hangzhou dish and the invention of tradition in Hangzhou tourism, see Liping Wang, "Tourism and Spatial Change," 116–19.

116. Pedith Chan, "In Search of the Southeast"; Gross, "Flights of Fancy," 135–42.

117. Madeleine Yue Dong, *Republican Beijing: The City and Its Histories* (Berkeley: University of California Press, 2003), 78–104.

118. She Guitang, "Zhongguo youlan shiye zhi huigu" [A review of Chinese tourism], *LZ* 17, no. 7 (July 1943): 8.

119. "Guangzhou shi xiangdao youlan guize" [Regulations of tour guides in Guangzhou], *Guangzhoushi zhengfu shizheng gongbao* [Guangzhou municipal gazette], no. 499 (1935): 10.

120. "Lüxingshe zuzhi erci youlan Huangshan tuan" [China Travel Service organizes a second tour to the Yellow Mountains], *Shenbao*, April 18, 1935.

121. "Kuaiji lüshi liangxiehui zhengqiu Huangshan lüxing tongzhi" [The Bar and Accountants' Associations soliciting members to tour Huangshan], *Shenbao*, March 19, 1936.

122. "Xuedoushan zhaodaisuo" [Xuedoushan Guesthouse], *LZ* 9, no. 9 (September 1935): 33.

123. "Xuedoushan zhaodaisuo"; "China Travel Service to Conduct Xmas Tour," *China Press*, December 5, 1935; "China Travel Service Tour to Visit Chiang's Birthplace," *China Press*, October 29, 1936.

124. Chen Cunren, "Huanan lüxing ji (shang)" [My travel in South China (part 1)], *LZ* 8, no. 5 (May 1934): 5–16.

125. Chen Cunren, "Huanan lüxing ji (xia)" [My travel in South China (part 2)], *LZ* 8, no. 6 (June 1934): 39–52.

126. Zhao Zheshi, "Huanan lüxing ji buzheng" [Corrections on "My travel in South China"], *LZ* 8, no. 9 (September 1934): 27–29.

127. "Zhongguo lüxingshe zuzhi youlantuan" [CTS organizes a tour club], *Shenbao*, August 4, 1935.

128. Yousheng lüxing tuan, *Yousheng lüxing tuan jianshi*, 12–13, 19.

129. "Lüxinggu baogao jingguo didian" [The tour department's report on tour destinations], *Yousheng*, no. 9 (1926): n.p.; "Huiyi jilu" [Meeting record] and "Lüxing huiwen" [The tourism news], *Yousheng*, no. 7 (1932): 3–4; "Bentuan san si yuefen lüxing tongji" [The statistics of the club's tours in March and April], *Yousheng* (May 1935): 2.

130. "Tongao, di eryi-wu hao" and "Tongao, di eryi-liu hao" [Announcements, no. 21.5 and no. 21.6], *Yousheng* (July 1932): 5.

131. "Bentuan wei jianqin chejia cheng tiedaobu shu" [The club's petition to the Ministry of Railways in order to cut the price for train tickets], *Yousheng lüxing tuan yuekan* [Unison Travel Club monthly] (April 1933): 48–50.

132. "Lüxing gaikuang" [Overview of travel], *Yousheng* (October 1934): 4.

133. For example, in July 1932, the Unison Travel Club chartered a ship to carry 450 tour participants to Mount Putuo. "Lüxing huiwen," 4.

134. Zhiqin, "Wuxie shan tansheng tu" [Travelogue and photographs of our adventure at Wuxie mountains], *Zhonghua* [China pictorial], no. 15 (1933): 10–12.

135. On the income level of middle-class families and tourism consumption in Republican Shanghai, see Gross, "Flights of Fancy," 121–23.

136. In addition to the North China tours, the Unison Travel Club also organized tours to Sichuan and South China. "Yousheng lüxing tuan Sichuan lüxing tonggao" [Sichuan tour announcement from the Unison Travel Club], *Shenbao*, August 14, 1935; "Youshengtuan faqi wusheng changtu lüxing" [The Unison Travel Club initiates a tour of five provinces], *Shenbao*, February 6, 1937.

137. "Yousheng lüxing tuan jinyao qishi" [An emergency announcement from the Unison Travel Club], *Shenbao*, September 25, 1931.

138. "Youshengtuan, huabei lüxing zhi xingcheng" [The itineraries of the Unison Club's North China tour], *Shenbao*, August 24, 1934.

139. "Xiezhu yousheng lüxing tuan juban diqi diba liangci huabei lüxing" [Assisting the Unison Travel Club's seventh and eighth North China tours], *Jinghu huhangyong tielu rikan* [Daily of the Shanghai-Nanjing and Shanghai-Hangzhou-Ningbo Railways], no. 1922 (1937): 132–33.

140. One Unison tour-goer compared different parts of the North China tour to a banquet with different courses. Whereas Qufu was like a cold appetizer and Mount Tai a hot stir-fried side dish, Beijing was definitely the main course. Zheng Nuofu, "Beiyou riji (xu)" [Journal of a North China tour (2)], *Yousheng*, no. 7 (1935): 12.

141. "Bentuan zongzhi" [The principle of this tour], in "Huabei lüxing zhuanhao," 1–3.

142. LaSelle Gilman, "Peiping Guides Cheered by Old City's Political Decline," *China Press*, August 29, 1935.

143. Lu Shifu, "Huabei lüxing sheying" [The travel photographs of North China], *Shidai* [Modern miscellany] 7, no. 4 (1934): 17.

144. Zheng Nuofu, "Beiyou riji" [Journal of a North China tour (1)], *Yousheng*, no. 6 (1935): 12.

145. Zheng Nuofu, "Beiyou riji (xu)," 11.
146. Zheng Nuofu, "Beiyou riji," 11.
147. Zheng Nuofu.
148. Xia Ying, "Huabei lüxing guilai" [Returning from a North China tour], *Shengli* [Vitality] 1, no. 1 (1936): 14.
149. Xia Ying, 14–15.
150. Zheng Nuofu, "Beiyou riji" [Journal of a North China tour (6)], *Yousheng*, no. 11 (1936): 11.
151. Margaret McClure, *The Wonder Country: Making New Zealand Tourism* (Auckland: Auckland University Press, 2004), vi; Horace M. Albright and Robert Cahn, *The Birth of the National Park Service: The Founding Years, 1913–1933* (Chicago: Howe Brothers, 1985); David Leheny, *The Rules of Play: National Identity and the Shaping of Japanese Leisure* (Ithaca, NY: Cornell University Press, 2003); Zuelow, *Making Ireland Irish*; Koenker, *Club Red*.

2. Travel as Narratives

1. Zhao Junhao, "Bianjishi shinian ji" [Ten years of my editorship], *LZ* 10, no. 1 (January 1936): 197.
2. On the modern Chinese travel literature, see Ping-hui Liao, "Travels in Modern China: From Zhang Taiyan to Gao Xingjian," in *The Oxford Handbook of Modern Chinese Literatures*, ed. Carlos Rojas and Andrea Bachner (Oxford: Oxford University Press, 2016), 39–51.
3. Timothy Brook, "Guides for Vexed Travelers: Route Books in the Ming and Ch'ing," *Ch'ing-shi wen-t'i* [Late imperial China] 4, no. 5 (1981): 32–76; Du Yongtao, *The Order of Places: Translocal Practices of the Huizhou Merchants in Late Imperial China* (Leiden: Brill, 2015), 201–37.
4. On the destination guidebooks by late imperial literati, see Tobie Meyer-Fong, "Seeing the Sights in Yangzhou from 1600 to the Present," in *When Images Speak: Visual Representation and Cultural Mapping in Modern China*, ed. Huang Ko-wu (Taipei: Academia Sinica, Institute of Modern History, 2004), 213–51; Siyen Fei, "Ways of Looking."
5. Catherine Yeh, *Shanghai Love: Courtesans, Intellectuals, and Entertainment Culture, 1850–1910* (Seattle: University of Washington Press, 2006), 306–16.
6. For a general format of *Shanghai zhinan*, see Shangwu yinshu guan [Commercial Press], ed., *Shanghai zhinan* [Guide to Shanghai] (Shanghai: Shangwu yinshu guan, 1909); Lin Zhen, ed., *Zengding Shanghai zhinan* [Expanded guide to Shanghai], 23rd ed. (Shanghai: Shangwu yinshu guan, 1930).
7. Shangwu yinshu guan, "Diwu juan: shisu youlan" [Volume 5: Eateries, lodgings, and touring], *Shanghai zhinan*, 21st ed. (Shanghai: Shangwu yinshu guan, 1925), 1–30.
8. Xu Ke, *Shiyong Beijing zhinan* [Practical guide to Beijing] (Shanghai: Shangwu yinshu guan, 1919); Xu Ke, *Zengding shiyong Beijing zhinan* [Expanded practical guide to Beijing] (Shanghai: Shangwu yinshu guan, 1926).
9. "Tiantan" [The temple of heaven], in Xu Ke, *Zengding shiyong Beijing zhinan*, chapter 9, 5.

10. For a few examples of railroad travel guides, see Cao Jinggao, ed., *Jingsui tielu lüxing zhinan* [Travel guide for the Beijing-Suiyuan Railroad] (Beijing: Jingsui tielu guanli ju, 1916); *Huning huhangyong tielu lüxing zhinan*; Jinpu tielu guanli ju zongwu chu biancha ke [The compilation department of the Tianjin-Pukou Railroad], ed., *Jinpu tielu lüxing zhinan* [Travel guide for the Tianjin-Pukou Railroad] (Tianjin: Jinpu tielu guanli ju, 1921).

11. Since national railroads in China were not integrated during the warlord era, the earliest travel guides for all national railroads were issued by private publishing houses. *Quanguo tielu lüxing zhinan* [National railroad travel guide] (Shanghai: Guangyi shuju, 1921); Jiaotong bu tielu lianyun shiwu chu [Division of rail traffic, Ministry of Transportation], ed., *Zhonghua guoyou tielu lüxing zhinan* [Travel guides for Chinese national railroads] (Beijing: Jinghua yinshu ju, 1922).

12. Shangwu yinshu guan, ed., *Zengding Zhongguo lüxing zhinan* [Expanded guide for travelers in China], 13th ed. (Shanghai: Shangwu yinshu guan, 1926).

13. For the organization and the front matter of Zhonghua book's travel guide, see Ge Suicheng, ed., *Quanguo duhui shangbu lüxing zhinan* [Travel guide to metropolitan cities and treaty ports nationwide] (Shanghai: Zhonghua shuju, 1926).

14. Shangwu yinshu guan, *Zengding Zhongguo lüxing zhinan*, 157.

15. Ge Suicheng, *Quanguo duhui shangbu lüxing zhinan*, 2:101.

16. Shangwu yinshu guan, *Zengding Zhongguo lüxing zhinan*, 157; Ge Suicheng, *Quanguo duhui shangbu lüxing zhinan*, 2:101.

17. Xu Ke, *Beidaihe zhinan*; Xu Ke, *Moganshan zhinan* [Guide to Mokanshan] (Shanghai: Shangwu yinshu guan, 1922); Xu Ke, *Jigongshan zhinan* [Guide to Chi Kung Shan] (Shanghai: Shangwu yinshu guan, 1922); Xu Ke, *Lushan zhinan* [Guide to Lü Shan] (Shanghai: Shangwu yinshu guan, 1922).

18. For an example of the itineraries organized by area, see Chen Yunzhang and Chen Xiachang, "Wu lu youcheng" [Itineraries for five routes], *Lushan daoyou* [Guide to Lushan], 4th ed. (Shanghai: Shangwu yinshu guan, 1925).

19. Xu Ke, *Moganshan zhinan*, 36–37.

20. Commercial Press published new and updated city and tourist guidebooks throughout the Nanjing Decade. Huanghai Sanren, *Huangshan zhinan* [Guide to Huangshan] (Shanghai: Shangwu yinshu guan, 1929); Xu Ke, *Zengding Moganshan zhinan* [Expanded guide to Mokanshan] (Shanghai: Shangwu yinshu guan, 1930); Zheng Bajia, *Fuzhou lüxing zhinan* [Travel guide to Fuzhou] (Shanghai: Shangwu yinshu guan, 1934).

21. For example, Commercial Press published a four-volume collection of new travelogues in 1915, which would go through four editions by 1924. Zhang Ying, ed., *Benguo xin youji* [The new travelogues of our country] (Shanghai: Shangwu yinshu guan, 1915).

22. Wang Wenru and Ling Guiqing, eds., *Xin youji huikan* [Collectanea of new travelogues] (Shanghai: Zhonghua shuju, 1921, 1928 [5th ed.], 1932 [6th ed.]); Wang Wenru and Yao Zhuxuan, eds., *Xin youji huikan xubian* [Collectanea of new travelogues: The sequel] (Shanghai: Zhonghua shuju, 1923, 1935 [4th ed.]).

23. Sun Qing, "'Xin' youji huibian yu jindai Zhongguo 'kongjian' biaoshu zhuanbian chutan" [A preliminary exploration of the collections of "new" travelogues and the shifts in the expressions of "space" in modern China], *Xin shixue* [New history], no. 11 (2019): 58–59.

24. In 1936, Zhonghua Books issued seven titles under this series. Ni Xiying, *Nanjing* (Shanghai: Zhonghua shuju, 1936); Ni Xiying, *Hangzhou* (Shanghai: Zhonghua shuju, 1936); Ni Xiying, *Jinan* (Shanghai: Zhonghua shuju, 1936); Ni Xiying, *Beiping* [Beijing] (Shanghai: Zhonghua shuju, 1936); Ni Xiying, *Qingdao* (Shanghai: Zhonghua shuju, 1936); Ni Xiying, *Xijing* [Xi'an] (Shanghai: Zhonghua shuju, 1936); Ni Xiying, *Guangzhou* (Shanghai: Zhonghua shuju, 1936). In 1938–39, two additional titles were added to the series. Ni Xiying, *Shanghai* (Shanghai: Zhonghua shuju, 1938); Ni Xiying, *Luoyang* (Shanghai: Zhonghua shuju, 1939).

25. Ni Xiying, *Qufu Taishan youji* [Travelogues of Qufu and Mount Taishan] (Shanghai: Zhonghua shuju, 1931); Ni Xiying, *Luoyang youji* [Travelogues of Luoyang] (Shanghai: Zhonghua shuju, 1935).

26. Ni Xiying, *Beiping*.

27. Ni Xiying, 33–34.

28. Ni Xiying, *Nanjing*, 52.

29. Ni Xiying, *Qingdao*, 14–16.

30. Ni Xiying, 16.

31. Ni Xiying, *Beiping*, 2–3.

32. Ni Xiying, *Xijing*, 3.

33. Rudy Koshar, "'What Ought to Be Seen': Tourists' Guidebooks and National Identities in Modern Germany and Europe," *Journal of Contemporary History* 33, no. 3 (July 1998): 339.

34. Shanghai shangchu yinhang lüxing bu [Travel Department of the Shanghai Commercial and Savings Bank], ed., *You Dian xu zhi* [Guide to Yunnan] (Shanghai: Shanghai shangchu yinhang lüxing bu, 1923); Shanghai shangchu yinhang lüxing bu, ed., *You Chuan xu zhi* [Notes on travel to Sichuan] (Shanghai: Shanghai shangchu yinhang lüxing bu, 1924); Li Zhenfu, ed., *Dongsansheng lüxing zhinan* [Travel guide to Manchuria] (Shanghai: Shanghai yinhang lüxing bu, 1926).

35. For a few examples of CTS guidebooks, see Zhongguo lüxingshe, ed., *Xizihu* [The West Lake] (Shanghai: Zhongguo lüxingshe, 1929); Zhongguo lüxingshe, ed., *Shoudu daoyou* [Tourists' guide to the capital] (Shanghai: Zhongguo lüxingshe, 1931); Zhongguo lüxingshe, ed., *Moganshan daoyou* [Tourists' guide to Moganshan] (Shanghai: Zhongguo lüxingshe, 1932); Beining tielu guanli ju [The Beijing-Mukden Railway], ed., *Beidaihe haibin daoyou* [Tourists' guide to Beidaihe] (Shanghai: Zhongguo lüxingshe, 1935).

36. Chen Guangfu, "Chuangban Zhongguo lüxingshe zishu," in Wu Jingyan, *Chen Guangfu yu Shanghai yinhang*, 224–31.

37. Li Butong, "Shijie geguo zhi lüxing shiye" [Tourism businesses around the world], *LZ* 4, no. 5 (May 1930): 81–87; Peigan, "Jianada lüxing shiye" [Tourism industry in Canada], *LZ* 4, no. 12 (December 1930): 61–66; Zhang Shuiqi, "Ouzhou Riben zhaoyin youke qingkuang" [Tourism promotion in Europe and Japan], *LZ* 5, no. 4 (April 1931): 15–18.

38. See, for example, Wang Huanwen, "Zhongguo jiying tichang lüxing shiye" [China should promote tourism business immediately], *LZ* 5, no. 11 (November 1931): 65–67.

39. Zhao Junhao, "Chen Guangfu xiansheng fangwen ji," 86.

40. Zhao Junhao, "Gu Shaochuan xiansheng fangwen ji" [An interview with Mr. Gu Shaochuan], *LZ* 9, no. 7 (July 1935): 80.

41. Zhao Junhao, "Pan Gongzhan xiansheng fangwen ji" [An interview with Mr. Pan Gongzhan], *LZ* 9, no. 5 (May 1935): 91.

42. Zhao Junhao, "Wang Rutang xiansheng fangwen ji" [An interview with Mr. Wang Rutang], *LZ* 9, no. 10 (October 1935): 66.

43. Xu Zhaofeng, "Benshe banli liuxuesheng fangyang shouxu zhi huiyi" [The recollection of our agency's assistance to study-abroad students], *LZ* 3, no. 8 (August 1929): 42.

44. For example, the special issue featured an article on Western etiquette to inform future *liuxuesheng*. "Xili zhaiyao" [A summary of Western etiquette], *LZ* 3, no. 8 (August 1928): 11.

45. Chen Xiangtao, "Zhongguo lüxingshe chuangshe zhaodaisuo zhi zhiqu" [The purports of the CTS guesthouses], *LZ* 9, no. 9 (September 1935): 6.

46. "Huangshan lüshe" [The Huangshan guesthouse], *LZ* 9, no. 9 (September 1935): 29–32.

47. "Chunyou zhuanhao" [The special number for spring], *LZ* 3, no. 3 (March 1929).

48. *LZ* 5, no. 7 (July 1931).

49. Ding Huikang, "Qingdao linzhua" [A glimpse of Qingdao], *LZ* 5, no. 7 (July 1931): 25–28.

50. Ding Huikang, 26.

51. *LZ* 3, no. 7 (July 1929).

52. *LZ* 4, no. 1 (January 1930).

53. Qin Lizhai, "Guling youcheng" [Itineraries for Guling], *LZ* 5, no. 6 (June 1931): 31–39.

54. Zhenju jushi, "Lushan zhen mianmu" [The true colors of Lushan], *LZ* 5, no. 6 (June 1931): 3–18.

55. *Zhonghua minguo quanguo tielu lüxing zhinan* [Travel guide to national railroads in the Republic of China] (n.p.: Tiadaobu lianyunchu, 1934).

56. "Yiyue yitan" [Monthly words], *LZ* 11, no. 7 (July 1937): n.p.

57. "Yiyue yitan."

58. *LZ* 11, no. 7 (July 1937): 7.

59. Yu Songhua, "Shengping zhi jianghu quwei (san)" [The fun of traveling the country (3)], *LZ* 8, no. 1 (January 1934): 2.

60. Huang Boqiao, "Daoyou yu aiguo" [Tourism and patriotism], *LZ* 10, no. 1 (January 1936): 3.

61. The prevalence of travel columns in many Republican-era publications can testify to this. For example, several popular magazines and journals published by Commercial Press, such as *Dongfang zazhi* (*Eastern Miscellany*) and *Xiaoshuo yuebao* (*Short Story Monthly*), featured travel columns and travelogues regularly. Newspaper supplements like *Chenbao fukan* (*Supplement to Chenbao*) in Beijing and *Shenbao yuekan* (*Shenbao Monthly*) in Shanghai ran travel-related news and travel writings. Even niche publications—from women's magazines to photography journals—contained travel-related contents.

62. Ma Guoliang, "Liangyou quanguo sheying lüxing tuan" [*Liangyou* Nationwide Photographic Tour], *Dazhong sheying* [Popular photography] 6 (1980): 28–29.

63. On *Liangyou*'s founding story, see Ma Guoliang, *Liangyou yijiu: yijia huabao yu yige shidai* [Remembering *Liangyou*: A pictorial journal and a historical period] (Taipei: Zhengzhong shuju, 2002), 10–14.

64. For the background of the Liangyou Company, especially its distribution network, see Wang Chuchu, "Appendix: Distributing *Liangyou*," *Liangyou: Kaleidoscopic Modernity and the Shanghai Global Metropolis, 1926–1945*, ed. Paul Pickowicz, Shen Kuiyi, and Zhang Yingjin (Leiden: Brill, 2013), 248–58.

65. Graham Smith, *Photography and Travel* (London: Reaktion Books, 2013), 10.

66. It even included two images, supposedly, of the Potala Temple in Lhasa. Shangwu yinshu guan, *Zhongguo mingsheng* [Views of China], 3rd ed. (Shanghai: Shangwu yinshu guan, 1912).

67. The twenty-two newer additions to the *Views of China* series featured the following destinations: Huangshan, Lushan, Putuoshan, West Lake, the Imperial Summer Villa, Taishan, Hengshan, the Confucius Cemetery, Yushan, Yandangshan, Tiantaishan, Hengshan, Wutaishan, Xishan, Panshan, Dafangshan, Huashan, Wuyishan, Tianmushan, Yuntaishan, Moganshan, and Laoshan. Rather than professional photographers, the authors of these albums, such as Huang Yanpei and Jiang Weiqiao, were avid travelers from an elite background. Shangwu yinshu guan, *Zhongguo mingsheng* [Views of China], series (Shanghai: Shangwu yinshu guan, 1914–26).

68. Published by the Shanghai branch of the Kodak Company, *Kodak Magazine* ran a column called "Kodak travelogue" from 1931 to 1937, featuring travel photographs of domestic destinations. *Keda zazhi* [Kodak magazine] (Shanghai: Keda gongsi, 1931–37).

69. Wu Liande, "Wei Liangyou sheying tuan fayan" [Speech for the *Liangyou* Nationwide Photographic Tour], *Liangyou* [Young companion], no. 69 (September 1932): 28.

70. Ye Gongchuo, "Duiyu Liangyou quanguo sheying lüxingtuan de ganxiang" [Some thoughts on the *Liangyou* Nationwide Photographic Tour], *Liangyou* [Young companion], no. 69 (September 1932): 27.

71. Liang Desuo, "Quanguo lieying ji" [Reports on the Nationwide Photographic Tour], *Liangyou* [Young companion], no. 70 (October 1932): 28; Wu Liande, *Zhonghua jingxiang*.

72. Liang Desuo, "Quanguo lieying ji," 30. Here, "the songs by Shangnü near the Qinhuai River" alludes to Tang poet Du Mu's (803–52) quatrain "Mooring on the River Qinhuai," whose last two lines read, "The singing girls know no vanquished kingdom's sadness / And still sing, across the river, 'Jade Flowers in the Rear Court.'" "Jade Flowers in the Rear Court" was a song composed at the court of the last emperor of the Chen dynasty, who also built the capital in Nanjing in the sixth century. The connotation of "the songs by Shangnü near the Qinhuai River," therefore, is one of sensual music sung in an age of decline.

73. Liang Desuo, "Quanguo lieying ji," 28–29.

74. On the literary and cultural importance of travel narratives in Chinese reportage in the Republican era, see Charles Laughlin, *Chinese Reportage: The Aesthetics of Historical Experience* (Durham, NC: Duke University Press, 2002), 37–74.

75. Liang Desuo, "Quanguo lieying ji zhi er: shenglin xunli" [Reports on the Nationwide Photographic Tour 2: A visit to the land of ancient sages], *Liangyou* [Young companion], no. 71 (November 1932): 7.

76. Liang Desuo, "Zhongyuan richeng zhailu" [Excerpts of our travels in Central China], *Liangyou* [Young companion], no. 74 (February 1933): n.p. More details about

the South Grotto Temple will be explored in the next chapter, where I turn my attention to travel in China's Northwest.

77. Strassberg, introduction to *Inscribed Landscapes*, 57.

78. "Dingxian cujin pingmin jiaoyü zhi sheshi" [Promotion of mass education in Tinghsien], *Liangyou* [Young companion], no. 72 (December 1932): n.p.

79. Dong, "Shanghai's *China Traveler*," 210–11.

80. Liang Desuo, "Quanguo lieying ji zhi er: shenglin xunli," 7.

81. Liang Desuo, "Xiang Yue jiaotong yu guoji wenti" [Transportation in Hunan and Guangdong and the problems of the nation], *Liangyou* [Young companion], no. 75 (March 1933): n.p.

82. Liang Desuo, "Xibei de yinxiang" [Impressions of the Northwest], *Liangyou* [Young companion], no. 73 (January 1933): n.p.

83. Liang Desuo, "Xiang Yue jiaotong yu guoji wenti," 13.

84. Wang Xiaoting, "Shiqu de Rehe" [The lost Jehol], *Dazhong huabao*, no. 8 (June 1934): 20–21.

85. This is the direct translation from the Chinese title (*Liangyou duzhe lüxing lieche*), although the English caption in the magazine called the column "The Liang You Travel Book."

86. Zhang Yuanhen, "Hetao zhong de mimi guo" [A secret country in the Hetao region], *Liangyou* [Young companion], no. 92 (August 1934): 6–7.

3. "Head to the Northwest"

1. Liu Bannong, "Zhi Zhou Zuoren de xin" [A letter to Zhou Zuoren], in *Fuqin Liu Bannong* [My father, Liu Bannong], by Liu Xiaohui (Shanghai: Shanghai renmin chubanshe, 2000), 223.

2. On the European invention of the "Silk Road" concept, see Tamara Chin, "The Invention of the Silk Road, 1877," *Critical Inquiry* 40, no. 1 (Autumn 2013): 194–219.

3. Ni Xiying, *Xijing*, 14.

4. Chin, "Invention of the Silk Road," 199–200, 206–8, 217.

5. Margarita Diaz-Andreu Garcia, *A World History of Nineteenth-Century Archaeology: Nationalism, Colonialism, and the Past* (Oxford: Oxford University Press, 2007), 192.

6. On the growing interest in Central Asia among Western Sinologists and Orientalists, see Fa-ti Fan, "Circulating Material Objects: The International Controversy over Antiquities and Fossils in Twentieth-Century China," in *The Circulation of Knowledge between Britain, India, and China: The Early-Modern World to the Twentieth Century*, ed. Bernard Lightman, Gordon McOuat, and Larry Stewart (Leiden: Brill, 2013), 209–36.

7. Chen Yinke, "Chen Yuan *Dunhuang jieyu lu* xu" [Preface to Chen Yuan's *The aftermath of the disaster in Dunhuang*], in *Dunhuang jieyu lu* [The aftermath of the disaster in Dunhuang], by Chen Yuan (Beijing: Guoli zhongyang yanjiu yuan lishi yuyan yanjiu suo, 1931), 1:1–2.

8. Valerie Hansen, *The Silk Road: A New History* (Oxford: Oxford University Press, 2012), 167.

9. On the history of archaeological raids made by foreign explorers in Dunhuang and other regions in northwestern China, see Peter Hopkirk, *Foreign Devils on the Silk Road: The Search for the Lost Cities and Treasures of Chinese Central Asia* (London: John Murray, 1980).

10. Rong Xinjiang, introduction to *Eighteen Lectures on Dunhuang*, trans. Imre Calambos (Leiden: Brill, 2013), 1–18.

11. Liu Bannong later compiled his copies into a three-volume book called *Dunhuang duosuo* [Miscellany from Dunhuang]. Liu Bannong, "*Dunhuang duosuo* xu" [Preface to *Miscellany from Dunhuang*], in *Dunhuang duosuo* [Miscellany from Dunhuang] (Beijing: Zhongyang yanjiu yuan lishi yuyan yanjiu suo, 1925), 12.

12. Lydia Liu, *Translingual Practice, Literature, National Culture, and Translated Modernity—China, 1900–1937* (Stanford, CA: Stanford University Press, 1995), 241.

13. Luo Zhitian, "Xifang xueshu fenlei yu minchu guoxue de xueke dingwei" [Western academic classification and subject orientation in national studies in the early Republican period], *Sichuan daxue xuebao* [Journal of Sichuan University] 38, no. 5 (May 2001): 75–82.

14. Traveling to postwar Europe in 1919, Liang Qichao, for example, was shocked by the devastation he witnessed there. In his travelogue, *Ouyou xinying lu* [Impressions of travel in Europe], he highlighted the corruption of the West and argued for using Chinese spiritualism as an antidote to Western materialism. Liang Qichao, *Ouyou xinying lu jielu* [Excerpts from *Impressions of travel in Europe*] (Shanghai: Zhonghua shuju, 1936). For analysis of Liang Qichao's transition, see Shu-mei Shih, *Lure of the Modern*, 160–66.

15. Upon seeing some manuscripts at Pelliot's residence in Beijing in 1909, Wang Guowei petitioned the Qing government to retrieve the remaining manuscripts from Dunhuang to be placed in the Capital Library (*Shoudu tushuguan*) in Beijing. Susan Whitfield, "The Dunhuang Manuscripts: From Cave to Computer," in *Books in Numbers: Seventy-Fifth Anniversary of the Harvard-Yenching Library; Conference Papers*, ed. Lucille Chia and W. L. Idema (Cambridge, MA: Harvard-Yenching Library, Harvard University, 2007), 121.

16. On the introduction of archaeology into China, see Rongyu Su, "The Reception of 'Archaeology' and 'Prehistory' and the Founding of Archaeology in Late Imperial China," in *Mapping Meanings: The Field of New Learning in Late Qing China*, ed. Michael Lackner and Natascha Vittinghoff (Leiden: Brill, 2004), 423–50.

17. Luo Zhitian, "Zouxiang guoxue yu shixue de saixiansheng" (Mr. Science as applied in national studies and history), *Jindaishi yanjiu* [Modern Chinese history studies], no. 3 (2000): 59–94.

18. Ge Zhaoguang, "From the Western Regions to the Eastern Sea: Formations, Methods and Problems in a New Historical World," *Here in "China" I Dwell: Reconstructing Historical Discourses of China for Our Time*, trans. Jesse Field and Qin Fang (Leiden: Brill, 2017), 172–78.

19. Fan, "Circulating Material Objects," 209–36.

20. The technique Warner used is called the "strappo technique," which he had learned from Daniel Thompson at Harvard. For more details of the method used by Warner, see Sanchita Balachandran, "Object Lessons: The Politics of Preservation and Museum Building in Western China in the Early Twentieth Century," *International Journal of Cultural Property*, no. 14 (2007): 13–14.

21. Balachandran, 14–18.

22. Lynne Warren, ed., *Encyclopedia of 20th Century Photography* (London: Routledge, 2006), 1: 269.

23. Chen Wanli, "Zixu" [Self-preface], *Xixing riji* [Journal of a westward journey] (Lanzhou: Gansu renmin chubanshe, 2000), 12.

24. Chen Wanli, *Xixing riji*, 83.

25. Chen Wanli, 79.

26. Chen Wanli, 113.

27. Susan Chan Egan, *A Latterday Confucian: Reminiscences of William Hung (1893–1980)* (Cambridge, MA: Harvard University Press, 1987), 115.

28. Chen Wanli, *Xixing riji*, 90.

29. Egan, *Latterday Confucian*, 112–16.

30. Chen Wanli, "Meidi touqie Dunhuang bihua de yinmou" [The US imperialists' plot to steal Dunhuang frescoes], *Wenwu* [Cultural relics], no. 1 (1951): n.p.

31. For a discussion of the tension between scientific nationalism and scientific universalism, see Fa-ti Fan, "How Did the Chinese Become Native? Science and the Search for National Origins in the May Fourth Era," in *Beyond the May Fourth Paradigm: In Search of Chinese Modernity*, ed. Kai-wing Chow et al. (Lanham, MD: Lexington Books, 2008), 183–208.

32. Chen Wanli, *Xixing riji*, 40.

33. Chen Wanli, 42–44.

34. Chen Wanli, 43–44.

35. Gu Jiegang, "Xuwen" [Preface], in Chen Wanli, *Xixing riji*, 7.

36. Chen first serialized his travel journal in the publication of Beijing University's National Studies Department. Shortly after that, his travel photographs of Dunhuang also appeared in magazines like *Guoxue* and *Liangyou*. His travel journal was also published as a book in 1926, while the Liangyou Company published a photo album of Dunhuang murals using his photographs. Chen Wanli, "Xixing riji" [Journal of a westward journey], *Beijing daxue yanjiu suo guoxue men zhoukan* [Weekly journal of the National Studies Department at Beijing University] 1, nos. 1–3, 5, 7, 9 (1925); Chen Wanli, "Dunhuang qianfodong huabi liangzhong" [Two murals in the Thousand Buddha caves in Dunhuang], *Guoxue* [National studies] 1, no. 2 (1926): n.p.; Chen Wanli, "Dunhuang qianfodong gubihua" [Ancient murals discovered at the Thousand Buddha caves in Dunhuang], *Liangyou* [Young companion], no. 37 (1929): 33; Chen Wanli, *Xixing riji* [Journal of a westward journey] (Beijing: Pushe, 1926); Chen Wanli, *Xichui bihua ji* [Murals on the western frontiers] (Shanghai: Liangyou tushu yinshua gongsi, 1928).

37. The most notable case was Shao Yuanchong's visit to Dunhuang in 1935. An early follower of Sun Yat-sen and the vice president of the Legislative Yuan in the Nationalist government, Shao led a government delegation to the Northwest in 1935 to pay respects to the mythical Yellow Emperor, whose mausoleum was located in Shaanxi Province. Shao's delegation then toured the northwestern region and visited Dunhuang. Gao Liangzuo, *Xibei suiyao ji* [An inspection tour of the Northwest] (Nanjing: Jianguo yuekan she, 1936), 211–41; Xu Shizhen and Shao Yuanchong, eds., *Xibei lansheng* [China's Northwest, a pictorial survey] (Nanjing: Zhengzhong shuju, 1936), 90–109.

38. For Zhang's artwork based on the Dunhuang murals, see National Palace Museum, *The Record of Mogao by Zhang Daqian* (Taipei: National Palace Museum, 1985).

39. Grace Shen, *Unearthing the Nation: Modern Geology and Nationalism in Republican China* (Chicago: University of Chicago Press, 2014), 47–72; Hsiao-pei Yen, "From Paleoanthropology in China to Chinese Paleoanthropology: Science, Imperialism and Nationalism in North China, 1920–1939," *History of Science* 53, no. 1 (2015): 23–28.

40. For more information about Hedin's three earlier expeditions in China (1893–97, 1899–1902, and 1906–8, respectively), see Sven Hedin, *My Life as an Explorer*, trans. Alfhild Huebsch (New York: Garden City, 1925).

41. "Beijing xueshu tuanti lianhe fandui Ruidian yuanzhengdui bing jiji yanjiu baocun guwu fa" [Beijing academic community objects to the Swedish expedition and actively looks for means to protect antiquities], *Chenbao* [Morning news], March 6, 1927.

42. Sven Hedin, *History of the Expedition in Asia 1927–1935*, trans. Donald Burton (Stockholm: Elanders boktryckeri aktiebolag, 1943), part 1, 25.

43. Hedin, part 1, 20.

44. Hedin, part 1, 52–54.

45. Hedin, part 1, 33.

46. Hedin, part 1, 19.

47. Hedin, part 1, 25.

48. Hedin, part 1, 26.

49. Hedin, part 1, 32. On Hedin's Pan-German nationalism and consequently his support for Nazi Germany, see Sarah K. Danielsson, *The Explorer's Roadmap to National-Socialism: Sven Hedin, Geography and the Path to Genocide* (Burlington, VT: Ashgate, 2012).

50. Hedin, *History of the Expedition*, part 1, 59.

51. Sven Hedin, *Across the Gobi Desert* (New York: E. P. Dutton, 1932), 7.

52. Xu Bingchang (Xusheng), *Xu Xusheng xiyou riji* [Xu Xusheng's journal of a westward journey] (Beijing: Zhongguo xueshu tuanti xiehui xibei kexue kaochatuan lishihui, 1930), 1:2–3.

53. Similarly, Yang Zhongjian, a vertebrate paleontologist who had joined the US-led Central Asiatic Expedition in 1928, articulated how the collaboration between Chinese and Americans in the expedition was largely based on mutual advantages. See Hsiao-pei Yen, "From Paleoanthropology in China," 38.

54. For a discussion of a similar strategy among Chinese geologists, see Shen, *Unearthing the Nation*, 109–44.

55. The associations made between the non-Han peoples' independence movements and Western and Japanese imperialist powers were not unfounded. On the different roles Japan and the Soviet Union played in independence movements in Mongolia, Manchuria, and Xinjiang, see Wang Ke, *Dongtujue sitan duli yundong 1930 niandai zhi 1940 niandai* [The East Turkestan independence movement in the 1930s–1940s] (Hong Kong: Chinese University of Hong Kong Press, 2013).

56. In a lecture at the Library of Congress, Kenneth Pomeranz, then American Historical Association president, insightfully suggested that historians should frame China's Republican era "not just as the construction of a nation but as a largely successful resistance to the 20th-century wave of decolonization." Kenneth Pomeranz, "Resisting Imperialism, Resisting Decolonization: Making China from the Ruins of the Qing, 1912–1949," Library of Congress, Washington, DC, July 23, 2013, https://www.loc.gov/item/webcast-6019/.

57. Justin M. Jacobs, *Xinjiang and the Modern Chinese State* (Seattle: University of Washington Press, 2016), 3–4.

58. Liang Qichao, for example, had celebrated Chinese emigrants from Fujian and Guangdong provinces to Southeast Asia as pioneers in China's "colonial" (*zhimin*) activities. Liang Qichao, "Zhongguo zhimin bada weiren zhuan" [Biographies of eight eminent Chinese colonizers], in *Liang Qichao quanji* [Liang Qichao anthology] (Beijing: Beijing chubanshe, 1997), 3:1368.

59. Among academics, sociologist Yu Tianxiu (also known as Tinn-Hugh Yu) was an early advocate of the idea of *zhibian*. Liu Ying and Wang Yingying, "Yu Tianxiu shehuixue sixiang tanxi" [An analysis of Yu Tianxiu's sociological thoughts], *Xibei daxue xuebao (zhexue shehui kexue ban)* [Journal of Northwest University (philosophy and social sciences edition)] 42, no. 5 (September 2012): 17–22.

60. Perry Johansson, *Saluting the Yellow Emperor: A Case of Swedish Sinography* (Leiden: Brill, 2012), 81–82.

61. Hedin, *History of the Expedition*, part 1, 76.

62. Xu Xusheng, *Xu Xusheng xiyou riji*, 1:n.p.

63. Sven Hedin, preface to *Riddles of the Gobi Desert* (London: Routledge, 1933), ix.

64. Xu Xusheng, *Xu Xusheng xiyou riji*, 1:1–2.

65. Gordon Stewart, "The Exploration of Central Asia," in *Reinterpreting Exploration: The West in the World*, ed. Dane Kennedy (Oxford: Oxford University Press, 2014), 195–213.

66. Xu Xusheng, *Xu Xusheng xiyou riji*.

67. Xu Xusheng, 1:46.

68. Xu Xusheng.

69. Xu Xusheng, 1:24.

70. Xu Xusheng, 2:5–6; Henning Haslund, *Men and Gods in Mongolia* (New York: Adventures Unlimited, 2000), 132–34.

71. Xu Xusheng, *Xu Xusheng xiyou riji*, 1:26–27.

72. Xu Xusheng, 1:35.

73. Xu Xusheng, 3:138.

74. Xu Xusheng, 1:122–24. By "Kozlov, Stein, or Warner," Xu was referring to the Russian explorer Pyotr Kuzmich Kozlov, the British archaeologist Aurel Stein, and the US scholar Langdon Warner.

75. Xu Xusheng, 3:32–34.

76. Xu Xusheng, 2:6–7.

77. Xu Xusheng, 2:7. When he published the travel journal in 1930, Xu added a note saying that he had later checked and found that there had been no such earthquake. Many of the objects Le Coq's expeditions had obtained in Xinjiang would be destroyed in 1945, when a bomb from the Allied forces hit the Ethnological Museum in Berlin. Herbert Härtel and Marianne Yaldiz, eds., *Along the Ancient Silk Routes: Central Asian Art from the West Berlin State Museums* (New York: Metropolitan Museum of Art, 1982), 37–38.

78. Huang Wenbi, *Huang Wenbi mengxin kaocha riji (1927–1930)* [Huang Wenbi's journal of the exploration in Inner Mongolia and Xinjiang (1927–1930)] (Beijing: Wenwu chubanshe, 1990), 13–19. For an early brief report on this discovery, see Huang Wenbi, "Xibei kexue kaochetuan zhi gongzuo ji qi zhongyao faxian" [Work of the scientific mission to the Northwest and its important discoveries], *Yanjing xuebao* [Yenching journal of Chinese studies], no. 8 (December 1930): 1610.

79. Xu Xusheng, *Xu Xusheng xiyou riji*, 1:26.

80. Xu Xusheng, 1:26–27.

81. Tjalling H. F. Halbertsma, *Early Christian Remains of Inner Mongolia* (Leiden: Brill, 2015), 81–89.

82. In many entries in his travel journal, Xu Xusheng mentioned reading the official dynastic histories. He paid special attention to the "Account of the Western Regions" ("Xiyu zhuan") and the "Treatise on Geography" ("Dili zhi") in various dynastic histories. Xu Xusheng, *Xu Xusheng xiyou riji*, 1:75.

83. For example, the title for Albert von Le Coq's account of the Turfan expeditions suggested German orientalism's interest in finding Hellenic influence in Central Asia. Suzanne Marchand, *German Orientalism in the Age of Empire: Religion, Race and Scholarship* (Cambridge: Cambridge University Press, 2010), 422.

84. Chin, "Invention of the Silk Road," 195.

85. Hedin, *Riddles of the Gobi Desert*, 317–19.

86. Christel Braae and Henning Haslund-Christensen, *Among Herders of Inner Mongolia: The Haslund-Christensen Collection at the National Museum of Denmark* (Aarhus: Aarhus University Press, 2017), 91–92.

87. Hedin, *Across the Gobi Desert*, 91.

88. Xu Xusheng, *Xu Xusheng xiyou riji*, vol. 1.

89. Xu Xusheng, 3:42. It possibly also referred to Burhan Shahidi. For a discussion of Yang Zengxin's use of Tatar expatriates such as Burhan during his rule of Xinjiang, see Jacobs, *Xinjiang and Modern Chinese State*, 31–32, 34.

90. Xu Xusheng, *Xu Xusheng xiyou riji*, 3:55.

91. For an analysis of the strategic manipulation of the politics of difference in non-Han peripheries, see Jacobs, *Xinjiang and the Modern Chinese State*.

92. Xu Xusheng, *Xu Xusheng xiyou riji*, 1:27.

93. In Hedin's travel writing, he was sometimes called "the Mohammedan 'King,'" signifying his Turkic background. Hedin, *Across the Gobi Desert*, 311.

94. Xu Xusheng, *Xu Xusheng xiyou riji*, 3:1.

95. Xu Xusheng, 3:1–2.

96. Xu Xusheng, 1:8–10.

97. Hedin, *History of the Expedition*, part 1, 244–45.

98. This German expedition was probably Emil Trinkler's 1927–28 exploration of Xinjiang. Emil Trinkler, "Explorations in the Eastern Karakoram and in the Western Kunlun," *Geographical Journal* 75, no. 6 (June 1930): 505–15.

99. Xu Xusheng, *Xu Xusheng xiyou riji*, 3:43.

100. Hedin, *Across the Gobi Desert*, 215.

101. Hedin, 78.

102. Xu Xusheng, *Xu Xusheng xiyou riji*, 1:39–41, 3:60–61.

103. Xu Xusheng, 1:29–30.

104. Huang Wenbi, *Huang Wenbi mengxin kaocha riji*, 24.

105. Xu Xusheng, *Xu Xusheng xiyou riji*, 1:30.

106. Tong Lam, *A Passion for Facts: Social Surveys and the Construction of the Chinese Nation-State, 1900–1949* (Berkeley: University of California Press, 2011).

107. Johannes Fabian, *Time and the Other: How Anthropology Makes Its Object* (New York: Columbia University Press, 1983).

108. Xu Xusheng, *Xu Xusheng xiyou riji*, 1:53–54.

109. Xu Xusheng, 1:54.

110. Xu Xusheng, 3:34.

111. Xu Xusheng, 1:47.

112. Xu Xusheng, 1:48.

113. Xu Xusheng, 1:48–49.

114. Tani Barlow, "Introduction: On 'Colonial Modernity,'" in *Formations of Colonial Modernity in East Asia*, ed. Tani Barlow (Durham, NC: Duke University Press, 1997), 1–20.

115. Even though Xu Xusheng did not compare China's Northwest and the American West in his travel journal, the parallels between the two regions were not lost to his contemporaries. When Sun Yat-sen stressed the benefits of building railways in Mongolia and Xinjiang in *The International Development of China*, he cited the examples of the "United States, Canada, Australia, and Argentina" as proof. *The International Development of China* (New York: Putnam, 1922), 23. In the opening statement for the journal *Xibei* (*The Northwest*), published by the Association of Cultural Promotion in the Northwest (*Xibei wenhua cujin hui*) in 1929, the author compared Chinese ignorance about the Northwest to European ignorance of the American West a century prior. "Fakan ci" [Foreword to the journal], *Xibei* [The Northwest] (March 1929): 1. In 1933, when Xue Guilun, a professor in mining engineering, embarked on a journey to the Northwest alongside a government delegation, he compared the importance of exploring the mineral resources in China's Northwest to similar endeavors in the American West. Xue Guilun, *Xibei shicha riji* [The journal of an inspection tour in the Northwest] (Shanghai: Shenbao, 1934), 1.

116. On the Qing statecraft writing about the Northwest, see Peter Perdue, "Embracing Victory, Effacing Defeat: Rewriting the Qing Frontier Campaigns," in *The Chinese State at the Borders*, ed. Diana Lary (Vancouver: UBC Press, 2007), 105–25. On the rising emphasis on "fieldwork" among Chinese intellectuals, see Shen, *Unearthing the Nation*, 47–72.

117. "Xibei kexue kaochatuan tekan" [A special issue on the scientific mission to the Northwest], *Shijie huabao* [World pictorial], January 27, 1929, n.p.; "Ji Xibei kexue kaochatuan" [On the scientific mission to the Northwest], *Guowen zhoubao* [Chinese weekly] 6, no. 6 (February 3, 1929): 1–9; "Xibei kexue kaochatuan kaocha jingguo" [The scientific research expedition to northwestern China], *Tuhua shibao* [Pictorial newspaper], February 3, 1929.

118. For example, after his return to Beijing in early 1929, Xu Xusheng gave talks at Qinghua University, Yenching University, Labor University (*Laodong daxue*), and L'Institut Franco-Chinoise (*Zhongfa daxue*) in Beijing, and at Communications University (*Jiaotong daxue*) in Shanghai. At least three of his talks were published in school journals. Xu Xusheng, "Xibei diaocha yuxian ji" [Mishaps in the scientific mission to the Northwest], *Yanjing daxue xiaokan* [Yenching University weekly], no. 22 (1929): n.p.; Xu Xusheng, "Jiangyan: Xinjiang zhi bokeda shan" [Speech: Mount Bogda in Xinjiang], *Zhongfa jiaoyujie* [Education in China and France], no. 25 (April 1929): 1–7; Xu Xusheng, "Xibei kaocha ji" [Exploration of the Northwest], *Jiaotong daxue rikan* [Communications University daily news], March 27, 1929.

119. The most notable example was the paleontologist Yang Zhongjian's travelogues based on his participation in the US-led Central Asiatic Expeditions in 1930 and the Croisière Jaune (or Yellow Expedition) sponsored by the Citroen car firm in

France in 1931. Yang Zhongjian, *Xibei de poumian* [The cross section of the Northwest] (Beiping: Dizhi tushuguan, 1932).

120. Pierre-Étienne Will, "Xi'an, 1900–1940: From Isolated Backwater to Resistance Center," in *New Narratives of Urban Space in Republican Chinese Cities*, ed. Billy K. L. So and Madeleine Zelin (Leiden: Brill, 2013), 227–30.

121. Jeremy Tai, "The Northwest Question: Capitalism in the Sands of Nationalist China," *Twentieth-Century China* 40, no. 3 (October 2015): 203.

122. According to Wang Zhaosheng, thirty-seven frontier research organizations were formed in the first two decades of the Republican period, and the majority were established in the 1930s. Wang Zhaosheng, *Xibei jianshe lun* [On building the Northwest] (Beijing: Qingnian chubanshe, 1943), 4. In the 1930s, Chinese witnessed more than a dozen new magazines dedicated to the Northwest. Zong Yumei, "20 shiji 30 niandai baokan meijie yu Xibei kaifa" [News media and the development of the Northwest in the 1930s], *Shixue yuekan* [Journal of history], no. 5 (2003): 54–58.

123. Made by the Shanghai Mingxing Film Company in 1934, the film *Go to the Northwest* (*Dao Xibei qu*) featured the famous movie star Xu Lai. Laikwan Pang, *Building a New China in Cinema: The Chinese Left-Wing Cinema Movement, 1932–1937* (Lanham, MD: Rowman & Littlefield, 2002), 45.

124. Shen Songqiao, "Jiangshan ruci duojiao—1930 niandai de Xibei lüxing shuxie yu guozu xiangxiang" [The nation being so beautiful: Travel writing of the Northwest and the national imagination], *Taida lishi xuebao* [Historical journal of Taiwan University], no. 37 (June 2006): 145–216; Zhihong Chen, "Romance with the Periphery: Chinese Travel Writing about the 'Frontier' during the Nanjing Decade (1927–1937)," *Journal of Tourism History* 8, no. 3 (2016): 215–38.

125. For the journalists' travel accounts about the Northwest, see Gu Zhizhong and Lu Yi, *Dao Qinghai qu* [Head to Qinghai] (Shanghai: Shangwu yinshu guan, 1934); Chen Gengya, *Xibei shicha ji* [Record of investigation in the Northwest] (Shanghai: Shenbao, 1936); Fan Changjiang, *Zhongguo de xibei jiao* [China's northwestern corner] (Tianjin: Dagong bao, 1936).

126. On the improvements in transportation, see Wang Yinqiao, *Xijing youlan zhinan* [Tourist guide to Xijing] (Xi'an: Tianjin Dagongbao Xi'an fenguan, 1936), 33–129; Ni Xiying, *Xijing*, 38–50. For examples of guidebooks published by government agencies, see Zhang Changgong, *Xijing shengji* [Scenic spots and historic relics in Xijing] (Xi'an: Shaanxi sheng diyi tushuguan, 1932); Chen Guangyao, *Xijing zhi xiankuang* [Contemporary conditions in Xijing] (Shanghai: Xijing choubei weiyuanhui, 1933).

127. Even though the title of the CTS guidebook suggests a broader remit, the content focuses on Shaanxi Province alone. Hu Shiyuan, *Xibei daoyou* [Travel guide to the Northwest] (Shanghai: Zhongguo lüxing she, 1935).

128. "Pingsui lu yaoxun" [Important news about the Ping-Sui line], *Tiedao gongbao*, no. 1040 (1934): 8.

129. "Pingsui lu yaoxun," 8.

130. Dickson Tong, "Tourism in the Northwest—a Neglected Opportunity," *China Weekly Review*, January 9, 1932.

131. Pingsui tielu guanli ju [The Ping-Sui Railway Administration], *Youlan zhi changdao* [Promoting tourism] (n.p.: Pingsui tielu guanli ju, 1935), 6.

132. Pingsui tielu guanli ju, 6–7.

133. Xiao Qian, "Pingsui xianshang (youji)" [On the Ping-Sui line (travelogue)], *Xinsheng zhoukan* [New life weekly] 1, no. 39 (1934): 804.

134. According to Bing Xin's travelogue, the Yenching professors on this tour were given the VIP treatment. The chartered train sent by the Ping-Sui line was the one reserved for railway managers going on inspection tours. Aside from a sleeper car, writing desks, and other amenities, the group also had a private chef on board. At Datong, Suiyuan, and other major stopovers, local Chinese authorities, usually military garrisons, would send automobiles to chauffeur the group to various tourist sites. Bing Xin, *Pingsui yanxian lüxing ji* [Travelogue of my journey along the Ping-Sui Railway] (n.p.: Pingsui tielu guanli ju, 1935).

135. Other members included the writer and literary critic Zheng Zhenduo, the historian Gu Jiegang, the economist Augusta Wagner, the sociologist Lei Jieqiong, the economist Chen Qitian, the archaeologist Rong Geng, and the photographer Zhao Cheng. After setting out for the Northwest on July 7, 1934, the group had to turn around and return to Beijing as a flood damaged the track near Datong. Weeks later, when they regrouped to depart westward again, the archaeologist Rong Geng took Augusta Wagner's spot, as Wagner had gone to the Beidaihe beach resort. Bing Xin, *Pingsui yanxian lüxing ji*.

136. Pingsui tielu guanli ju, *Youlan zhi changdao*, 8–9.

137. Gu Jiegang, *Wang Tongchun kaifa hetao ji* [The history of Wang Tongchun's efforts in developing the Hetao area] (n.p.: Pingsui tielu guanli ju, 1935); Wu Wenzao, *Menggu bao* [Mongolian yurts] (n.p.: Pingsui tielu guanli ju, 1935); Lei Jieqiong, *Pingsui yanxian zhi tianzhu jiaohui* [The Catholic missions along the Ping-Sui Railway] (n.p.: Pingsui tielu guanli ju, 1935).

138. Gu Jiegang, *Wang Tongchun kaifa hetao ji*.

139. Bing Xin, *Pingsui yanxian lüxing ji*, 1.

140. Bing Xin.

141. Bing Xin's travel diary, published by the Ping-Sui Railway in 1935, was so popular that the Beixin Bookstore in Shanghai published the same book under a different title in the same year and soon issued a second edition a few months later. Another travelogue in the same series was a collection of essays on famous sites in the Northwest penned by Zheng Zhenduo, Rong Geng, and Jiang Endian. Bing Xin, *Bing Xin youji* [Bing Xin's travelogue] (Shanghai: Beixin shuju, 1935); Zheng Zhenduo, Rong Geng, and Jiang Endian, *Xibei shengji* [Great sites in the Northwest] (n.p.: Pingsui tielu guanli ju, 1935).

142. Bing Xin, *Pingsui yanxian lüxing ji*, 19; Zheng Zhenduo, "Yungang shiku" [The Yungang Grottoes], in *Xibei shengji*, 7.

143. McDonald, *Placing Empire*, 59–62.

144. Bing Xin, *Pingsui yanxian lüxing ji*, 24–27; Xiao Qian, "Pingsui xianshang (youji)," *Xinsheng zhoukan* 1, no. 40 (1934): 826–27.

145. For the state-building efforts in Suiyuan in the early twentieth century, see Justin Tighe, *Constructing Suiyuan: The Politics of Northwestern Territory and Development in Early Twentieth-Century China* (Boston: Brill, 2005).

146. Bing Xin, *Pingsui yanxian lüxing ji*, 37.

147. Xiao Qian, "Pingsui xianshang (youji)," *Xinsheng zhoukan* 1, no. 41 (1934): 844.

148. For example, Pingsui tielu guanli ju, *Youlan zhi changdao*, 3; Fang Shan, "Qingzhong" [Green tumulus], *LZ* 9, no. 5 (May 1935): 63–65; Zhao Shuyong, "Pingsui

yanxian daoyou zhilüe" [Guide to the sites along the Ping-Sui Railway line], *LZ* 9, no. 9 (September 1935): 97–99; Baoshi, "Suiyuan youji" [Travelogue of Suiyuan], *Keda zazhi* 7, no. 4 (1936): 7–9.

149. Zheng Zhenduo, "Zhaojun mu" [The Zhaojun tomb], in *Xibei shengji*, 46–50.

150. Tighe, *Constructing Suiyuan*, 222, especially n. 10.

151. Tighe, 218–55.

152. Bing Xin, *Pingsui yanxian lüxing ji*, 53–54.

153. Bing Xin, 56.

154. Huang Yubin, "Dao Bailingmiao qu" [Go to Bailingmiao] *LZ* 9, no. 6 (June 1935): 15.

155. Huang Fensheng, *Bailingmiao xunli* [March to Bailingmiao] (Shanghai: Shangwu yinshu guan, 1936), 48.

156. Huang Fensheng, 57.

157. Fang Shan, "Yungang shiku si youji" [Travelogue of the Yungang Grottoes], *Keda zazhi* 5, no. 3 (1934): 7–9; Baoshi, "Suiyuan youji," 7–9.

158. For example, a group of more than forty people participated in an exploratory tour of Shaanxi's industrial areas in 1932. Zhao Renqiao and Meng Shi, "Canjia Xibei shiye kaochatuan zhi jingguo ji qi gongzuo" [Processes and work of the group exploring the industries of the Northwest], *Qinghua zhoukan* [Qinghua University weekly] 38, no. 5 (1932): 52–54.

159. "Longhai lu Huayin Lintong liangzhan piwei guonei laihui youlan zhandian" [The Huayin and Lintong stops on the Longhai line are designated as "tourism round-trip" stops], *Jinghu huhangyong tielu rikan*, no. 1389 (1935):143.

160. For example, the Shanghai journalists Gu Zhizhong, Lu Yi, and two other colleagues went to Qinghai via Shaanxi and Gansu in 1933. Gu Zhizhong and Lu Yi, *Dao Qinghai qu*.

161. Hu Shiyuan, *Xibei daoyou*, 33.

162. Hu Shiyuan, 33–35.

163. Xingchu [Shen Xingchu] and Juexue, "Xibei youcheng jilue" [A brief itinerary of the Northwest], *Zhonghua*, no. 27 (1934): n.p.

164. Bian Wenjian, "Xibei kaocha ji" [An exploration of the Northwest], *LZ* 7, no. 5 (May 1933): 57–67.

165. Bian Wenjian, 64.

166. Shu Yongkang, "Xixing riji" [Journal of a westward journey], *LZ* 7, no. 9 (September 1933): 47–53; Shu Yongkang, "Xixing riji (er)" [Journal of a westward journey (2)], *LZ* 7, no. 10 (October 1933): 31–48; Shu Yongkang, "Xixing riji (san)" [Journal of a westward journey (3)], *LZ* 7, no. 11 (November 1933): 47–62.

167. Shu Yongkang, "Xixing riji (er)," 37.

168. On Bian and Shu's essays in *China Traveler*, see Dong, "Shanghai's *China Traveler*," 217–19.

169. For a discussion of this practice in commercial guidebooks in Europe, see Nicholas Parsons, *Worth the Detour: A History of the Guidebook* (Stroud: History Press, 2007), 177–252.

170. Zhang Henshui, "Xiyou xiaoji (yi-shiyi)" [Brief account of a westward journey (1–11)], *LZ* 8, no. 9, and *LZ* 9, no. 7 (September 1934–July 1935).

171. For a discussion of tourism and the development of Republican Beijing, see Dong, *Republican Beijing*, 78–104.

172. Zhang Henshui, "Xiyou xiaoji (yi)," *LZ* 8, no. 9 (September 1934): 8.
173. Zhang Henshui, "Xiyou xiaoji (er)," *LZ* 8, no. 10 (October 1934): 36–38.
174. Zhang Henshui, "Xiyou xiaoji (san)," *LZ* 8, no. 11 (November 1934): 69.
175. Zhang Henshui, "Xiyou xiaoji (liu)," *LZ* 9, no. 2 (February 1935): 65–66.
176. Zhang Henshui, "Xiyou xiaoji (qi)," *LZ* 9, no. 3 (March 1935): 51–56.
177. Zhang Henshui, 55–56.
178. Zhang Henshui, "Xiyou xiaoji (ba)," *LZ* 9, no. 4 (April 1935): 33–34.
179. Zhang Henshui, "Xiyou xiaoji (shiyi)," *LZ* 9, no. 7 (July 1935): 60.
180. Zhang Henshui, 62–64.
181. Zhang Henshui, 64–65.
182. See for example, Duomo, "Luoyang youji" [Travelogue of Luoyang], *Keda zazhi* 6, no. 9 (1935): 7–9; Wu Yishou, "Lanzhou youji" [Travelogue of Lanzhou], *Keda zazhi* 6, no. 4 (1935): 7–9.
183. Zhang Henshui, "Xiyou xiaoji (shiyi)," 61.
184. Zhang Henshui, 58.
185. Zhang Henshui, 3.
186. For some examples, see Ying Xing, "Yungang shiku" [The Yungang Grottoes], *Dongfang zazhi* 27, no. 2 (1930): 91–104; Liang Desuo, "Xibei de yinxiang"; Yang Lingde, "Cong Suiyuan dao Bailingmiao" [From Suiyuan to Bailingmiao], *Shenbao yuekan* [Shenbao monthly] 3, no. 8 (1934): n.p.
187. For example, see Shu Yongkang, "Dao Xibei qu" [Head to the Northwest], *Sheying huabao* [Photography pictorial] 9, no. 33 (1933): 16–17; Zhang Zhihong, "Dao Xibei qu" [Head to the Northwest], *Libailiu* [Saturday], no. 584 (1934): n.p.; Huang Huaju, "Dao Xibei qu" [Go to Northwest], *Sanmin huakan* [Pictorial of the Three Principles of the People], no. 8 (1934): 2.
188. Xiao Qian, "Pingsui xianshang (youji)," *Xinsheng zhoukan* 1, no. 41 (1934): 845.
189. Bai Dizhou, "Ji Liu Bannong xiansheng zhi bingyin" [On the cause of the illness of Mr. Liu Bannong], *Guoyu zhoukan* [National language weekly] 6, no. 151 (August 1934): 43–44.
190. Peter Perdue, "Identifying China's Northwest, for Nation and Empire," in *Locating China: Space, Place, and Popular Culture*, ed. Jing Wang (New York: Routledge, 2005), 98; Joseph Esherick, "How the Qing Became China," in *Empire to Nation: Historical Perspectives on the Making of the Modern World*, ed. Joseph Esherick, Hasan Kayali, and Eric Van Young (Lanham, MD: Rowman & Littlefield, 2006), 229.

4. Facilitating the Exodus

1. *LZ* 12, no. 11 (November 1938).
2. Sun Fuxi, "Xi'nan shi jianguo de tianyuan" [The Southwest is the field for state building], *LZ* 12, no. 11 (November 1938): 5.
3. On the estimated figures of wartime refugees during the War of Resistance against Japan, see Parks Coble, *China's War Reporters* (Cambridge, MA: Harvard University Press, 2015), 80–81.
4. Rana Mitter, *Forgotten Ally: China's World War II, 1937–1945* (Boston: Houghton Mifflin Harcourt, 2013), 12.

5. Diana Lary, introduction to *China at War: Regions of China, 1937–1945*, ed. Stephen R. MacKinnon, Diana Lary, and Ezra F. Vogel (Stanford, CA: Stanford University Press, 2007), 11.

6. The negotiation of the opening up of Chongqing as a treaty port was a long process and extended from 1878 to 1891. Danke Li, *Echoes of Chongqing: Women in Wartime China* (Urbana: University of Illinois Press, 2009), 188n38.

7. On the early history of the steamer navigation to Chongqing, see Deng Shaoqin, ed., *Jindai Chuanjiang hangyun jianshi* [A brief history of shipping on Chuanjiang] (Chongqing: Chongqing difang shi ziliao zu, 1982), 93–96.

8. Shanghai shangchu yinhang lüxing bu, *You Chuan xu zhi*, 2; Zheng Bicheng, *Sichuan daoyou* [Guide to Sichuan] (Shanghai: Zhongguo lüxing she, 1935), 7.

9. Zheng Bicheng, *Sichuan daoyou*.

10. For example, it took one author twenty-four days to travel from Beijing to Chengdu in 1928–29. Zhang Qinshi, "Lü Shu dao shang" [On the way to Sichuan], *LZ* 3, no. 1 (January 1929): 25.

11. "Hangkong yu gonglu" [Aviation and highways], *Liangyou* [Young companion], no. 78 (1933): 2–3; Zheng Bicheng, *Sichuan daoyou*, 7.

12. Lai Yanyu, ed., *Guangxi youli xuzhi* [Notes on travel to Guangxi] (Nanning: Guangxi yinshu chang, 1935), 2–3.

13. Qian Hua, "Guilin shanshui" [Guilin's scenery], *LZ* 10, no. 1 (January 1936): 58.

14. Kan Zonghua and Zheng Houbang, "Xi'nan hangkong gongsi de xingshuai" [The rise and decline of Southwest Aviation], in *Zhonghua wenshi ziliao wenku* [Library of Chinese historical data], ed. Quanguo zhengxie wenshi ziliao weiyuanhui [Committee for research on historical data] (Beijing: Zhongguo wenshi chubanshe, 1996), 13:707–12.

15. In the travel guide to Yunnan published by the Travel Department of Shanghai Bank, this travel route via Vietnam was the only suggested route. Shanghai shangchu yinhang lüxing bu, *You Dian xu zhi*.

16. Zhang Yunfen, "Cong Shanghai dao Yunnan fu" [From Shanghai to Kunming], *LZ* 8, no. 1 (January 1934): 60–61.

17. Shanghai shangchu yinhang lüxing bu, *You Dian xu zhi*, 1.

18. Ding Wenjiang, *Youji erzhong* [Two travel accounts] (Shenyang: Liaoning jiaoyu chubanshe, 1998), 4–6.

19. Zhang Yunfen, "Cong Shanghai dao Yunnan fu," 62.

20. Ding Wenjiang, *Youji erzhong*, 6–15.

21. Ding Wenjiang, 6–7.

22. On the history of *Maxiangyue*, see Si Jiaozhi, "Xi'nan minjian yunshuye 'Maxiangyue'" ["Maxiangyue": The popular transportation service in the Southwest], in Quanguo zhengxie wenshi ziliao weiyuanhui, *Zhonghua wenshi ziliao wenku*, 13:971–84.

23. Zhonguo lüxingshe, ed., *Qianyou jilüe* [Travel guide of Guizhou] (Shanghai: Zhongguo lüxingshe, 1934), 7.

24. Shanghai shangchu yinhang lüxing bu, *You Dian xu zhi*; Shanghai shangchu yinhang lüxing bu, *You Chuan xu zhi*.

25. Reinhardt, *Navigating Semi-Colonialism*, 242.

26. Zheng Jianfeng, "You Xi'an yan gonglu zhi Chongqing" [From Xi'an to Chongqing via motorways], *LZ* 11, no. 6 (June 1937): 17–19.

27. Hu Shiquan, "Jingdian gonglu zhoulan tuan suizheng ji" [Tour along the Jingdian highway], *LZ* 11, no. 6 (June 1937): 21–38.

28. Hu Shiquan, 23.

29. James Hargett, *Stairway to Heaven: A Journey to the Summit of Mount Emei* (Albany: State University of New York Press, 2006), 183–85.

30. For discussions of the Three Gorges and Mount Emei in classic Chinese literature, see Corey Byrnes, *Fixing Landscape: A Techno-Poetic History of China's Three Gorges* (New York: Columbia University Press, 2018), 25–87; Hargett, *Stairway to Heaven.*

31. Byrnes, *Fixing Landscape*, 2.

32. Shanghai shangchu yinhang lüxing bu, *You Chuan xu zhi*, cover, 24–31.

33. Wang Xiaoting, "Sanxia tianxian" [The Three Gorges as a natural barrier], *Dazhong huabao*, no. 1 (1933): 14–15.

34. Yun Yintang, "Wushan guo yan ji" [A brief view of the Wu mountain], *LZ* 7, no. 2 (February 1933): 1–6; "Chuanyou suojian: sanxia tianxian" [Views of Sichuan: The natural barrier of the Three Gorges], *Shenbao yuekan* 4, no. 12 (December 1935): n.p.

35. "Guilin shanshui jia tianxia" [Guilin's scenery is the best under heaven], *Liangyou* [Young companion], no. 77 (1933): 23–24; Shao Yuxiang, "Guangxi shanshui" [Guilin scenery], *LZ* 10, no. 1 (January 1936): n.p.

36. "Liangyou duzhe lüxing lieche: Yunnan Kunming" [*Liangyou* readers' travel train: Kunming, Yunnan], *Liangyou* [Young companion], no. 105 (September 1935): 24–25.

37. "Liangyou duzhe lüxing lieche," 24–25.

38. Gao Bochen, "Sanxia daoguan" [My trip to the Three Gorges], *LZ* 10, no. 1 (January 1936): 103.

39. For the rise of the Minsheng Company in the late 1920s and 1930s, see Reinhardt, *Navigating Semi-Colonialism*, 237–44, 253–93.

40. Gao Bochen, "Sanxia daoguan," 103.

41. "Chuanjiang lüxingshe kaimu le" [The opening of the Chuangjiang Travel Service], *Xingcha* [Xingcha weekly], no. 48 (1931): 1.

42. Chongqing Zhongguo yinhang [Bank of China, Chongqing Branch], ed., *Emei shan* [Mount Emei] (Chongqing: Zhongguo yinhang, 1935), 226–29.

43. Gao Bochen, "Sanxia daoguan," 104.

44. "Yousheng lüxingtuan Sichuan lüxing tonggao."

45. *Yousheng lüxingtuan dishijie zhengqiu dahui tekan* [Special issue for the tenth recruiting conference] (Shanghai: Yousheng lüxingtuan, 1935), n. p.

46. "Huanan lüxing, Yousheng lüxingtuan tonggao" [South China tour, a notice for the Unison Travel Club], *Shenbao*, April 3, 1937.

47. Tang Weibin, "Zhonglü ershisan nian" [The twenty-three years of the China Travel Service], *LZ* 20, no. 1 (January 1946): 97.

48. Within two months of the outbreak of the war, *China Traveler* registered with the Shanghai Municipal Council of the International Settlement. In 1938, it moved its main operations, such as article collection, advertising, and distribution, to Hong Kong. Tang Weibin, "Zhonglü ershisan nian," 96; "Lüxing zazhi zengshe huanan zong faxing jinyao qishi" [A breaking announcement from *China Traveler*: Adding a main distribution office in South China], *LZ* 12, no. 6 (June 1938): n.p.

49. Shortly after the fighting in Shanghai broke out, the Shanghai Bank first set up a joint committee for the Shanghai Bank, CTS, and the Daye Trading Company

in Hankou, one of the tri-cities of Wuhan. When Wuhan was captured in 1938, these operations were transferred to Hong Kong. A year later, the headquarters of CTS relocated back to Shanghai. Pan Taifeng, "Kangri zhanzheng shiqi de Zhongguo lüxingshe," 912–13.

50. Ulrich Theobald, "Southwest China: Local Conditions and Economic Trajectories," in *Southwestern China in a Regional and Global Perspective (c.1600–1911): Metals, Transport, Trade and Society*, ed. Ulrich Theobald and Jin Cao (Leiden: Brill, 2018), 1–8; Bin Yang, *Between Winds and Clouds: The Making of Yunnan* (New York: Columbia University Press, 2009), 23–72.

51. Jin Ye, "You Guangzhouwan zhuandao" [Transits through Guangzhouwan], *LZ* 13, no. 12 (December 1939): 7–11.

52. These locales in Southeast Asia became so important that *China Traveler* published a special issue on Manila in 1941. Another special issue on Rangoon was announced, although it was never published. "Manila zhuanhao" [Special issue on Manila], *LZ* 15, no. 11 (November 1941).

53. "Zhongguo lüxingshe Xigong fenshe kaimu" [The opening of the Saigon branch of CTS], *LZ* 15, no. 1 (January 1941): 5; Pan Taifeng, "Kangri zhanzheng shiqi de Zhongguo lüxingshe," 907–28. In August 1940, *China Traveler* also published a special issue on Malaysia. *LZ* 14, no. 8 (August 1940).

54. Pan Taifeng, "Kangri zhanzheng shiqi de Zhongguo lüxingshe," 912.

55. Fu Hua Chen, *Between East and West: Life on the Burma Road, the Tibetan Highway, the Ho Chi Minh Trail, and in the United States* (Niwot: University Press of Colorado, 1996), 80.

56. Parks M. Coble, *Chinese Capitalists in Japan's New Order: The Occupied Lower Yangzi, 1937–1945* (Berkeley: University of California Press, 2003), 29.

57. Zhou Xiuquan, "Dianmian daozhong" [On the Burma Road], *LZ* 14, no. 1 (January 1940): 51–53; Xu Yifang, "Wo canjia Dianmian gonglu xiujiangongcheng de jingguo" [My participation in the construction of the Burma Road], *Zhonghua wenshi ziliao wenku*, 13:740–46.

58. Wang Jianshi, "Zhongyin gonglu quancheng xing" [Travel experiences along the entire Sino-India motorway], *LZ* 19, no. 1 (January 1945): 62–65.

59. He Yongji, "Yindu jiyou" [Travelogue of India], *LZ* 17, no. 9 (September 1943): 100–102.

60. ZLZD, Q 368-1-37.

61. In July 1937, the Nationalist government called a meeting in Nanjing, at which the Executive Yuan, the Military Affairs Commission, the National Economic Council, the Ministry of Military Affairs, the Ministry of Transportation, the Ministry of Railroads, and the provincial governments of Sichuan, Yunnan, Guizhou, and Hunan gathered to discuss "ways to improve highway transportation in the Southwest." Establishing a Southwest Provinces Joint Transportation Committee, participants hoped to break down the barriers among the southwestern provinces and integrate the unconnected transportation system in the Southwest. Lin Bing, "Xi'nan gonglu: Zhuliu duan" [The Southwest Motorways: Between Guiyang and Liuzhou], *LZ* 13, no. 5 (May 1939): 21–28; "Xi'nan gesheng gonglu piaojia biao" and "Xi'nan gesheng gonglu lianyun lichen biao" [The prices and distances of through travel on the Southwest Motorways], *LZ* 13, no. 5 (May 1939): 67–68; Lin Bing,

"Xi'nan gonglu: Xiangqian duan" [The Southwest Motorways: Between Changsha and Guiyang], *LZ* 12, no. 11 (November 1938): 47–53.

62. The system consisted of four main parts: the Hunan-Guizhou line (*Xiangqian xian*), the Sichuan-Guizhou line (*Chuanqian xian*), the Yunnan-Guizhou line (*Dianqian xian*), and the Guangxi-Guizhou line (*Guiqian xian*). Xue Cishen, preface to *Sannian lai zhi xi'nan gonglu* [The Southwest Motorways over the past three years], ed. Ding Songchang (China: s.n., 1940), 1.

63. Yang Rennong, "Chuandian gonglu jixing" [Travel notes on the Chuandian highway], *LZ* 14, no. 7 (July 1940): 15–17.

64. "Zhongguo lüxingshe jingban gedi zhaodaisuo ji shitang" [The guesthouses and dining halls run by CTS], *LZ* 17, no. 3 (March 1943): n.p.; Pan Taifeng, "Kangri zhanzheng shiqi de Zhongguo lüxingshe," 909–11.

65. "Gedi ziying ji chengban zhaodaisuo" [CTS guesthouses in various locations], *LZ* 14, no. 6 [June 1940]: n.p.

66. "Fu Dian xingcheng" [Itineraries to Yunnan], *LZ* 12, no. 2 (February 1938): 18.

67. Zhao Junhao, *Kunming daoyou* [Guidebook to Kunming] (Shanghai: China Travel Service, 1939), 51–62.

68. She Guitang, "Yuenan zhitan" [A summary of Vietnam], *LZ* 14, no. 6 (June 1940): 51–54; She Guitang, "Dianmian gonglu jixing" [A travel record of the Yunnan-Burma Road], *LZ* 14, no. 12 (December 1940): 5–9.

69. Keith Schoppa, *In a Sea of Bitterness: Refugees during the Sino-Japanese War* (Cambridge, MA: Harvard University Press, 2011), 285.

70. *Lianda* was formed in Changsha in November 1937 from a merging of the top three universities in China: Beijing University, Qinghua University, and Nankai University. When the war spread to central and southern China in 1938, the government decided to relocate the school to Kunming. While a team of more than two hundred largely male students led by a dozen faculty members traveled on foot from Changsha to Yunnan, the majority of the faculty and students retreated via Vietnam. For an account of the retreat on foot, see Qian Nengxin, *Xi'nan sanqian wubai li* [3,500 *li* to the Southwest] (Changsha: Shangwu yinshu guan, 1939). For the history of *Lianda*, see John Israel, *Lianda: A Chinese University in War and Revolution* (Stanford, CA: Stanford University Press, 1999).

71. Deng Hanjun, "Yue Xiang Gui lücheng suoji" [Fragmented recollections of my journey through Guangdong, Hunan, and Guangxi Provinces], *LZ* 12, no. 11 (November 1938): 37.

72. Sa Kongliao, *Cong Xianggang dao Xinjiang* [From Hong Kong to Xinjiang] (Yinchuan: Ningxia chubanshe, 2000), 1.

73. A close friend of the well-known leftist journalist Zou Taofen, Du Zhongyuan became a columnist for Zou's widely read magazine *Life Weekly*, and then went on to become editor of its successor journal, *New Life Weekly* (*Xinsheng zhoukan*), in which he wrote an essay called "On Emperors" leading to Japanese protest and his arrest and imprisonment by the Nationalist government. After the war broke out, Du became a war correspondent and traveled through central and northwestern China to promote an anti-Japanese united front. In late 1937, he journeyed to Xinjiang to meet with Sheng Shicai, the pro-Soviet warlord in Xinjiang. Urged by Sheng to help build a new Xinjiang, Du returned to Hong Kong to enlist other progressives.

For Du's life, see Zhang Baoyu, ed., *Du Zhongyuan* [Biography of Du Zhongyuan] (Urumqi: Xinjiang daxue chubanshe, 1987), 1–44.

74. Sa Kongliao, *Cong Xianggang dao Xinjiang*, 1–2.

75. Sa Kongliao.

76. The rumor was not baseless speculation. When Sa Kongliao was on his way to Vietnam in March 1939, the Japanese military was in the middle of its attack on Hainan Island and preparing for its advance into French Indochina. In fact, Sa recorded that when his ship was crossing the Qiongzhou strait, he saw Japanese navy fleets close by. Sa Kongliao, 4, 7.

77. Pan Taifeng, "Kangri zhanzheng shiqi de Zhongguo lüxingshe," 915.

78. Qian Huipu, "Hudian jixing" [Travel notes on my trip from Shanghai to Yunnan], *LZ* 12, no. 12 (December 1938): 7.

79. Sa Kongliao, *Cong Xianggang dao Xinjiang*, 8–13.

80. Sa Kongliao, 5–6, 9–10, 19–20.

81. Sa Kongliao, 15–16.

82. Li Changzhi, "Xi'nan jixing" [Travel notes on my journey to the Southwest], *LZ* 12, no. 11 (November 1938): 16.

83. Sa Kongliao, *Cong Xianggang dao Xinjiang*, 3.

84. Sa Kongliao, 18.

85. Pan Enlin, introduction to *Xi'nan lansheng* [Scenic beauty in Southwest China], ed. Zhao Junhao (Shanghai: Zhongguo lüxingshe, 1940), n.p.

86. According to the introduction to the second edition of *Xi'nan lansheng* in 1940, both domestic readers and overseas markets had welcomed the album, and the copies of the first edition sold out so fast that CTS had to print a second edition within a year. Pan Enlin, introduction, n.p.

87. Sa Kongliao, *Cong Xianggang dao Xinjiang*, 30.

88. Sa Kongliao, 42.

89. Tang Weibin, "Zhonglü ershisan nian," 98.

90. Qian Zhitong, "Pingyi zhaodaisuo chengli jingguo" [The founding of the Pingyi Guesthouse], *Lüguang* 1, no. 2 (February 1940): 9.

91. Tang Weibin, "Zhonglü ershisan nian," 98.

92. Danke Li, *Echoes of Chongqing*, 193n17.

93. Sa Kongliao, *Cong Xianggang dao Xinjiang*, 52.

94. "Xi'nan zhuanhao shuhou" [Postscript to the special issue on the Southwest], *LZ* 12, no. 11 (November 1938): 109.

95. Sun Fuxi, "Xi'nan shi jianguo de tianyuan," 5–7.

96. See Kenneth J. Ruoff, *Imperial Japan at Its Zenith: The Wartime Celebration of the Empire's 2600th Anniversary* (Ithaca, NY: Cornell University Press, 2010); McDonald, *Placing Empire*.

97. "Xi'nan zhuanhao shuhou," 109.

98. Sun Peigan, "Tan zhanshi youlan shiye" [On wartime tourism], *LZ* 15, no. 7 (July 1941): 91.

99. "Xi'nan zhuanhao shuhou," 109.

100. Li Qiyu, "Kunming fengguang" [The sights of Kunming], *LZ* 12, no. 1 (January 1938): 23.

101. Aside from the aforementioned *Lianda*, half a dozen institutes of the National Academy of Beiping and several branches of the Academia Sinica also moved to Kunming during wartime. Israel, *Lianda*, 195.

102. Sha Ou, "Shanguang shuise de Kunming" [Scenic Kunming], *LZ* 13, no. 1 (January 1939): 69.

103. Li Qiyu, "Kunming fengguang," 24.

104. Li Qiyu, 28.

105. Sha Ou, "Shanguang shuise de Kunming," 69.

106. Zhao Junhao, *Kunming daoyou* [Guidebook to Kunming], ed. Huang Lisheng and Ge Mo'an (Kunming: Zhongguo lüxingshe, 1944).

107. Li Qiyu, "Kunming fengguang," 28–32; Shuai Yucang, "Kunming manji" [A travelogue of Kunming], *LZ* 13, no. 8 (August 1939): 25–31; Huang Zhuoqiu, "Kunming fengjing xian" [Scenery of Kunming], *LZ* 15, no. 2 (February 1941): 13–19.

108. Li Qiyu, "Kunming fengguang," 26.

109. Ban Gong, "Kunming de chaguan" [The teahouses of Kunming], *LZ* 13, no. 7 (July 1939): 27–30.

110. Liu Zhiping, "Kunming huanhu xing ji" [Around-the-lake tour in Kunming], *LZ* 14, no. 6 (June 1940): 3–17.

111. Zhang Henshui, "Chongqing lügan lu" [Impression of my sojourn in Chongqing], *LZ* 13, no. 1 (January 1939): 49.

112. Shen Tao, "Yi Chongqing" [Reminiscing about Chongqing], *LZ* 15, no. 8 (August 1941): 55.

113. Zhang Henshui, "Chongqing lügan lu"; Gao Shaocong, "Chongqing suoji" [Notes on Chongqing], *LZ* 14, no. 4 (April 1940): 3; Shen Tao, "Yi Chongqing," 55–56.

114. Zhang Henshui, "Chongqing lügan lu xupian" [Impression of my sojourn in Chongqing (continued)], *LZ* 14, no. 1 (January 1940): 33.

115. Gu Mengwu, "Xianhua zhanshi shoudu" [On the wartime capital], *LZ* 13, no. 11 (November 1939): 9–10.

116. Gu Mengwu, 10; Sihong, "Chongqing shenghuo pianduan" [Snippets of Chongqing life], *LZ* 14, no. 4 (April 1940): 9–10; Shen Tao, "Yi Chongqing," 56–57.

117. Gu Mengwu, "Xianhua zhanshi shoudu," 10–11; Gao Shaocong, "Peidu Chongqing sumiao" [Sketches of the wartime capital Chongqing], *LZ* 15, no. 2 (February 1941): 6.

118. Tang Weibin, "Zhonglü ershisan nian," 100.

119. Qiu Peihao, "Xianhua Chongqing nanquan" [A chat about the South Hot Spring in Chongqing], *LZ* 13, no. 10 (October 1939): 23–33; Yi Shengbo, "Chongqing de fengjing qu: Nan wenquan" [The scenic areas of Chongqing: The South Hot Spring], *LZ* 17, no. 4 (April 1943): 13–16.

120. Zhang Henshui, "Chongqing lügan lu," 50.

121. Qiu Peihao, "Xianhua Chongqing nanquan," 31.

122. Shao Zuping, "Chengdu mingsheng fangwen ji" [Visiting Chengdu's famous attractions], *LZ* 14, no. 4 (April 1940): 27–35.

123. Tang Weibin, "Zhonglü ershisan nian," 99–100.

124. Shao Zuping, "Chengdu mingsheng fangwen ji"; Guo Zhusong, "Chengdu shengji daoyou" [Guide to the famous sites in Chengdu], *LZ* 16, no. 12 (December 1942): 19–26.

125. The Wuhou Temple was first built in the Tang dynasty to commemorate Zhuge Liang, the loyal and ingenious military strategist from the second century who had supported the mission to reunify China and restore the Han dynasty. Du Fu's thatched cottage was associated with the celebrated Tang poet Du Fu, who

sought refuge in Shu (today's Sichuan) during the An Shi rebellion. And the River-Viewing Pavilion was built in memory of a famous Tang-dynasty poet and courtesan Xue Tao, whose father took the family from Chang'an to Sichuan shortly before he passed away.

126. Yi Junzuo, "Jincheng qi ri ji (shang)" [Seven days in Chengdu (1)], *LZ* 14, no. 5 (May 1940): 3–14.

127. Yi Junzuo, "Jincheng qi ri ji (xia)" [Seven days in Chengdu (2)], *LZ* 14, no. 6 (June 1940): 19–28.

128. Chen Qi, "Ji Guilin zhi xing" [My trip to Guilin], *LZ* 12, no. 5 (May 1938): 13–20; Chen Qi, "Ji Guilin zhi xing (xia)" [My trip to Guilin (continued)], *LZ* 12, no. 6 (June 1938): 55–60. Wang Hongdao, "Guilin xiaozhu" [Sojourn in Guilin], *LZ* 13, no. 11 (November 1939): 20.

129. Deng Hanjun, "Yue Xiang Gui lücheng suoji," 42–44.

130. Deng Hanjun, 43.

131. Wan Feng, "Cong Guilin dao Yangshuo" [From Guilin to Yangshuo], *LZ* 16, no. 8 (August 1942): 13–14.

132. Sha Ou, "Guiyang yipie" [A glimpse of Guiyang], *LZ* 12, no. 12 (December 1938): 14.

133. Sha Ou, 13–15; Gu Jungu, "Guiyang zaxie" [A miscellany of Guiyang] *LZ* 13, no. 3 (March 1939): 5–12.

134. Liu Tizhi, "Liangyou duzhe lüxing lieche: Guizhou fengjing mei" [*Liangyou* readers' travel train: Beautiful Guizhou], *Liangyou* [Young companion], no. 102 (June 1935): 20–21.

135. Sha Ou, "Guiyang yipie," 15; Gu Jungu, "Guiyang zaxie," 9–10.

136. Li Changzhi, "Xi'nan jixing," 16.

137. "Qianjin zhong zhi Guiyang" [Kweiyang in war color], *Liangyou* [Young companion], no. 142 (May 1939): 28.

138. "Xi'nan fengjing xian" [The scenic line in the Southwest], *LZ* 12, no. 10 (October 1938): n.p.

139. Sha Ou, "Shanguang shuise de Kunming," 69–70.

140. For an example of the print media's coverage of the non-Han people in the Southwest, see Yajun Mo, "The New Frontier: Zhuang Xueben and Xikang Province," in *Chinese History in Geographical Perspective*, ed. Yongtao Du and Jeff Kyong-McClain (Lanham, MD: Lexington Books, 2013), 121–40.

141. Bertram Gordon, *War Tourism: Second World War France from Defeat and Occupation to the Creation of Heritage* (Ithaca, NY: Cornell University Press, 2018), 7.

5. Between Empire and Nation-State

1. Naoko Shimazu, "Colonial Encounters: Japanese Travel Writing on Colonial Taiwan," in *Refracted Modernity: Visual Culture and Identity in Colonial Taiwan*, ed. Yūko Kikuchi (Honolulu: University of Hawai'i Press, 2007), 21–37; Ruoff, *Imperial Japan at Its Zenith*, 129–47; McDonald, *Placing Empire*.

2. Feuerwerker, "Economic Trends, 1912–1949," 95.

3. For the history of Han migration to Manchuria, see Thomas R. Gottschang and Diana Lary, *Swallows and Settlers: The Great Migration from North China to Manchuria*

(Ann Arbor: Center for Chinese Studies, University of Michigan, 2000); Rana Mitter, *The Manchurian Myth: Nationalism, Resistance, and Collaboration in Modern China* (Berkeley: University of California Press, 2000), 20–23.

4. Everard Cotes, *Signs and Portents in the Far East* (New York: G. P. Putnam, 1907), 141–43.

5. "China-Japan Circular and Overland Tour," *Peking Daily News*, September 30, 1915; "China-Japan Railway Conference: Celebrations in Tokyo," *Peking Gazette*, April 10, 1917.

6. *Zhongri tielu zhouyou zhinan* [Guide to railways in China and Japan] (Tianjin: Shangwu yinshu guan, 1916); Thomas Cook Ltd., *Chinese Government Railways: Travellers' Notes and Through Booking Arrangements, Return Tourist Tickets, Circular Tours in China, Chosen and Japan* (Shanghai: Thomas Cook & Son, 1918).

7. Thomas Cook Ltd., *Chinese Government Railways*, 40–41.

8. *Zhongri tielu zhouyou zhinan*, 19–22.

9. Louise Young, *Japan's Total Empire: Manchuria and the Culture of Wartime Imperialism* (Berkeley: University of California Press, 1998).

10. Eliza Scidmore, "Mukden, the Manchu Home, and Its Great Art Museum," *National Geographic Magazine* 21, no. 4 (1910): 289.

11. Emily G. Kemp, *The Face of Manchuria, Korea, and Russian Turkestan* (New York: Duffield, 1911).

12. Thomas Cook Ltd., *Cook's Handbook for Tourists to Peking*; Thomas Cook Ltd., *Peking, North China, South Manchuria and Korea*, 4th ed. (London: Thomas Cook & Son, 1920); Claudius Madrolle, *North-Eastern China: Manchuria, Mongolia, Vladivostock, Korea* (Paris: Hachette, 1912).

13. Thomas Cook Ltd., *Cook's Handbook for Tourists to Peking*, 75.

14. Thomas Cook Ltd., 77–78; Claudius Madrolle, *North-Eastern China*, 16–18.

15. Scidmore, "Mukden, the Manchu Home," 305–6.

16. Kemp, *Face of Manchuria*, 22.

17. Cotes, *Signs and Portents*, 133–39.

18. Thomas Cook Ltd., *Cook's Handbook for Tourists to Peking*, 83–85.

19. Kemp, *Face of Manchuria*, 4.

20. Kemp, 8.

21. Frederick Simpich, "Manchuria, Promised Land of Asia," *National Geographic Magazine* 56, no. 4 (October 1929): 379.

22. Koyūkai [Friends of Shanghai Association], *Tōa dōbun shoin daigakushi* [The history of the East Asia Common Culture Academy] (Tokyo: Koyūkai, 1955), 1–42.

23. In January 1915, Japan's Okuma government presented the "Twenty-One Demands" to the Yuan Shikai regime, pressing for China's assent to Japan's takeover of the German leased lands in Shandong, the strengthening of its control in Manchuria, and the establishment of a Japanese sphere of influence in the southeastern province of Fujian. These harsh terms and Yuan's appeasements to Japan caused a huge backlash and triggered the rise of anti-Japanese sentiments among Chinese.

24. Koyūkai, ed., *Shanhai Tōa dōbun shoin dairyokō kiroku: jitsuroku Chūgoku tōsaki* [Shanghai East Asia Common Culture Academy's records on Grand Travels: Authentic records of personal investigations in China] (Tokyo: Shin Jinbutsu ōraisha, 1991), 153–58.

25. Koyūkai, 161, 165.

26. Li Zhenfu, *Dongsansheng lüxing zhinan*, 1.

27. Li Zhenfu, 55, 67, 75.

28. "Zhongyang junxiao dongbei kaochatuan di Ha'erbin chezhan sheying" [A picture of the arrival of the Central Military Academy investigation group at Harbin station], *Zhongyang huakan* [Central pictorial], no. 7 (September 1929): 8.

29. Jiang Zunyou, "Dongbei shiye kaochatuan yu yanda nongxi" [The industry investigation tour to the Northeast and the Department of Agriculture at Yenching University], *Yanda nongxun* [Agricultural bulletin of Yenching] 2, no. 11 (February 1929): 1–2; Yang Zhongjian, *Xibei de poumian*, 51–71.

30. Lu Zuofu's travelogue about his Manchurian trip was serialized in a few publications in 1930–31. It was later published as a book. Lu Zuofu, *Dongbei youji* [Travelogue of the Northeast], 2nd ed. (Chengdu: Chengdu shuju, 1931).

31. Zhichao, "Lüxing xinxun: Susheng dangbu rexin tichang dongbei lüxing" [Travel newsletter: The Nationalist Party in Jiangsu Province enthusiastically promotes travel to the Northeast], *Yousheng lüxing yuekan* [Unison Travel magazine] 5, no. 3 (March 1930): 53–54.

32. "Tiedaobu xunling" [Instructions from the Ministry of Railways], *Tiedao gongbao*, no. 48 (March 1930): 37–40.

33. "Benlu dashi ji" [Big events with the railway], *Shenhai tielu yuekan* [Shenyang-Hailong Railway monthly] 1, no. 1 (1929), 1.

34. Miao Wenhua, *Beiling zhilüe* [Brief record of the North Mausoleum] (Shenyang: Beiling gongyuan guanlichu, 1929).

35. "Zi Yaohua jun lüxing dongbei" [Mr. Zi Yaohua's visit to the Northeast], *Haiguang* 2, no. 4 (April 1930): 5–6.

36. For some examples, see Tu Zheyin, "Cong Shanghai dao Harbin" [From Shanghai to Harbin], *LZ* 2, no. 1 (January 1928): 15–17; Wuyong, "Ha'erbin yipie" [A glimpse of Harbin], *LZ* 3, no. 2 (February 1929): 19–23; Zhao Junhao, "Dongbei jihen ji (yi)" [Travelogue of the Northeast (1)], *LZ* 3, no. 7 (July 1929): 53–66; Yang Weibei, "Dalian caifeng lu" [Travel record of Dalian], *LZ* 4, no. 12 (December 1930): 11–13.

37. Liu Yazhi, ed., *Bao Ri zuijin zhi jingji qinlüe yu dongbei* [An exposé of the recent economic invasion of the Japanese in the Northeast] (Shanghai: Dongbei yanjiu she, n.d.), 83–85.

38. SSCYD, Q275-1-433.

39. "Zhongguo lüxingshe baogao Pingshen tongche jingguo shishi" [China Travel Service report on the process of the through traffic arrangement on the Beijing-Mukden Railway], *Zhonghang yuekan* [Bank of China monthly] 9, no. 3 (September 1934): 134–36.

40. SSCYD, Q275-1-819.

41. SSCYD, Q275-1-433.

42. "Pingshen tongche zhongtu chucha" [An accident midway along the Beijing-Mukden train line], *Meizhou pinglun* [Weekly review], no. 124 (1934): 26–27.

43. For an example, see Leng Xiang, "Cong Zhang Shuiqi lianxiang qi" [The Zhang Shuiqi connection], *Xinghua* [Xinghua weekly] 1, no. 6 (1936): n.p.

44. SSCYD, Q275-1-819; Pan Taifeng, "Youguan 'Dongfang lüxingshe' de shiliao" [Historical data on the Oriental Travel Bureau], in Wu Jingyan, *Chen Guangfu yu Shanghai yinhang*, 222.

45. Tu Zheyin, "Ha'erbin yipie" [A glimpse of Harbin], *Liangyou* [Young companion], no. 39 (1929): 6; "Liaoji fengguang" [The scenery of Liaoning and Jilin], *Liangyou* [Young companion], no. 63 (1931): 38.

46. Ma Hetian, *Dongbei kaocha ji* [Travel notes on investigations in the Northeast] (Nanjing: Xin yaxiya xuehui, 1934).

47. First serialized in the official magazine of the Minsheng Shipping Company, *Xingcha*, in 1931, Lu Zuofu's travel journal was also published in *Qingnian shijie zazhi* (*Youth World Journal*) and *Doubao* in 1932.

48. Zhao Junhao, "Dongbei jihen ji (yi)," 57.

49. Zhao Junhao, 58, 64, 65.

50. Zhao Junhao, 65; Zhao Junhao, "Dongbei jihen ji (er)" [Travelogue of the Northeast (2)], *LZ* 3, no. 8 (August 1929): 33; Zhao Junhao, "Dongbei jihen ji (si)" [Travelogue of the Northeast (4)], *LZ* 3, no. 11 (November 1929): 36.

51. Lu Zuofu, *Dongbei youji*, 38.

52. Xu Hongtao, *Xinan dongbei* [The Southwest and the Northeast] (Hangzhou: Dafeng she, 1935), 86–87.

53. Zhao Junhao, "Dongbei jihen ji (yi)," 61.

54. Ma Hetian, *Dongbei kaocha ji*, 137–38.

55. Wang Yuting, *Dongbei yinxiang ji* [Impressions of the Northeast] (Shanghai: Shixian she, 1933), 79, 95.

56. Zhao Junhao, "Dongbei jihen ji (si)," 37.

57. Lu Zuofu, *Dongbei youji*, 64.

58. Zhao Junhao, "Dongbei jihen ji (si)," 38.

59. Wang Yuting, *Dongbei yinxiang ji*, 99; Zhao Junhao, "Dongbei jihen ji (wu)" [Travelogue of the Northeast (5)], *LZ* 3, no. 12 (December 1929): 25.

60. Xu Hongtao, *Xinan dongbei*, 54, 58.

61. Xu Hongtao, 59–60.

62. Lu Zuofu, *Dongbei youji*, 67.

63. Zhao Junhao, "Dongbei jihen ji (yi)," 60; Lu Zuofu, *Dongbei youji*, 22–23.

64. Xu Hongtao, *Xinan dongbei*, 93; Wang Yuting, *Dongbei yinxiang ji*, 137; Ma Hetian, *Dongbei kaocha ji*, 131.

65. Lu Zuofu, *Dongbei youji*, 29.

66. Xu Hongtao, *Xinan dongbei*, 98.

67. Ma Hetian, *Dongbei kaocha ji*, 136.

68. Ma Hetian, 12; Xu Hongtao, *Xinan dongbei*, 11–12.

69. Zhao Junhao, "Dongbei jihen ji (si)," 39.

70. Zhao Junhao, "Dongbei jihen ji (yi)," 64.

71. Zhao Junhao, 61.

72. Lu Zuofu, *Dongbei youji*, 44–45.

73. Zhao Junhao, "Dongbei jihen ji (yi)," 64–65; Zhao Junhao, "Dongbei jihen ji (si)," 34.

74. Ma Hetian, *Dongbei kaocha ji*, 139, 142–143.

75. Lu Zuofu, *Dongbei youji*, 32.

76. Zhao Junhao, "Dongbei jihen ji (yi)," 66.

77. Zhao Junhao, "Dongbei jihen ji (si)," 35.

78. Zhao Junhao, "Dongbei jihen ji (si)."

79. Zhao Junhao, 35.

80. For example, *Liangyou* featured the border between Manchuria and Russia in its regular travel column, *Liangyou* Readers' Travel Train, in 1936. Han Qinghui, "Ri'e zhanzheng de diyi zhanxian" [The front line of the Russo-Japanese War] *Liangyou* [Young companion], no. 114 (1936): 34–35.

81. Emma Jinhua Teng, *Taiwan's Imagined Geography: Chinese Colonial Travel Writing and Pictures, 1683–1893* (Cambridge, MA: Harvard University Asia Center, 2004).

82. Andrew Morris, "Taiwan's History: An Introduction," in *The Minor Arts of Daily Life: Popular Culture in Taiwan*, ed. Andrew Morris, David Jordan, and Marc Moskowitz (Honolulu: University of Hawai'i Press, 2004), 3–32.

83. For a discussion of the concept of *wangguo*, see Rebecca E. Karl, *Staging the World: Chinese Nationalism at the Turn of the Twentieth Century* (Durham, NC: Duke University Press, 2002), 15–16; Teng, *Taiwan's Imagined Geography*, 238.

84. Teng, *Taiwan's Imagined Geography*, 247.

85. Edward Owen Rutter, *Through Formosa: An Account of Japan's Island Colony* (London: T. F. Unwin, 1923), 15.

86. When the British novelist and travel writer Edward Owen Rutter and his wife journeyed through Taiwan in 1921, a Japanese civil servant sent by the governor-general of Taiwan accompanied them. Rutter, *Through Formosa*, 15–16. Alice Ballantine Kirjassoff, the contributor of the only essay on Taiwan published in *National Geographic* before 1945, was the wife of Max David Kirjassoff, the US consul to Taipei in 1916–19. Alice Ballantine Kirjassoff, "Formosa the Beautiful," *National Geographic Magazine* 37, no. 3 (March 1920): 246–92.

87. As the president of the University of Wichita, Harold W. Foght was invited by the Imperial Department of Education in Japan to lecture there in 1928. He and his wife, Alice Robbins Foght, published a book about their journey, which included chapters about their travels in Taiwan. Harold Foght and Alice Foght, *Unfathomed Japan: A Travel Tale in the Highways and Byways of Japan and Formosa* (New York: Macmillan, 1928). For an anthropologist's account of Taiwan, see, for example, Janet B. Montgomery, *Among the Head-Hunters of Formosa* (London: T. F. Unwin, 1922).

88. Cangjiang (Liang Qichao), "You Tai di'er xin" [The second letter from my trip to Taiwan], *Guofengbao* [National trends] 2, no. 7 (1911): 4.

89. Jiangsu shengzhang gongshu shiye ke [The department of industry, Jiangsu governor's office], ed., *Canguan Taiwan quanye gongjin hui baogaoshu* [The report on our visit to the Taiwan Industrial Exhibition] (Shanghai: Jiangsu shengzhang gongshu shiye ke, 1917). Wang Yang, "Taiyou riji" [Travel journal of Taiwan], in *Taiwan shicha baogao shu* [Travel report of Taiwan], 2nd ed. (Shanghai: Zhonghua shuju, 1928).

90. The Japanese authorities became suspicious of Huang's group and sent military police to question them when they arrived in Takao (Kaohsiung). Coincidentally, Huang would later serve briefly as the mayor of Kaohsiung in 1949. Huang Qiang, *Taiwan biefu hongxue lu* [Travelogue of Taiwan] (Hong Kong: Shangwu yinshu guan, 1928), 90.

91. Jiang Yong, "Taiwan banyue ji" [Travelogue of my half-month trip to Taiwan], *LZ* 4, no. 8 (August 1930): 1–18.

92. Even though Jiang Kanghu was better known as the politician who founded China's first socialist party and later an official in Wang Jingwei's puppet government during the Second Sino-Japanese War, his academic career in North America, which

included a professorship at the University of California at Berkeley in 1914–20 and a stint as a Chinese consultant for the Library of Congress, provided him with international recognition and greater geographical mobility in the 1930s. Jiang Kanghu, *Taiyou zhuiji* [Taiwan travelogue] (Shanghai: Zhonghua shuju, 1935), 4.

93. For an itinerary, see Japan Tourist Bureau, ed., "A Suggested Itinerary for Taiwan Travel," in *Ryotei to hiyō gaisan Taishō 12-nen* (Tokyo: Japan Tourist Bureau, 1923), 287–95, cited in McDonald, *Placing Empire*, 60. On the recommended itineraries from British and US publishers, see Basil Hall Chamberlain and W. B. Mason, *A Handbook for Travellers in Japan (Including Formosa)*, 9th ed. (London: John Murray, 1913), 526–32; Thomas Philip Terry, *Terry's Japanese Empire* (Boston: Houghton Mifflin, 1928), 761–91. For an example of the Chinese travel experience, see Jiang Yong, "Taiwan banyue ji," 1–18.

94. Terry, *Terry's Japanese Empire*, 762; Japanese Department Railways and Frederic De Garis, *The Hot Springs of Japan (and the Principal Cold Springs) Including Chosen (Korea) Taiwan (Formosa) South Manchuria* (Tokyo: s.n., 1922), 428.

95. Kihinkai [Welcome Society], *A Guide-Book for Tourists in Japan*, 5th ed. (Tokyo: s.n., 1910), 205.

96. Foght and Foght, *Unfathomed Japan*, 332.

97. Kirjassoff, "Formosa the Beautiful," 265–72.

98. McDonald, *Placing Empire*, 64–65.

99. Cangjiang [Liang Qichao], "You Tai diyi xin" [The first letter from my trip to Taiwan], *Guofengbao* 2, no. 7 (1911): 1–3.

100. Shi Jingchen, "Kunying riji" [Kunying diary], in *Taiwan youji* [Taiwan travelogues], ed. Taiwan Bank (Taipei: Taiwan yinhang, 1960), 41.

101. Qiu Wenluan et al., *Taiwan lüxing ji* [Travel accounts of Taiwan] (Taipei: Taiwan Yinhang, 1965); Shao Yaonian, "Taiwan lüxing zaji" [Miscellany of my travel to Taiwan], *Guangdong nonglin yuebao* [Canton agriculture and forest monthly] 2, no. 2 (1917): 109–12; Xu Shizhuang, "Shu Taiwan lüxing" [My travels to Taiwan], *Xuesheng* [Students] 8, no. 12 (1921): 67–68.

102. "Guanlan gongjinhui riji" [Travel journal of visiting the industrial exhibition], *Shiye huibao* [Reports on industries] 1, no. 3 (1916): 16.

103. Huang Qiang, *Taiwan biefu hongxue lu*, 66.

104. Cangjiang [Liang Qichao], "You Tai diliu xin" [The sixth letter from my trip to Taiwan], *Guofengbao* 2, no. 9 (1911): 1.

105. Jiang Yong, "Taiwan banyue ji," 3–4, 8.

106. Jiang Yong, 2.

107. Foght and Foght, *Unfathomed Japan*, 332; Kirjassoff, "Formosa the Beautiful," 272.

108. Huang Qiang, *Taiwan biefu hongxue lu*, 130.

109. Wang Yang, "Taiyou riji," 168.

110. Teng, *Taiwan's Imagined Geography*, 122–48.

111. Kirjassoff, "Formosa the Beautiful," 287; Foght and Foght, *Unfathomed Japan*, 403–4.

112. Foght and Foght, *Unfathomed Japan*, 396.

113. Terry, *Terry's Japanese Empire*, 762.

114. Kirjassoff, "Formosa the Beautiful," 275, 283–85, 287.

115. Huang Qiang, *Taiwan biefu hongxue lu*, 122.

116. Huang Qiang, 130.

117. Huang Qiang, 134–42.

118. Kirjassoff, "Formosa the Beautiful," 283. For a similar judgement, see Foght and Foght, *Unfathomed Japan*, 333.

119. Rutter, *Through Formosa*, 89–90.

120. Rutter, 93, 224; Foght and Foght, *Unfathomed Japan*, 348, 411.

121. Huang Qiang, *Taiwan biefu hongxue lu*, 80.

122. Cangjiang [Liang Qichao], "You Tai disan xin" [The third letter from my trip to Taiwan], *Guofengbao* 2, no. 7 (1911): 6.

123. Jiang Yong, "Taiwan banyue ji," 2.

124. Shi Jingchen, "Kunying riji," 50–51.

125. This analogy was prevalent in early twentieth-century China. For example, in his poem "Song of Seven Sons" written in 1925, Wen Yiduo compared the seven places taken over by foreign powers from China—Macao, Hong Kong, Taiwan, Weihaiwei, Canton Bay, Kowloon, and Dalian and Lüshun—to seven children forced to leave their mother. With seven verses devoted to each of them, they all end with the same cry: "Mother! I want to come home, Mother!"

126. Jiang Kanghu, *Taiyou zhuiji*, 75.

127. Jiang Yong, "Taiwan banyue ji," 2.

128. Wang Yang, "Taiyou riji," 183.

129. McDonald, *Placing Empire*, 135–59.

130. Wang Yang, "Taiyou riji," 188.

131. E. Patricia Tsurumi, *Japanese Colonial Education in Taiwan, 1895–1945* (Cambridge, MA: Harvard University Press, 1978), 45.

132. Jiang Yong, "Taiwan banyue ji," 4.

133. Jiang Yong.

134. Jiang Kanghu, *Taiyou zhuiji*, 18–24.

135. Jiang Kanghu, 19.

136. Jiang Kanghu, 10.

137. "Lüxing zazhi zhengwen teji" [Special column for *China Traveler*], *LZ* 20, no. 1 (January 1946): 1.

138. Wang Huimin, ed., *Xin dongbei zhinan* [Guidebook to the new Northeast] (Chongqing: Shangwu yinshu guan, 1946). Advertisement of CTS facilities, *LZ* 21, no. 2 (February 1947): n.p.

139. Tang Weibin, "Shengli hou zhi Zhongguo lüxingshe" [The China Travel Service after victory], *LZ* 22, no. 4 (April 1948): 81–83.

140. Tang Weibin, 82–83.

141. Tang Weibin, 83.

142. Steven Phillips, *Between Assimilation and Independence: The Taiwanese Encounter with Nationalist China, 1945–50* (Stanford, CA: Stanford University Press, 2003), 64–88.

143. For some examples of Taiwan travelogues in *China Traveler*, see Zhu Mei, "Dong Taiwan huanyou ji" [A tour of eastern Taiwan], *LZ* 21, no. 6 (June 1947): 21–25; Zhang Shichao, "Dong Taiwan lücheng" [An itinerary of eastern Taiwan], *LZ* 23, no. 3 (March 1949): 23–29.

144. Chen Qiying, *Taiwan lansheng* [Scenic beauty in Taiwan] (Shanghai: Zhongguo lüxingshe, 1948).

145. After Japan's defeat, the Taiwan branch of the Japan Travel Bureau was taken over by the Taiwan Travel Service (*Taiwan lüxingshe*) and continued to provide travel services in Taiwan. Some of the earliest postwar guidebooks on Taiwan were published by this agency. Taiwan lüxingshe [Taiwan Travel Service], ed., *Taiwan fengjing* [Taiwanese scenery] (Taipei: Taiwan lüxingshe, 1946); Taiwan lüxingshe, ed., *Taiwan lüxing zhinan* [Guidebook to Taiwan] (Taipei: Taiwan lüxingshe, 1947). As for group tours to Taiwan, the Unison Travel Club organized trips during this era. Xu Renhan, "Taiyou guan'gan" [My impression of Taiwan], *LZ* 23, no. 1 (January 1949): 65.

146. Huang Yingzhe, *Qu Ribenhua; zai Zhongguohua: zhanhou Taiwan wenhua chongjian, 1945–1947* [De-Japanification, re-Sinification: Cultural reconstruction in postwar Taiwan] (Taipei: Maitian, 2007).

147. Chen Qiying, "Taiwan huanyou ji" [A circumnavigation of Taiwan], *LZ* 20, no. 9 (September 1946): 5.

148. Wu Chenyi, "Taiwan youxue" [Some travel impressions of Taiwan], *LZ* 22, no. 9 (September 1948): 21.

149. Chen Qiying, "Taiwan huanyou ji," 5; Wu Shenyi, "Taiwan youxue," 21.

150. Wu Chenyi, "Taiwan youxue," 26.

151. Chen Qiying, "Taiwan huanyou ji (yi)," 1.

152. Chen Qiying.

153. Cai Yumen, "Taiwan jihen ji" [Records of my travels in Taiwan], *LZ* 21, no. 4 (April 1947): 44; Guo Zhusong, "Zoufang Taiwan (shangpian)" [A visit to Taiwan (part 1)], *LZ* 21, no. 10 (October 1947): 4.

154. Kaiming, "Taiwan de lunkuo" [An outline of Taiwan], *LZ* 20, no. 2 (February 1946): 53; Dequn, "Taiwan yiyue" [A month in Taiwan], *LZ* 20, no. 5 (May 1946): 51.

155. Kaiming, "Taiwan de lunkuo," 53; Dequn, "Taiwan yiyue," 51.

156. Chen Qiying, *Taiwan lansheng*, 2.

157. Kaiming, "Taiwan de lunkuo," 55.

158. Chen Qiying, "Taiwan huanyou ji (yi)," 8.

159. Ju Xiaoming, "Taiwan dizhi," *LZ* 20, no. 1 (January 1946): 13.

160. Chen Qiying, "'Taiwan huanyou ji (er)" [A circumnavigation of Taiwan (2)], *LZ* 20, no. 10 (October 1946): 21.

161. Guo Zhusong, "Zoufang Taiwan (xiapian)" [Visiting Taiwan (2)], *LZ* 21, no. 11 (November 1947): 30.

162. Guo Zhusong; Cai Yumen, "Taiwan jihen ji," 43.

163. Chen Qiying, "Taiwan huanyou ji (er)," 22.

164. Yufeng, "Jiadao nongyin zhong: youshang Taibei shi" [A tour of Taipei in the shade of trees], *LZ* 20, no. 7 (July 1946): 43–45.

165. Guo Zhusong, "Zoufang Taiwan (xiapian)," 29.

166. Guo Zhusong.

167. Guo Zhusong.

168. Dequn, "Taiwan yiyue," 52.

169. Guo Zhusong, "Zoufang Taiwan (shangpian)," 9.

170. Guo Zhusong, 29, 32.

171. Kaiming, "Taiwan de lunkuo," 56; Chen Qiying, "Taiwan huanyou ji (er)," 27.

172. For a discussion of the effect of the selection of the "Eight Views and Twelve Famous Sites" of Taiwan, see Chiang Chu-shan, *Daoyu fushihui: Rizhi Taiwan de*

dazhong shenghuo [An ukiyo-e of an island: Ordinary life in Taiwan under Japanese rule] (Taipei: Weilan wenhua chubanshe, 2014), 165–74.

173. Cai Yumen, "Taiwan jihen ji," 43–49.

174. When Jiang Yong visited the Sun-Moon Lake in 1929, he noted that there was no railway or motorway for the last stretch of the trip and travelers needed to travel by handcars on a track to reach the lake. Jiang Yong, "Taiwan banyue ji," 15.

175. Zhang Qiqu, "Taiwan you cheng ji (xiapian)" [Touring Taiwan (part 2)], *LZ* 22, no. 4 (April 1948): 39.

176. For the travel essays in *China Traveler* focusing entirely on the lake, see He Mingxian, "'Riyuetan' fengjing xian" [The scenery of the Sun-Moon Lake], *LZ* 20, no. 11 (November 1946): 17–21; Wu Silan, "Yuanye you riyuetan" [Touring the Sun-Moon Lake], *LZ* 23, no. 4 (April 1949): 46–47.

177. He Mingxian, "'Riyuetan' fengjing xian," 17; Wu Silan, "Yuanye you riyuetan," 46; Zhang Shichao, "Taizhong xingjiao" [A trip to Taizhong], *LZ* 22, no. 7 (July 1948): 56.

178. He Mingxian, "'Riyuetan' fengjing xian," 17; Guo Zhusong, "Zoufang Taiwan (shangpian)," 10.

179. He Mingxian, "'Riyuetan' fengjing xian," 17.

180. Wu Chenyi, "Taiwan youxue," 25.

181. Zhang Qiqu, "Taiwan you cheng ji," 38.

182. Chen Qiying, "Taiwan huanyou ji (er)," 22.

183. For a discussion of the term Takasago-zoku, see McDonald, *Placing Empire*, 122.

184. Chen Qiying, "Taiwan huanyou ji (er)," 22.

185. Zhang Qiqu, "Taiwan you cheng ji," 38.

186. He Mingxian, "'Riyuetan' fengjing xian," 20.

187. Zhang Qiqu, "Taiwan you cheng ji," 38.

188. Xu Renhan, "Taiyou guan'gan," 67.

189. Guo Zhusong, "Zoufang Taiwan (shangpian)," 12.

190. Guo Zhusong, "Zoufang Taiwan (xiapian)," 30.

191. Guo Zhusong, "Zoufang Taiwan (shangpian)," 11.

192. Xu Yinxiang, "Dong Taiwan xinjiang" [Travel in eastern Taiwan], *LZ* 23, no. 7 (July 1949): 53–54.

Conclusion

1. "Lüxing shiye guowai xuanchuan diyi sheng" [Our first overseas promotional mission of the tourism industry], *LZ* 22, no. 3 (March 1948): n.p.

2. Shelley Baranowski et al., "Discussion: Tourism and Empire," *Journal of Tourism History* 7, no. 1–2 (2015): 100–130.

3. *The Constitutional Compact, i.e., the Amended Provisional Constitution of the Republic of China* (Peking: Daily News, May 1, 1914), chapter 1, article 3.

4. Anne E. Gorsuch and Diane P. Koenker, eds. *Turizm: The Russian and East European Tourist Under Capitalism and Socialism* (Ithaca, NY: Cornell University Press, 2006); Koenker, *Club Red*.

5. Andrea Colantonio and Robert B. Potter, *Urban Tourism and Development in the Socialist State: Havana during the "Special Period"* (Aldershot, UK: Ashgate, 2006), 27–28.

6. She Guitang, "Lüxing xiangdao" [Tour guide], *LZ* 24, no. 6 (June 1950): 16–17.

7. In Hong Kong, the CTS branch was taken over by the People's Liberation Army in 1952, while Chen Guangfu reregistered CTS in Taiwan in the early 1950s. Zheng Yijun, *Chuanqi rensheng: ji Cai Fujiu zouguo de lu* [Legendary life: Cai Fujiu's path] (Hong Kong: Haifeng chubanshe, 1997), 208–43.

8. For a definition of "tourist diplomacy," see David Airey and King Chong, *Tourism in China: Policy and Development Since 1949* (New York: Routledge, 2011), 26.

9. For example, in a travel pictorial published in Hong Kong in the mid-1950s, tourist sites widely celebrated in the Republican era, such as the Great Wall and the Forbidden City in Beijing, the Central Square in Dalian, the Great Goose Pagoda in Xi'an, and the Bund in Shanghai, were juxtaposed with new constructions of the socialist state. *Zhongguo lüxing huace* [Traveling in China] (Hong Kong: Jingji daobao she, 1956).

10. Zhang Yuanfeng, "Xihu shi shuyu laodong renmin de" [West Lake belongs to the working people], *LZ* 25, no. 6 (June 1951): 23–26.

11. Pal Nyiri, *Scenic Spots: Chinese Tourism, the State, and Cultural Authority* (Seattle: University of Washington Press, 2006), 3–25.

12. Ryan and Gu, *Tourism in China*, 18–29.

13. Zhang Yuanfeng, *Dao Xinjiang qu* [Going to Xinjiang] (Shanghai: Zhongguo lüxingshe, 1952); Wei Lihai, *Menggu renmin gongheguo lüxing ji* [Travelogue of the People's Republic of Mongolia] (Shanghai: Zhongguo lüxingshe, 1952); Mingzhe, *Neimeng sanji* [Sketches of Inner Mongolia] (Shanghai: Zhongguo lüxingshe, 1952); Wei Lihai, *Zhandou zai kenteshan shang* [Battles on Mount Khentii] (Shanghai: Zhongguo lüxingshe, 1953); Su Lan, *Kangzang suijun ji* [My journey with the army to Tibet] (Shanghai: Zhongguo lüxingshe, 1953).

14. For an example, see "Xin Xinjiang" [New Xinjiang] and "Gaoyuan shang meili de gucheng Lasa" [Lhasa, the beautiful ancient town on the plateau], in *Zhongguo lüxing huace*, 38–41.

15. Nyiri, *Scenic Spots*, 15.

16. On the case of mainland China's outbound tourism to Taiwan, see Yingzhi Guo et al., "Tourism and Reconciliation between Mainland China and Taiwan," *Tourism Management* 27, no. 5 (2006): 997–1005.

Bibliography

Archival Sources

Shanghai shangye chuxu yinhang dang'an [The archives of Shanghai Commercial and Saving Bank], 1915–50. Shanghai Municipal Archives.

Zhongguo lüxingshe zongshe dang'an [The archives of the Head Office of China Travel Service], 1923–54. Shanghai Municipal Archives.

Newspaper and Periodicals

Chinese

Beijing daxue yanjiu suo guoxue men zhoukan [Weekly journal of the National Studies Department at Beijing University]

Chenbao [Morning news]

Chenbao fukan [Supplement to *Chenbao* newspaper]

Dazhong huabao [The cosmopolitan]

Dongfang zazhi [Eastern miscellany]

Guangdong nonglin yuebao [Canton agriculture and forest monthly]

Guangzhoushi zhengfu shizheng gongbao [Guangzhou municipal gazette]

Guofengbao [National trends]

Guowen zhoubao [Kuowen weekly]

Guoxue [National studies]

Guoyu zhoukan [National language weekly]

Haiguang [Internal journal of Shanghai Bank]

Jiangtong daxue rikan [Communications University daily news]

Jiaotong gongbao [Bulletin for transportation regulations]

Jiaoyu zhoubao [Education weekly]

Jinghu huhangyong tielu rikan [Daily of the Shanghai-Nanjing and Shanghai-Hangzhou-Ningbo Railways]

Jingwu congbao [Jingwu journal]

Jingwu huabao [Jingwu pictorial]

Keda zazhi [Kodak magazine]

Liangyou [Young companion]

Libailiu [Saturday]

Lüguang [Light of travel]

Lüxing zazhi [China traveler]
Meizhou pinglun [Weekly review]
Neizheng gongbao [Bulletins of the Ministry of the Interior]
Qinghua zhoukan [Qinghua University weekly]
Qingnian jinbu [Youth progress]
Qingnian shijie zazhi [Youth world journal]
Sanmin huakan [Pictorial of the Three Principles of the People]
Shenbao [Shanghai news]
Shenbao yuekan [Shenbao monthly]
Shengli [Vitality]
Shenhai tielu yuekan [Shenyang-Hailong Railway monthly]
Sheying huabao [Photography pictorial]
Shidai [Modern miscellany]
Shijie huabao [World pictorial]
Shiye huibao [Reports on industries]
Tiedao banyue kan [Railway bimonthly]
Tiedao gongbao [Railway bulletins]
Tielu gongbao: Huning Huhangyong xian [Railway gazetteer: Shanghai-Nanjing and Shanghai-Hangzhou-Ningbo lines]
Tielu yuekan: Beining xian [Railway monthly: The Beijing-Mukden Railway]
Tielu zazhi [Railway magazine]
Tuhua shibao [Pictorial newspaper]
Wenwu [Cultural relics]
Xiaoshuo yuebao [Short story monthly]
Xibei [The Northwest]
Xihu bolanhui rikan tekan [Special daily for West Lake Exposition]
Xingcha [Xingcha weekly]
Xinsheng zhoukan [New life weekly]
Xiuxiang xiaoshuo [Fiction illustrated]
Xuesheng [Students]
Yanda nongxun [Agricultural bulletin of Yenching]
Yanjing daxue xiaokan [Yenching University weekly]
Yinhang yuekan [Banker's magazine]
Yousheng [Unison]
Yousheng lüxing tuan yuekan [Unison Travel Club monthly]
Yuehua [Crescent China]
Zhongfa jiaoyujie [Education in China and France]
Zhonghang yuekan [Bank of China monthly]
Zhonghua [China pictorial]

Zhonghua gongchengshi xuehui huibao [Journal of associations of Chinese engineers]
Zhongyang huakan [Central pictorial]

English

China Press
China Weekly Review
Chinese Students' Monthly
Peking Daily News
Peking Gazette

Books and Articles

Airey, David, and King Chong. *Tourism in China: Policy and Development Since 1949*. New York: Routledge, 2011.

Albright, Horace M., and Robert Cahn. *The Birth of the National Park Service: The Founding Years, 1913–1933*. Chicago: Howe Brothers, 1985.

Anderson, Benedict. *Imagined Communities: Reflections on the Origin and Spread of Nationalism*. New York: Verso, 1991.

Bai Dizhou. "Ji Liu Bannong xiansheng zhi bingyin" [On the cause of the illness of Mr. Liu Bannong]. *Guoyu zhoukan* [National language weekly] 6, no. 151 (August 1934): 43–44.

Balachandran, Sanchita. "Object Lessons: The Politics of Preservation and Museum Building in Western China in the Early Twentieth Century." *International Journal of Cultural Property*, no. 14 (2007): 1–32.

Ban Gong. "Kunming de chaguan" [The teahouses of Kunming]. *LZ* 13, no. 7 (July 1939): 27–30.

"Banli xiangshan youlantuan" [The tour to Fragrant Mountain]. *Tiedao banyue* 1, no. 13 (1936): 63.

Baoshi. "Suiyuan youji" [Travelogue of Suiyuan]. *Keda zazhi* 7, no. 4 (1936): 7–9.

Baranowski, Shelley. *Strength through Joy: Consumerism and Mass Tourism in the Third Reich*. New York: Cambridge University Press, 2004.

Baranowski, Shelley, Christopher Endy, Waleed Hazbun, Stephanie Malia Hom, Gordon Pirie, Trevor Simmons, and Eric G. E. Zuelow. "Discussion: Tourism and Empire." *Journal of Tourism History* 7, no. 1–2 (2015): 100–130.

Barlow, Tani. "Introduction: On 'Colonial Modernity.'" In *Formations of Colonial Modernity in East Asia*, edited by Tani Barlow, 1–20. Durham, NC: Duke University Press, 1997.

Barrento, António. "Going Modern: The Tourist Experience at the Seaside and Hill Resorts in Late Qing and Republican China." *Modern Asian Studies* 52, no. 4 (2018): 1089–133.

——. "On the Move: Tourist Culture in China, 1895–1949." DPhil diss., SOAS, University of London, 2012.

"Beijing xueshu tuanti lianhe fandui Ruidian yuanzhengdui bing jiji yanjiu baocun guwu fa" [Beijing academic community objects to the Swedish expedition and actively looks for means to protect antiquities]. *Chenbao*, March 6, 1927.

Beining tielu guanli ju [The Beijing-Mukden Railway], ed. *Beidaihe haibin daoyou* [Tourists' guide to Beidaihe]. Shanghai: Zhongguo lüxingshe, 1935.

"Ben jie zhengqiu hui zhangcheng" [The rules of this recruiting conference]. *Yousheng* (September 1925): n.p.

"Benlu dashi ji" [Big events with the railway]. *Shenhai tielu yuekan* 1, no. 1 (1929): 1–3.

"Bentuan san si yuefen lüxing tongji" [The statistics of the club's tours in March and April]. *Yousheng* (May 1935): 2.

"Bentuan wei jianqin chejia cheng tiedaobu shu" [The club's petition to the Ministry of Railways in order to cut the price for train tickets]. *Yousheng lüxing tuan yuekan* (April 1933): 48–50.

"Bentuan zongzhi" [The principle of this tour]. In "Huabei lüxing zhuanhao" [Special issue on the North China tour]. *Yousheng* (August 1931): 1–3.

Bian Wenjian. "Xibei kaocha ji" [An exploration of the Northwest]. *LZ* 7, no. 5 (May 1933): 57–67.

Bing Xin. *Bing Xin youji* [Bing Xin's travelogue]. Shanghai: Beixin shuju, 1935.

——. *Pingsui yanxian lüxing ji* [Travelogue of my journey along the Ping-Sui Railway]. N.p.: Pingsui tielu guanli ju, 1935.

Borocz, Jozsef. "Travel-Capitalism: The Structure of Europe and the Advent of the Tourist." *Comparative Studies in Society and History* 34, no. 4 (1992): 708–41.

Braae, Christel, and Henning Haslund-Christensen, *Among Herders of Inner Mongolia: The Haslund-Christensen Collection at the National Museum of Denmark*. Aarhus: Aarhus University Press, 2017.

Brook, Timothy. "Communications and Commerce," in *The Cambridge History of China*. Vol. 8, *The Ming Dynasty, 1368–1644, Part 2*, edited by D. Twichett and F. W. Mote, 619–39. Cambridge: Cambridge University Press, 1998.

——. "Guides for Vexed Travelers: Route Books in the Ming and Ch'ing." *Ch'ing-shi wen-t'i* [Late imperial China] 4, no. 5 (1981): 32–76.

Byrnes, Corey. *Fixing Landscape: A Techno-Poetic History of China's Three Gorges*. New York: Columbia University Press, 2018.

Cai Yumen. "Taiwan jihen ji" [Records of my travels in Taiwan]. *LZ* 21, no. 4 (April 1947): 43–49.

Callahan, William, "The Cartography of National Humiliation and the Emergence of China's Geobody." *Public Culture* 21, no. 1 (2009): 141–73.

Cao Jinggao, ed., *Jingsui tielu lüxing zhinan* [Travel guide for the Beijing-Suiyuan Railroad]. Beijing: Jingsui tielu guanli ju, 1916.

Chamberlain, Basil Hall, and W. B. Mason, *A Handbook for Travellers in Japan (Including Formosa)*. 9th ed. London: John Murray, 1913.

Chan, Pedith. "In Search of the Southeast: Tourism, Nationalism, and Scenic Landscape in Republican China." *Twentieth-Century China* 43, no. 3 (2018): 207–31.

Chatterjee, Partha. *The Nation and Its Fragments: Colonial and Postcolonial Histories*. Princeton, NJ: Princeton University Press, 1993.

Chen, Fu Hua. *Between East and West: Life on the Burma Road, the Tibetan Highway, the Ho Chi Minh Trail, and in the United States*. Niwot: University Press of Colorado, 1996.

Chen Cunren. "Huanan lüxing ji" [My travel in South China]. *LZ* 8, nos. 5, 6 (May, June 1934): 5–16; 39–52.

Chen Gengya. *Xibei shicha ji* [Record of investigation in the Northwest]. Shanghai: Shenbao, 1936.

Chen Guangfu. *Chen Guangfu riji* [Chen Guangfu's diary]. Shanghai: Shanghai shudian, 2002.

——. *Chen Guangfu xiansheng yanlun ji* [A collection of Mr. Chen Guangfu's speeches]. Taipei: Shanghai shangye chuxu yinhang, 1970.

——. "Fakan ci" [Inaugural preface]. *LZ* 1, no. 1 (Spring 1927): 1.

Chen Guangyao. *Xijing zhi xiankuang* [Contemporary conditions in Xijing]. (Shanghai: Xijing choubei weiyuanhui, 1933.

Chen Qi, "Ji Guilin zhi xing" [My trip to Guilin]. *LZ* 12, nos. 5, 6 (May, June 1938): 13–20; 55–60.

Chen Qiying. "Taiwan huanyou ji" [A circumnavigation of Taiwan]. *LZ* 20, nos. 9–10 (September, October 1946): 5–9, 21–28.

——. *Taiwan lansheng* [Scenic beauty in Taiwan]. Shanghai: Zhongguo lüxingshe, 1948.

Chen Wanli. "Dunhuang qianfodong gu bihua" [Ancient murals discovered at the Thousand Buddha caves in Dunhuang]. *Liangyou*, no. 37 (1929): 33.

——. "Dunhuang qianfodong huabi liangzhong" [Two murals in the Thousand Buddha caves in Dunhuang]. *Guoxue* 1, no. 2 (1926): n.p.

——. "Meidi touqie Dunhuang bihua de yinmou" [The US imperialists' plot to steal Dunhuang frescoes]. *Wenwu*, no. 1 (1951): n.p.

——. *Xichui bihua ji* [Murals on the western frontiers]. Shanghai: Liangyou tushu yinshua gongsi, 1928.

——. "Xixing riji" [Journal of a westward journey]. *Beijing daxue yanjiu suo guoxue men zhoukan* 1, nos. 1–3, 5, 7, 9 (1925).

——. *Xixing riji* [Journal of a westward journey]. Beijing: Pushe, 1926.

——. *Xixing riji* [Journal of a westward journey]. Lanzhou: Gansu renmin chubanshe, 2000.

Chen Xiangtao [Chen, C.]. "China's Appeal for the Tourist." *China Press*, October 10, 1936.

——. "Zhongguo lüxingshe chuangshe zhaodaisuo zhi zhiqu" [The purports of the CTS guesthouses]. *LZ* 9, no. 9 (September 1935): 5–6.

Chen, Zhihong, "Romance with the Periphery: Chinese Travel Writing about the 'Frontier' during the Nanjing Decade (1927–1937)." *Journal of Tourism History* 8, no. 3 (2016): 215–38.

Chen Yinke. "Chen Yuan *Dunhuang jieyu lu* xu" [Preface to Chen Yuan's *The aftermath of the disaster in Dunhuang*]. In *Dunhuang jieyu lu* [The aftermath of the disaster in Dunhuang], by Chen Yuan, 1:1–2. Beijing: Guoli zhongyang yanjiu yuan lishi yuyan yanjiu suo, 1931.

Chen Yunzhang and Chen Xiachang. *Lushan daoyou* [Guide to Lushan]. 4th ed. Shanghai: Shangwu yinshu guan, 1925.

Chiang Chu-shan. *Daoyu fushihui: Rizhi Taiwan de dazhong shenghuo* [An ukiyo-e of an island: Ordinary life in Taiwan under Japanese rule]. Taipei: Weilan wenhua chubanshe, 2014.

Chin, C. P. "China Travel Service Takes Lead in Promoting Comfort." *China Press*, May 28, 1933.

Chin, Tamara. "The Invention of the Silk Road, 1877." *Critical Inquiry* 40, no. 1 (Autumn 2013): 194–219.

"China-Japan Circular and Overland Tour." *Peking Daily News*, September 30, 1915.

"China-Japan Railway Conference: Celebrations in Tokyo." *Peking Gazette*, April 10, 1917.

"China Travel Service Plans Olympic Tour." *China Press*, May 28, 1936.

"China Travel Service to Conduct Xmas Tour." *China Press*, December 5, 1935.

"China Travel Service to Link Hankow and Canton by Motors." *China Press*, June 3, 1933.

"China Travel Service Tour to Visit Chiang's Birthplace." *China Press*, October 29, 1936.

Chinese Eastern Railway. *North Manchuria and the Chinese Eastern Railway*. Harbin: C.E.R. Printing Office, 1924.

Chongqing Zhongguo yinhang [Bank of China, Chongqing Branch], ed. *Emei shan* [Mount Emei]. Chongqing: Zhongguo yinhang, 1935.

"Chuanjiang lüxingshe kaimu le" [The opening of the Chuangjiang Travel Service]. *Xingcha*, no. 48 (1931): n.p.

"Chuanyou suojian (yi): sanxia tianxian" [Views of Sichuan (part 1): The natural barrier of the Three Gorges]. *Shenbao yuekan* 4, no. 12 (December 1935): n.p.

"Chunyou zhuanhao" [The special number for spring] *LZ* 3, no. 3 (March 1929).

Clark, J. D. *Sketches in and around Shanghai*. Shanghai: Shanghai Mercury, 1894.

Coble, Parks M. *China's War Reporters*. Cambridge, MA: Harvard University Press, 2015.

——. *Chinese Capitalists in Japan's New Order: The Occupied Lower Yangzi, 1937–1945*. Berkeley: University of California Press, 2003.

Colantonio, Andrea, and Robert B. Potter. *Urban Tourism and Development in the Socialist State: Havana during the "Special Period."* Aldershot, UK: Ashgate, 2006.

Cook, Thomas. *Letters from the Sea and from Foreign Lands: Descriptive of a Tour round the World*. London: Thomas Cook & Son, 1873.

Cotes, Everard. *Signs and Portents in the Far East*. New York: G. P. Putnam, 1907.

Culp, Robert. *Articulating Citizenship: Civic Education and Student Politics in Southeastern China, 1912–1940*. Cambridge, MA: Harvard University Asia Center, 2007.

Danielsson, Sarah K. *The Explorer's Roadmap to National-Socialism: Sven Hedin, Geography and the Path to Genocide*. Burlington, VT: Ashgate, 2012.

Deng Hanjun. "Yue Xiang Gui lücheng suoji" [Fragmented recollections of my journey through Guangdong, Hunan, and Guangxi Provinces]. *LZ* 12, no. 11 (November 1938): 37–45.

Deng Shaoqin, ed. *Jindai Chuanjiang hangyun jianzhi* [A brief history of shipping on Chuanjiang]. Chongqing: Chongqing difang shi ziliao zu, 1982.

Dequn. "Taiwan yiyue" [A month in Taiwan]. *LZ* 20, no. 5 (May 1946): 51–52.

Dikötter, Frank. *The Age of Openness*. Hong Kong: Hong Kong University Press, 2008.

Ding Huikang. "Qingdao linzhua" [A glimpse of Qingdao]. *LZ* 5, no. 7 (July 1931): 25–28.

Ding Songchang, ed. *Sannian lai zhi xi'nan gonglu* [The Southwest Motorways over the past three years]. China: s.n., 1940.

Ding Wenjiang. *Youji erzhong* [Two travel accounts]. Shenyang: Liaoning jiaoyu chubanshe, 1998.

"Dingxian cujin pingmin jiaoyü zhi sheshi" [Promotion of mass education in Tinghsien]. *Liangyou*, no. 72 (December 1932): n.p.

Dong, Madeleine Yue. *Republican Beijing: The City and Its Histories*. Berkeley: University of California Press, 2003.

——. "Shanghai's *China Traveler*." In *Everyday Modernity in China*, edited by Madeleine Yue Dong and Joshua L. Goldstein, 195–226. Seattle: University of Washington Press, 2005.

Dott, Brian. *Identity Reflections: Pilgrimages to Mount Tai in Late Imperial China*. Cambridge, MA: Harvard University Asia Center, 2005.

Du, Yongtao. *The Order of Places: Translocal Practices of the Huizhou Merchants in Late Imperial China*. Leiden: Brill, 2015.

Duomo. "Luoyang youji" [Travelogue of Luoyang]. *Keda zazhi* 6, no. 9 (1935): 7–9.

Egan, Susan Chan. *A Latterday Confucian: Reminiscences of William Hung (1893–1980)*. Cambridge, MA: Harvard University Press, 1987.

Epelde, Kathleen R. "Travel Guidebooks to India: A Century and a Half of Orientalism." PhD diss., University of Wollongong, 2004.

Esherick, Joseph. "How the Qing Became China." In *Empire to Nation: Historical Perspectives on the Making of the Modern World*, edited by Joseph Esherick, Hasan Kayali, and Eric Van Young, 229–59. Lanham, MD: Rowman & Littlefield, 2006.

Fabian, Johannes. *Time and the Other: How Anthropology Makes Its Object*. New York: Columbia University Press, 1983.

Fairbank, John King, and Merle Goldman. *China: A New History*. Cambridge, MA: Harvard University Press, 2006.

"Fakan ci" [Foreword to the journal]. *Xibei*, no. 1 (March 1929): 1–2.

Fan, Fa-ti. "Circulating Material Objects: The International Controversy over Antiquities and Fossils in Twentieth-Century China." In *The Circulation of Knowledge between Britain, India, and China: The Early-Modern World to the Twentieth Century*, edited by Bernard Lightman, Gordon McOuat, and Larry Stewart, 209–36. Leiden: Brill, 2013.

——. "How Did the Chinese Become Native? Science and the Search for National Origins in the May Fourth Era." In *Beyond the May Fourth Paradigm: In Search of Chinese Modernity*, edited by Kai-wing Chow, Tze-ki Hon, Hung-Yok Ip, and Don Price, 183–208. Lanham, MD: Lexington Books, 2008.

Fan Changjiang. *Zhongguo de xibei jiao* [China's northwestern corner]. Tianjin: Dagong bao, 1936.

Fan Dimin. "Jinnian de jige guanchao tuan" [This year's tidal bore tours]. *Xihu bolanhui rikan tekan*, no. 11 (1929): 3.

Fang Shan. "Qingzhong" [Green tumulus]. *LZ* 9, no. 5 (May 1935): 63–65.

——. "Yungang shiku si youji" [Travelogue of the Yungang Grottoes]. *Keda zazhi* 5, no. 3 (1934): 7–9.

Fei, Siyen. "Ways of Looking: The Creation and Social Use of Urban Guidebooks in Sixteenth- and Seventeenth-Century China." *Urban History* 37, no. 2 (2010): 226–48.

Fernsebner, Susan. "When the Local is the Global: Case Studies in Early Twentieth-Century Chinese Exposition Projects." In *Expanding Nationalisms at World's Fairs*, edited by David Raizan and Ethan Robby, 173–94. London: Routledge, 2018.

Feuerwerker, Albert. "Economic Trends, 1912–1949." In *The Cambridge History of China*. Vol. 12, *Republican China, 1912–1949*, edited by John K. Fairbank and Denis C. Twitchett, 28–127. Cambridge: Cambridge University Press, 1983.

Field, Andrew. *Shanghai's Dancing World: Cabaret Culture and Urban Politics, 1919–1954*. Hong Kong: Chinese University Press, 2010.

Foght, Harold, and Alice Foght. *Unfathomed Japan: A Travel Tale in the Highways and Byways of Japan and Formosa*. New York: Macmillan, 1928.

Fong, Grace. "Reconfiguring Time, Space, and Subjectivity: Lü Bicheng's Travel Writings on Mount Lu." In *Different Worlds of Discourse: Transformations of Gender and Genre in Late Qing and Early Republican China*, edited by Nanxiu Qian, Grace Fong, and Richard J. Smith, 87–114. Leiden: Brill, 2008.

"Fu Dian xingcheng" [Itineraries to Yunnan]. *LZ* 12, no. 2 (February 1938): 18.

"Fuchun jiang luxingtuan ji" [About the Fuchun River tour]. *LZ* 4, no. 5 (May 1930): 53–58.

Furlough, Ellen. "Making Mass Vacations: Tourism and Consumer Culture in France, 1930s to 1970s." *Comparative Studies in Society and History* 40, no. 2 (April 1998): 247–86.

Gao Bochen. "Sanxia daoguan" [My trip to the Three Gorges]. *LZ* 10, no. 1 (January 1936): 103–12.

Gao Liangzuo. *Xibei suiyao ji* [An inspection tour of the Northwest]. Nanjing: Jianguo yuekan she, 1936.

Gao Shaocong. "Chongqing suoji" [Notes on Chongqing]. *LZ* 14, no. 4 (April 1940): 3–6.

——. "Peidu Chongqing sumiao" [Sketches of the wartime capital Chongqing]. *LZ* 15, no. 2 (February 1941): 3–11.

Garcia, Margarita Diaz-Andreu. *A World History of Nineteenth-Century Archaeology: Nationalism, Colonialism, and the Past*. Oxford: Oxford University Press, 2007.

Ge Suicheng, ed. *Quanguo duhui shangbu lüxing zhinan* [A travel guide to metropolitan cities and treaty ports nationwide]. Shanghai: Zhonghua shuju, 1926.

Ge Zhaoguang. "From the Western Regions to the Eastern Sea: Formations, Methods and Problems in a New Historical World." In *Here in "China" I Dwell: Reconstructing Historical Discourses of China for Our Time*. Translated by Jesse Field and Qin Fang, 172–78. Leiden: Brill, 2017.

"Gedi ziying ji chengban zhaodaisuo" [CTS guesthouses in various locations]. *LZ* 14, no. 6 (June 1940): n.p.

Gellner, Ernest. *Nations and Nationalism*. Ithaca, NY: Cornell University Press, 1983.

Gerth, Karl. *China Made: Consumer Culture and the Creation of the Nation*. Cambridge, MA: Harvard University Press, 2003.

Gilman, LaSelle. "Peiping Guides Cheered by Old City's Political Decline." *China Press*, August 29, 1935.

Goodman, Bryna, and David Goodman, eds. *Twentieth-Century Colonialism and China: Localities, the Everyday, and the World*. Abingdon, UK: Routledge, 2012.

Gordon, Bertram. *War Tourism: Second World War France from Defeat and Occupation to the Creation of Heritage*. Ithaca, NY: Cornell University Press, 2018.

Gorsuch, Anne E., and Diane P. Koenker, eds. *Turizm: The Russian and East European Tourist under Capitalism and Socialism* (Ithaca, NY: Cornell University Press, 2006).

Gottschang, Thomas R., and Diana Lary, eds. *Swallows and Settlers: The Great Migration from North China to Manchuria*. Ann Arbor: Center for Chinese Studies, University of Michigan, 2000.

Gross, Miriam. "Flights of Fancy from a Sedan Chair: Marketing Tourism in Republican China, 1927–1937." *Twentieth-Century China* 36, no. 2 (2011): 119–47.

"Group Tours Gaining Favor among Travel-Minded Chinese." *China Press*, July 8, 1936.

Gu Jiegang. *Wang Tongchun kaifa hetao ji* [The history of Wang Tongchun's efforts in developing the Hetao area]. N.p.: Pingsui tielu guanli ju, 1935.

——. "Xuwen" [Preface]. In *Xixing riji*, by Chen Wanli, 7. Lanzhou: Gansu renmin chubanshe, 2000.

Gu Jungu. "Guiyang zaxie" [A miscellany of Guiyang]. *LZ* 13, no. 3 (March 1939): 5–12.

Gu Mengwu. "Xianhua zhanshi shoudu" [On the wartime capital]. *LZ* 13, no. 11 (November 1939): 9–11.

Gu Zhizhong and Lu Yi, *Dao Qinghai qu* [Head to Qinghai]. Shanghai: Shangwu yinshu guan, 1934.

"Guangrong zhi ye" [Customer letters]. *LZ* 11, no. 3 (March 1937): n.p.

"Guangzhou shi xiangdao youlan guize" [Regulations of tour guides in Guangzhou]. *Guangzhoushi zhengfu shizheng gongbao*, no. 499 (1935): 10.

"Guanlan gongjinhui riji" [Travel journal of visiting the industrial exhibition]. *Shiye huibao* 1, no. 3 (1916): 14–18.

"Guilin shanshui jia tianxia" [Guilin's scenery is the best under heaven]. *Liangyou*, no. 77 (1933): 23–24.

Guo, Yingzhi, Samuel Seongseop Kim, Dallen J. Timothy, and Kuo-Ching Wang. "Tourism and Reconciliation between Mainland China and Taiwan." *Tourism Management* 27, no. 5 (2006): 997–1005.

Guo Zhusong. "Chengdu shengji daoyou" [Guide to the famous sites in Chengdu]. *LZ* 16, no. 12 (December 1942): 19–26.

——. "Zoufang Taiwan" [Visiting Taiwan]. *LZ* 21, nos. 10, 11 (October, November 1947): 1–12; 29–36.

Guomin zhengfu jiaotongbu [Ministry of Transportation and Communication]. "Lüxingye zhuce zanxing tiaoli" [The provisional regulations regarding registration in the tourism industry]. *Jiaotong gongbao* 1, no. 2 (1927): 69–71.

Halbertsma, Tjalling H. F. *Early Christian Remains of Inner Mongolia*. Leiden: Brill, 2015.

Hamilton, Jill. *Thomas Cook: The Holiday Maker*. Stroud, UK: Sutton, 2005.

Han Qinghui. "Ri'e zhanzheng de diyi zhanxian" [The front line of the Russo-Japanese War]. *Liangyou*, no. 114 (1936): 34–35.

"Hangkong yu gonglu" [Aviation and highways]. *Liangyou*, no. 78 (1933): 2–3.

Hansen, Valerie. *The Silk Road: A New History*. Oxford: Oxford University Press, 2012.

Hargett, James. *Jade Mountains and Cinnabar Pools: The History of Travel Literature in Imperial China*. Seattle: University of Washington Press, 2018.

——. *Stairway to Heaven: A Journey to the Summit of Mount Emei*. Albany: State University of New York Press, 2006.

Harrison, Henrietta. *The Making of the Republican Citizen: Political Ceremonies and Symbols in China, 1911–1929*. Oxford: Oxford University Press, 2000.

Härtel, Herbert, and Marianne Yaldiz, eds. *Along the Ancient Silk Routes: Central Asian Art from the West Berlin State Museums*. New York: Metropolitan Museum of Art, 1982.

Haslund, Henning. *Men and Gods in Mongolia*. New York: Adventures Unlimited Press, 2000.

He Mingxian. "'Riyuetan' fengjing xian" [The scenery of the Sun-Moon Lake]. *LZ* 20, no. 11 (November 1946): 17–21.

He Yongji. "Yindu jiyou" [Travelogue of India]. *LZ* 17, no. 9 (September 1943): 100–102.

Hedin, Sven. *Across the Gobi Desert*. New York: E. P. Dutton, 1932.

——. *History of the Expedition in Asia 1927–1935*. Part 1. Translated by Donald Burton. Stockholm: Elanders boktryckeri aktiebolag, 1943.

——. *My Life as an Explorer*. Translated by Alfhild Huebsch. New York: Garden City, 1925.

——. *Riddles of the Gobi Desert*. London: Routledge, 1933.

Hobsbawm, Eric. *The Age of Capital, 1848–1875*. London: Abacus, 1995.

Hopkirk, Peter. *Foreign Devils on the Silk Road: The Search for the Lost Cities and Treasures of Chinese Central Asia*. London: John Murray, 1980.

Hu Shiquan. "Jingdian gonglu zhoulan tuan suizheng ji" [Tour along the Jingdian highway]. *LZ* 11, nos. 6, 8 (June, August 1937): 21–38, 53–62.

Hu Shiyuan. *Xibei daoyou* [Travel guide to the Northwest]. Shanghai: Zhongguo lüxingshe, 1935.

"Huanan lüxing, Yousheng lüxingtuan tonggao" [South China tour, a notice for the Unison Travel Club]. *Shenbao*, April 3, 1937.

Huang Boqiao. "Daoyou yu aiguo" [Tourism and patriotism]. *LZ* 10, no. 1 (January 1936): 3–4.

Huang Fensheng. *Bailingmiao xunli* [March to Bailingmiao]. Shanghai: Shangwu yinshu guan, 1936.

Huang Huaju. "Dao Xibei qu" [Go to Northwest]. *Sanmin huakan*, no. 8 (1934): 2.

Huang Lisheng and Ge Mo'an, eds. *Kunming daoyou* [Guidebook to Kunming]. Kunming: Zhongguo lüxingshe, 1944.

Huang Qiang. *Taiwan biefu hongxue lu* [Travelogue of Taiwan]. Hong Kong: Shangwu yinshu guan, 1928.

Huang Weiqing. "Chongdao guilai" [Returning from Chongming Island]. *Jingwu huabao* 2, no. 10 (1929): 3.

Huang Wenbi. *Huang Wenbi mengxin kaocha riji (1927–1930)* [Huang Wenbi's journal of the exploration in Inner Mongolia and Xinjiang (1927–1930)]. Beijing: Wenwu chubanshe, 1990.

——. "Xibei kexue kaochetuan zhi gongzuo ji qi zhongyao faxian" [Work of the scientific mission to the Northwest and its important discoveries]. *Yanjing xuebao* [Yenching journal of Chinese studies], no. 8 (December 1930): 1610–14.

Huang Yingzhe. *Qu Ribenhua; zai Zhongguohua: zhanhou Taiwan wenhua chongjian, 1945–1947* [De-Japanification, re-Sinification: Cultural reconstruction in postwar Taiwan]. Taipei: Maitian, 2007.

Huang Yubin. "Dao Bailingmiao qu" [Go to Bailingmiao]. *LZ* 9, no. 6 (June 1935): 13–25.

Huang Zhuoqiu. "Kunming fengjing xian" [Scenery of Kunming]. *LZ* 15, no. 2 (February 1941): 13–19.

Huanghai Sanren. *Huangshan zhinan* [Guide to Huangshan]. Shanghai: Shangwu yinshu guan, 1929.

"Huangshan lüshe" [The Huangshan guesthouse]. *LZ* 9, no. 9 (September 1935): 29–32.

"Huiyi jilu" [Meeting record]. *Yousheng*, no. 7 (1932): 3–4.

Huning huhangyong lianglu biancha ke [The compilation department of the Shanghai-Nanjing and Shanghai-Hangzhou-Ningbo Railroads], ed. *Huning huhangyong tielu lüxing zhinan* [Travel guide for the Shanghai-Nanjing and Shanghai-Hangzhou-Ningbo Railroads]. Shanghai: Huning huhangyong tielu guanli ju, 1918.

Israel, John. *Lianda: A Chinese University in War and Revolution*. Stanford, CA: Stanford University, 1999.

Jacobs, Justin M. *Xinjiang and the Modern Chinese State*. Seattle: University of Washington Press, 2016.

Japanese Department Railways and Frederic De Garis. *The Hot Springs of Japan (and the Principal Cold Springs) Including Chosen (Korea) Taiwan (Formosa) South Manchuria*. Tokyo: s.n., 1922.

Ji, Zhaojin. *A History of Modern Shanghai Banking: The Rise and Decline of China's Finance Capitalism*. Armonk, NY: M. E. Sharpe, 2003.

"Ji Xibei kexue kaochatuan" [On the scientific mission to the Northwest]. *Guowen zhoubao* 6, no. 6 (February 3, 1929): 1–9.

Jiang Huaiqing. "Canguan shijie yundonghui qu" [Let's go tour the Olympic Games]. *LZ* 10, no. 3 (March 1936): 119–23.

Jiang Kanghu. *Taiyou zhuiji* [Taiwan travelogue]. Shanghai: Zhonghua shuju, 1935.

Jiang Weiqiao. "Moganshan jiyou" [Travelogue of Moganshan]. *Xiaoshuo yuebao* 11, no. 9 (1920): 1–3.

——. "Taishan jiyou" [Travelogue of Mount Tai]. *Xiaoshuo yuebao* 6, no. 10 (1915): 1–3.

——. "Wuyue yu sida mingshan" [Five sacred and four famous mountains]. *LZ* 10, no. 1 (January 1936): 49.

Jiang Yong. "Taiwan banyue ji" [Travelogue of my half-month trip to Taiwan]. *LZ* 4, no. 8 (August 1930): 1–18.

Jiang Zunyou. "Dongbei shiye kaochatuan yu yanda nongxi" [The industry investigation tour to the Northeast and the Department of Agriculture at Yenching University]. *Yanda nongxun* 2, no. 11 (February 1929): 1–2.

Jiangsu shengzhang gongshu shiye ke [The department of industry, Jiangsu governor's office], ed. *Canguan Taiwan quanye gongjin hui baogaoshu* [The report on our visit to the Taiwan Industrial Exhibition]. Shanghai: Jiangsu shengzhang gongshu shiye ke, 1917.

Jiaotong bu tielu lianyun shiwu chu, ed. *Zhonghua guoyou tielu lüxing zhinan* [Travel guides for Chinese national railroads]. Beijing: Jinghua yinshu ju, 1922.

"Jiaru Huabei lüxing tuanyuan xingshi yilan biao" [The name list of the tour members to North China]. In "Huabei lüxing zhuanhao" [Special issue on the North China tour], *Yousheng* (August 1931): 5–8.

Jin Ye. "You Guangzhouwan zhuandao" [Transits through Guangzhouwan]. *LZ* 13, no. 12 (December 1939): 7–11.

"Jinpu Jiaoji lianglu huitong juban qingdao youlantuan" [The Jinpu and Jiaoji Railways organize Qingdao tour]. *Tiedao gongbao*, no. 1457 (1936): 13.

Jinpu tielu guanli ju zongwu chu biancha ke [The compilation department of the Tianjin-Pukou Railroad], ed. *Jinpu tielu lüxing zhinan* [Travel guide for the Tianjin-Pukou Railroad]. Tianjin: Jinpu tielu guanli ju, 1921.

Johansson, Perry. *Saluting the Yellow Emperor: A Case of Swedish Sinography*. Leiden: Brill, 2012.

Ju Xiaoming. "Taiwan dizhi." *LZ* 20, no. 1 (January 1946): 13–18.

"Juban Beiping Qingdao liangdi lüxing youlantuan" [Tours between Beijing and Qingdao]. *Tielu banyue kan* 2, no. 9 (1937): 150–51.

Kaiming. "Taiwan de lunkuo" [An outline of Taiwan]. *LZ* 20, no. 2 (February 1946): 53–56.

Kan Zonghua and Zheng Houbang. "Xi'nan hangkong gongsi de xingshuai" [The rise and decline of Southwest Aviation]. In *Zhonghua wenshi ziliao wenku* [Library of Chinese historical data], edited by Wenshi ziliao yanjiu weiyuanhui [Research committee of historical materials], 13:707–12. Beijing: Zhongguo wenshi chubanshe, 1996.

Karl, Rebecca E. *Staging the World: Chinese Nationalism at the Turn of the Twentieth Century*. Durham, NC: Duke University Press, 2002.

Kemp, Emily G. *The Face of Manchuria, Korea, and Russian Turkestan*. New York: Duffield, 1911.

Kerr, Douglas, and Julia Kuehn, eds., *A Century of Travels in China: Critical Essays on Travel Writing from the 1840s to the 1940s*. Hong Kong: Hong Kong University Press, 2007.

Kihinkai. *A Guide-Book for Tourists in Japan*. 5th ed. Tokyo: s.n., 1910.

Kirjassoff, Alice Ballantine. "Formosa the Beautiful." *National Geographic Magazine* 37, no. 3 (March 1920): 246–92.

Koenker, Diane P. *Club Red: Vacation Travel and the Soviet Dream*. Ithaca, NY: Cornell University Press, 2013.

Köll, Elisabeth. *Railroads and the Transformation of China*. Cambridge, MA: Harvard University Press, 2019.

Koshar, Rudy. "'What Ought to Be Seen': Tourists' Guidebooks and National Identities in Modern Germany and Europe." *Journal of Contemporary History* 33, no. 3 (July 1998): 323–40.

Koyūkai [Friends of Shanghai Association]. *Tōa dōbun shoin daigakushi* [The history of the East Asia Common Culture Academy]. Tokyo: Koyūkai, 1955.

——. ed. *Shanhai Tōa dōbun shoin dairyokō kiroku: jitsuroku Chūgoku tōsaki* [Shanghai East Asia Common Culture Academy's records on Grand Travels: Authentic records of personal investigations in China]. Tokyo: Shin Jinbutsu ōraisha, 1991.

"Kuaiji lüshi liangxiehui zhengqiu Huangshan lüxing tongzhi" [The Bar and Accountants' Associations soliciting members to tour Huangshan]. *Shenbao*, March 19, 1936.

Lai Yanyu, ed. *Guangxi youli xuzhi* [Notes on travel to Guangxi]. Nanning: Guangxi yinshu chang, 1935.

Lam, Tong. *A Passion for Facts: Social Surveys and the Construction of the Chinese Nation-State, 1900–1949*. Berkeley: University of California Press, 2011.

Lary, Diana. Introduction to *China at War: Regions of China, 1937–1945*, edited by Stephen R. MacKinnon, Diana Lary, and Ezra F. Vogel, 1–14. Stanford, CA: Stanford University Press, 2007.

Laughlin, Charles A. *Chinese Reportage: The Aesthetics of Historical Experience*. Durham, NC: Duke University Press, 2002.

Lee, Leo Ou-fan. *Shanghai Modern: The Flowering of a New Urban Culture in China, 1930–1945*. Cambridge, MA: Harvard University Press, 1999.

Leheny, David R. *The Rules of Play: National Identity and the Shaping of Japanese Leisure*. Ithaca, NY: Cornell University Press, 2003.

Lei Jieqiong. *Pingsui yanxian zhi tianzhu jiaohui* [The Catholic missions along the Ping-Sui Railway]. N.p.: Pingsui tielu guanli ju, 1935.

Leng Xiang. "Cong Zhang Shuiqi lianxiang qi" [The Zhang Shuiqi connection]. *Xinghua* 1, no. 6 (1936): n.p.

Li, Danke. *Echoes of Chongqing: Women in Wartime China*. Urbana: University of Illinois Press, 2009.

Li Baorong. "Zhongguo wuda bishu qu" [Five largest summer resorts in China]. *LZ* 7, no. 7 (July 1933): 109–13.

Li Butong. "Shijie geguo zhi lüxing shiye" [Tourism businesses around the world]. *LZ* 4, no. 5 (May 1930): 81–87.

Li Changzhi. "Xi'nan jixing" [Travel notes on my journey to the Southwest]. *LZ* 12, no. 11 (November 1938): 9–20.

Li Qiyu. "Kunming fengguang" [The sights of Kunming]. *LZ* 12, no. 1 (January 1938): 23–32.

Li Zhenfu, ed. *Dongsansheng lüxing zhinan* [Travel guide to Manchuria]. Shanghai: Shanghai yinhang lüxingbu, 1926.

Liang, Y. T. "Open Letter to Travel Service." *China Press*, January 10, 1937.

Liang Desuo. "Quanguo lieying ji" [Reports on the Nationwide Photographic Tour]. *Liangyou*, no. 70 (October 1932): 28–29.

——. "Quanguo lieying ji zhi er: shenglin xunli" [Reports on the Nationwide Photographic Tour 2: A visit to the land of ancient sages]. *Liangyou*, no. 71 (November 1932): 6–7, 17.

——. "Xiang Yue jiaotong yu guoji wenti" [Transportation in Hunan and Guangdong and the problems of the nation]. *Liangyou*, no. 75 (March 1933): n.p.

——. "Xibei de yinxiang" [Impressions of the Northwest]. *Liangyou*, no. 73 (January 1933): n.p.

——. "Zhongyuan richeng zhailu" [Excerpts of our travels in Central China]. *Liangyou*, no. 74 (February 1933): n.p.

Liang Qichao. *Ouyou xinying lu jielu* [Excerpts from impressions of travel in Europe]. Shanghai: Zhonghua shuju, 1936.

——. "Zhongguo zhimin bada weiren zhuan" [Biographies of eight eminent Chinese colonizers]. In *Liang Qichao quanji* [Liang Qichao anthology], 3:1366–68. Beijing: Beijing chubanshe, 1997.

—— [Cangjiang, pseud.]. "You Tai diyi xin" [The first letter from my trip to Taiwan]. *Guofengbao* 2, no. 7 (1911): 1–3.

—— [Cangjiang, pseud.]. "You Tai di'er xin" [The second letter from my trip to Taiwan]. *Guofengbao* 2, no. 7 (1911): 3–4.

—— [Cangjiang, pseud.]. "You Tai disan xin" [The third letter from my trip to Taiwan]. *Guofengbao* 2, no. 7 (1911): 4–6.

—— [Cangjiang, pseud.]. "You Tai diliu xin" [The sixth letter from my trip to Taiwan]. *Guofengbao* 2, no. 9 (1911): 1–6.

"Liangyou duzhe lüxing lieche: Yunnan Kunming" [*Liangyou* readers' travel train: Kunming, Yunnan]. *Liangyou*, no. 105 (September 1935): 24–25.

"Liangyou sheying tuan" [The *Liangyou* photographic tour]. *Liangyou*, no. 70 (October 1932): 22.

Liao, Ping-hui. "Travels in Modern China: From Zhang Taiyan to Gao Xingjian." In *The Oxford Handbook of Modern Chinese Literatures*, edited by Carlos Rojas and Andrea Bachner, 39–51. Oxford: Oxford University Press, 2016.

"Liaoji fengguang" [The scenery of Liaoning and Jilin]. *Liangyou*, no. 63 (1931): 38.

Lin Bing. "Xi'nan gonglu: Xiangqian duan" [The Southwest Motorways: Between Changsha and Guiyang]. *LZ* 12, no. 11 (November 1938): 47–53.

——. "Xi'nan gonglu: Zhuliu duan" [The Southwest Motorways: Between Guiyang and Liuzhou], *LZ* 13, no. 5 (May 1939): 21–28.

Lin Zhen, ed. *Zengding Shanghai zhinan* [Expanded guide to Shanghai]. 23rd ed. Shanghai: Shangwu yinshu guan, 1930.

Liu, Lydia. *Translingual Practice, Literature, National Culture, and Translated Modernity—China, 1900–1937*. Stanford, CA: Stanford University Press, 1995.

Liu Bannong [Liu Fu]. *Dunhuang duosuo* [Miscellany from Dunhuang]. Beijing: Zhongyang yanjiu yuan lishi yuyan yanjiu suo, 1925.

Liu Tizhi. "Liangyou duzhe lüxing lieche: Guizhou fengjing mei" [*Liangyou* readers' travel train: Beautiful Guizhou]. *Liangyou*, no. 102 (June 1935): 20–21.

Liu Xiaohui. *Fuqin Liu Bannong* [My father, Liu Bannong]. Shanghai: Shanghai renmin chubanshe, 2000.

Liu Yazhi, ed. *Bao Ri zuijin zhi jingji qinlüe yu dongbei* [An exposé of the recent economic invasion of the Japanese in the Northeast]. Shanghai: Dongbei yanjiu she, n.d.

Liu Ying and Wang Yingying. "Yu Tianxiu shehuixue sixiang tanxi" [An analysis of Yu Tianxiu's sociological thoughts]. *Xibei daxue xuebao (zhexue shehui kexue ban)* [Journal of Northwest University (philosophy and social sciences edition)] 42, no. 5 (September 2012): 17–22.

Liu Zhiping. "Kunming huanhu xing ji" [Around-the-lake tour in Kunming]. *LZ* 14, no. 6 (June 1940): 3–17.

"Longhai lu Huayin Lintong liangzhan piwei guonei laihui youlan zhandian" [The Huayin and Lintong stops on the Longhai line are designated as "tourism round-trip" stops]. *Jinghu huhangyong tielu rikan*, no. 1389 (1935):143.

Lu Shifu. "Huabei lüxing sheying" [The travel photographs of North China]. *Shidai* 7, no. 4 (1934): 16–17.

Lu Weipin. "Fuchun jiang youji" [Travelogue of Fuchun River]. *LZ* 5, no. 6 (June 1931): 55–60.

Lu Zuofu. *Dongbei youji* [Travelogue of the Northeast]. 2nd ed. Chengdu: Chengdu shuju, 1931.

Luo Zhitian. "Xifang xueshu fenlei yu minchu guoxue de xueke dingwei" [Western academic classification and subject orientation in national studies in the early Republican period]. *Sichuan daxue xuebao* [Journal of Sichuan University] 38, no. 5 (May 2001): 75–82.

——. "Zouxiang guoxue yu shixue de saixiansheng" [Mr. Science as applied in national studies and history]. *Jindaishi yanjiu* [Modern Chinese history studies, no. 3 (2000): 59–94.

Lüsheng. "Chiren shuomeng ji: Di shi'er hui" [Chapter 12: A fool's tale of his dream]. *Xiuxiang xiaoshuo*, no. 30 (1904): 2–3.

"Lüxing gaikuang" [Overview of travel]. *Yousheng* (October 1934): 4.

"Lüxing huiwen" [The tourism news]. *Yousheng*, no. 7 (1932): 4–5.

"Lüxing shiye guowai xuanchuan diyi sheng" [Our first overseas promotional mission of the tourism industry]. *LZ* 22, no. 3 (March 1948): n.p.

"Lüxing zazhi zengshe huanan zong faxing jinyao qishi" [A breaking announcement from *China Traveler*: Adding a main distribution office in South China]. *LZ* 12, no. 6 (June 1938): n.p.

"Lüxing zazhi zhengwen teji" [Special column for *China Traveler*]. *LZ* 20, no. 1 (January 1946): 1.

"Lüxingbu zhengyou yundong jianzhang" [The general rule of membership at the travel club]. *Jingwu congbao* 2, no. 7 (1936): 5.

"Lüxinggu baogao jingguo didian" [The tour department's report on tour destinations]. *Yousheng*, no. 9 (1926): n.p.

"Lüxingshe zuzhi erci youlan Huangshan tuan" [China Travel Service organizes a second tour to the Yellow Mountains]. *Shenbao*, April 18, 1935.

Ma Guoliang. "Liangyou quanguo sheying lüxing tuan" [*Liangyou* Nationwide Photographic Tour]. *Dazhong sheying* [Popular photography], no. 6 (1980): 28–29.

——. *Liangyou yijiu: yijia huabao yu yige shidai* [Remembering *Liangyou*: A pictorial journal and a historical period]. Taipei: Zhengzhong shuju, 2002.

Ma Hetian. *Dongbei kaocha ji* [Travel notes on investigations in the Northeast]. Nanjing: Xin yaxiya xuehui, 1934.

Madrolle, Claudius. *North-Eastern China: Manchuria, Mongolia, Vladivostock, Korea*. Paris: Hachette, 1912.

"Manila zhuanhao" [Special issue on Manila]. *LZ* 15, no. 11 (November 1941).

Marchand, Suzanne. *German Orientalism in the Age of Empire: Religion, Race and Scholarship*. Cambridge: Cambridge University Press, 2010.

McClure, Margaret. *The Wonder Country: Making New Zealand Tourism*. Auckland: Auckland University Press, 2004.

McDonald, Kate. *Placing Empire: Travel and the Social Imagination in Imperial Japan*. Berkeley: University of California Press, 2017.

Meyer-Fong, Tobie. "Seeing the Sights in Yangzhou from 1600 to the Present." In *When Images Speak: Visual Representation and Cultural Mapping in Modern China*, edited by Huang Ko-wu, 213–51. Taipei: Academia Sinica, Institute of Modern History, 2004.

Miao Wenhua. *Beiling zhilüe* [Brief record of the North Mausoleum]. Shenyang: Beiling gongyuan guanlichu, 1929.

Ming. "Zhongguo de chaojin zhe" [Chinese pilgrims]. *Yuehua* 4, no. 4 (1932): 22.

"Mingsheng guji guwu baocun tiaoli" [The regulation of protecting scenic spots, historical sites, and historical relics]. *Neizheng gongbao* 1, no. 6 (1928): 13–16.

Mingzhe. *Neimeng sanji* [Sketches of Inner Mongolia]. Shanghai: Zhongguo lüxingshe, 1952.

Mitter, Rana. *Forgotten Ally: China's World War II, 1937–1945*. Boston: Houghton Mifflin Harcourt, 2013.

——. *The Manchurian Myth: Nationalism, Resistance, and Collaboration in Modern China*. Berkeley: University of California Press, 2000.

Mo, Yajun. "The New Frontier: Zhuang Xueben and Xikang Province." In *Chinese History in Geographical Perspective*, edited by Yongtao Du and Jeff Kyong-McClain, 121–40. Lanham, MD: Lexington Books, 2013.

"Moganshan bishu zhi lianyun guizhang" [Regulations of the through transportation to Moganshan]. *Zhonghua gongchengshi xuehui huibao* 7, no. 4 (1920): 1–2.

"Moganshan tielu lüguan" [Mokanshan Railway Hotel]. *Tielu gongbao: Huning Huhangyong xian*, no. 41 (1921): n.p.

Montgomery, Janet B. *Among the Head-Hunters of Formosa*. London: T. F. Unwin, 1922.

Morris, Andrew. "Taiwan's History: An Introduction." In *The Minor Arts of Daily Life: Popular Culture in Taiwan*, edited by David Jordan and Marc Moskowitz, 3–32. Honolulu: University of Hawai'i Press, 2004.

"Nanjing shoudu fandian" [The Metropolitan Hotel in Nanjing]. *LZ* 9, no. 9 (September 1935): 7–13.

"Nanxiang qima jingsai" [The horse race in Nanxiang]. *Yousheng* (May 1935): n.p.

"Nanyang qundao zhuanhao" [Special issue on Malaysia]. *LZ* 14, no. 8 (August 1940).

Naquin, Susan, and Chün-fang Yü, eds. *Pilgrims and Sacred Sites in China*. Berkeley: University of California Press, 1992.

National Palace Museum. *The Record of Mogao by Zhang Daqian*. Taipei: National Palace Museum, 1985.

Ni Xiying. *Beiping*. Shanghai: Zhonghua shuju, 1936.

——. *Guangzhou*. Shanghai: Zhonghua shuju, 1936.

——. *Hangzhou*. Shanghai: Zhonghua shuju, 1936.

——. *Jinan*. Shanghai: Zhonghua shuju, 1936.

——. *Luoyang*. Shanghai: Zhonghua shuju, 1939.

——. *Luoyang youji* [Travelogues of Luoyang]. Shanghai: Zhonghua shuju, 1935.

——. *Nanjing*. Shanghai: Zhonghua shuju, 1936.

——. *Qingdao*. Shanghai: Zhonghua shuju, 1936.

——. *Qufu Taishan youji* [Travelogues of Qufu and Mount Taishan]. Shanghai: Zhonghua shuju, 1931.

——. *Shanghai*. Shanghai: Zhonghua shuju, 1938.

——. *Xijing*. Shanghai: Zhonghua shuju, 1936.

Nie Baozhang and Zhu Yingui, eds. *Zhongguo jindai hangyunshi ziliao, 1895–1927* [Historical materials on modern Chinese shipping, 1895–1927]. Beijing: Zhongguo shehui kexue chubanshe, 2002.

Nield, Robert. *China's Foreign Places: The Foreign Presence in China in the Treaty Port Era, 1840–1943*. Hong Kong: Hong Kong University Press, 2015.

"Nippon Tour Plans Are Set." *China Press*, March 27, 1936.

Noack, Christian. "Building Tourism in One Country? The Sovietization of Vacationing, 1917–41." In *Touring beyond the Nation: A Transnational Approach to European Tourism History*, edited by Eric G. E. Zuelow, 171–93. Farnham, UK: Ashgate, 2011.

Nyiri, Pal. *Scenic Spots: Chinese Tourism, the State, and Cultural Authority*. Seattle: University of Washington Press, 2006.

"Outbound Tourism—Travel (Million US Dollars)." Knoema, World Data Atlas, accessed December 29, 2020. https://knoema.com/atlas/topics/Tourism/Outbound-Tourism-Indicators/Outbound-tourism-travel.

"Overseas Chinese Tourists Make 1st Tour of Shanghai." *China Press*, October 16, 1934.

"Paiding zhaoliao guanchao zhuanche zhi renyuan" [Personnel for the special trains for viewing tidal waves]. *Shenbao*, October 8, 1919.

Pan Enlin. Introduction to *Xi'nan lansheng* [Scenic beauty in Southwest China], edited by Zhao Junhao, n.p. Shanghai: Zhongguo lüxingshe, 1940.

Pan Taifeng. "Kangri zhanzheng shiqi de zhongguo lüxingshe" [The China Travel Service during the Anti-Japanese War]. In *Wenshi ziliao xuanji* [Selected historical materials], edited by Wenshi ziliao yanjiu weiyuanhui [Research committee of historical materials], 117:168–206. Beijing: Zhongguo wenshi chubanshe, 1989.

Pang, Laikwan. *Building a New China in Cinema: The Chinese Left-Wing Cinema Movement, 1932–1937*. Lanham, MD: Rowman & Littlefield, 2002.

Park Kyung Seok. "Minguo shiqi Shanghai de yousheng lüxing tuan he 'xiuxian lüxing'" [The Unison Travel Club in Republican Shanghai and "leisure travel"]. *Minguo yanjiu* [Studies of Republican China], no. 18 (2010): 246–61.

Parsons, Nicholas. *Worth the Detour: A History of the Guidebook*. Stroud: History Press, 2007.

Peigan. "Jianada lüxing shiye" [Tourism industry in Canada]. *LZ* 4, no. 12 (December 1930): 61–66.

Peking Daily News. *The Constitutional Compact, i.e., the Amended Provisional Constitution of the Republic of China*. Peking: Daily News, May 1, 1914.

Perdue, Peter. "Embracing Victory, Effacing Defeat: Rewriting the Qing Frontier Campaigns." In *The Chinese State at the Borders*, edited by Diana Lary, 105–25. Vancouver: UBC Press, 2007.

——. "Identifying China's Northwest, for Nation and Empire." In *Locating China: Space, Place, and Popular Culture*, edited by Jing Wang, 94–114. New York: Routledge, 2005.

Phillips, Steven. *Between Assimilation and Independence: The Taiwanese Encounter with Nationalist China, 1945–50*. Stanford, CA: Stanford University Press, 2003.

"Pingshen tongche zhongtu chucha" [An accident midway along the Beijing-Mukden train line]. *Meizhou pinglun*, no. 124 (1934): 26–27.

"Pingsui lu yaoxun" [Important news about the Ping-Sui line]. *Tiedao gongbao*, no. 1040 (1934): 8.

Pingsui tielu guanli ju [The Ping-Sui Railway Administration]. *Youlan zhi changdao* [Promoting tourism]. N.p.: Pingsui tielu guanli ju, 1935.

Pomeranz, Kenneth. "Resisting Imperialism, Resisting Decolonization: Making China from the Ruins of the Qing, 1912–1949." Lecture at the Library of Congress, Washington, DC, July 23, 2013. https://www.loc.gov/item/webcast-6019/.

Pratt, Mary Louise. *Imperial Eyes: Travel Writing and Transculturation*. 2nd ed. Abingdon, UK: Routledge, 2008.

Qian Hua. "Guilin shanshui" [Guilin's scenery]. *LZ* 10, no. 1 (January 1936): 58–59.

Qian Huipu. "Hudian jixing" [Travel notes on my trip from Shanghai to Yunnan]. *LZ* 12, no. 12 (December 1938): 7–12.

Qian Nengxin. *Xi'nan sanqian wubai li* [3,500 *li* to the Southwest]. Changsha: Shangwu yinshu guan, 1939.

Qian Zhitong. "Pingyi zhaodaisuo chengli jingguo" [The founding of the Pingyi Guesthouse]. *Lüguang* 1, no. 2 (February 1940): 9.

"Qianjin zhong zhi Guiyang" [Kweiyang in war color]. *Liangyou*, no. 142 (May 1939): 28–29.

Qin Lizhai. "Guling youcheng" [Itineraries for Guling]. *LZ* 5, no. 6 (June 1931): 31–39.

Qiu Peihao. "Xianhua Chongqing nanquan" [A chat about the South Hot Spring in Chongqing]. *LZ* 13, no. 10 (October 1939): 23–33.

Qiu Wenluan, Liu Fanzheng, and Xie Mingke. *Taiwan lüxing ji* [Travel accounts of Taiwan]. Taipei: Taiwan Yinhang, 1965.

Quanguo tielu lüxing zhinan [National railroad travel guide]. Shanghai: Guangyi shuju, 1921.

Reinhardt, Anne. *Navigating Semi-Colonialism: Shipping, Sovereignty, and Nation-Building in China, 1860–1937*. Cambridge, MA: Harvard University Asia Center, 2018.

——. "Treaty Ports as Shipping Infrastructure." In *Treaty Ports in Modern China: Law, Land and Power*, edited by Robert Bickers and Isabella Jackson, 101–20. London: Routledge, 2016.

Rogaski, Ruth. *Hygienic Modernity: Meanings of Health and Disease in Treaty-Port China*. Berkeley: University of California Press, 2004.

Rong Xinjiang. *Eighteen Lectures on Dunhuang*. Translated by Imre Calambos. Leiden: Brill, 2013.

Ruoff, Kenneth. *Imperial Japan at Its Zenith: The Wartime Celebration of the Empire's 2600th Anniversary*. Ithaca, NY: Cornell University Press, 2010.

Rutter, Edward Owen. *Through Formosa: An Account of Japan's Island Colony*. London: T. F. Unwin, 1923.

Ryan, Chris, and Gu Huimin, eds. *Tourism in China: Destination, Cultures and Communities*. New York: Routledge, 2009.

Sa Kongliao. *Cong Xianggang Dao Xinjiang* [From Hong Kong to Xinjiang]. Yinchuan: Ningxia chubanshe, 2000.

Schivelbusch, Wolfgang. *The Railway Journey: The Industrialization of Time and Space in the Nineteenth Century*. Berkeley: University of California Press, 2014.

Schoppa, Keith. *In a Sea of Bitterness: Refugees during the Sino-Japanese War*. Cambridge, MA: Harvard University Press, 2011.

Scidmore, Eliza. "Mukden, the Manchu Home, and Its Great Art Museum." *National Geographic Magazine* 21, no. 4 (1910): 289–320.

Sha Ou. "Guiyang yipie" [A glimpse of Guiyang]. *LZ* 12, no. 12 (December 1938): 13–15.

——. "Shanguang shuise de Kunming" [Scenic Kunming]. *LZ* 13, no. 1 (January 1939): 69–70.

Shaffer, Marguerite S. *See America First: Tourism and National Identity, 1880–1940*. Washington, DC: Smithsonian Institution Press, 2001.

Sheehan, Brett. "Webs and Hierarchies: Banks and Bankers in Motion, 1900–1950." In *Cities in Motion: Interior, Coast, and Diaspora in Transnational China*, edited by Sherman Cochran and David Strand, 81–105. Berkeley: Institute of East Asian Studies, University of California, 2007.

Shanghai shangchu yinhang lüxing bu [Travel Department of the Shanghai Commercial and Savings Bank], ed. *You Chuan xu zhi* [Guide to Sichuan]. Shanghai: Shanghai shangchu yinhang lüxing bu, 1924.

——, ed. *You Dian xu zhi* [Guide to Yunnan]. Shanghai: Shanghai shangchu yinhang lüxing bu, 1923.

"Shanghai shangye chuxu yinhang she lüxingbu" [Shanghai Commercial and Savings Bank adds a Travel Department]. *Shenbao* [Shanghai news], August 12, 1923.

"Shanghai yinhang lüxing bu zhi fada" [The development of the Travel Department at Shanghai Bank]. *Yinhang yuekan* 4, no. 4 (1924): 12–13.

Shangwu yinshu guan [Commercial Press], ed. *Shanghai zhinan* [Guide to Shanghai]. Shanghai: Shangwu yinshu guan, 1909.

——, ed. *Shiyong Beijign zhinan* [Practical guide to Beijing]. Shanghai: Shangwu yinshu guan, 1919.

——, ed. *Zengding Shanghai zhinan* [Expanded guide to Shanghai]. 21st ed. Shanghai: Shangwu yinshu guan, 1925.

——, ed. *Zengding Zhongguo lüxing zhinan* [Expanded guide for travelers in China]. 13th ed. Shanghai: Shangwu yinshu guan, 1926.

——, ed. *Zhongguo lüxing zhinan* [Guide for travelers in China]. Shanghai: Shangwu yinshu guan, 1912.

——, ed. *Zhongguo mingsheng* [Views of China]. 3rd ed. Shanghai: Shangwu yinshu guan, 1912.

——, ed. *Zhongguo mingsheng* [Views of China]. Series, 22 vols. Shanghai: Shangwu yinshu guan, 1914–26.

Shao Yaonian. "Taiwan lüxing zaji" [Miscellany of my travel to Taiwan]. *Guangdong nonglin yuebao* 2, no. 2 (1917): 109–12.

Shao Yuxiang. "Guangxi shanshui" [Guilin scenery]. *LZ* 10, no. 1 (January 1936): n.p.

Shao Zuping. "Chengdu mingsheng fangwen ji" [Visiting Chengdu's famous attractions]. *LZ* 14, no. 4 (April 1940): 27–35.

She Guitang. "Dianmian gonglu jixing" [A travel record of the Yunnan-Burma Road]. *LZ* 14, no. 12 (December 1940): 5–9.

——. "Lüxing xiangdao" [Tour guide]. *LZ* 24, no. 6 (June 1950): 16–17.

——. "Yuenan zhitan" [A summary of Vietnam]. *LZ* 14, no. 6 (June 1940): 51–54.

——. "Zhongguo youlan shiye zhi huigu" [A review of Chinese tourism]. *LZ* 17, no. 7 (July 1943): 5–10.

Shen, Grace. *Unearthing the Nation: Modern Geology and Nationalism in Republican China*. Chicago: University of Chicago Press, 2014.

Shen Songqiao. "Jiangshan ruci duojiao—1930 niandai de Xibei lüxing shuxie yu guozu xiangxiang" [The nation being so beautiful: Travel writing of the Northwest and the national imagination]. *Taida lishi xuebao* [Historical journal of Taiwan University], no. 37 (June 2006): 145–216.

Shen Tao. "Yi Chongqing" [Reminiscing about Chongqing]. *LZ* 15, no. 8 (August 1941): 55–58.

Shi Jingchen. "Kunying riji" [Kunying diary]. In *Taiwan youji* [Taiwan travelogues], edited by Taiwan Bank, 41–68. Taipei: Taiwan yinhang, 1960.

Shih, Shu-mei. *The Lure of the Modern: Writing Modernism in Semicolonial China, 1917–1937*. Berkeley: University of California Press, 2001.

Shimazu, Naoko. "Colonial Encounters: Japanese Travel Writing on Colonial Taiwan." In *Refracted Modernity: Visual Culture and Identity in Colonial Taiwan*, edited by Yūko Kikuchi, 21–37. Honolulu: University of Hawai'i Press, 2007.

"Shouhui jigong shandi lingyi zuwu bishu zhangcheng" [Reclaiming Jigongshan and discussions about leasing out the summer resort]. *Dongfang zazhi* 5, no. 11 (1908): 20–23.

Shu Yongkang. "Dao Xibei qu" [Head to the Northwest]. *Sheying huabao* 9, no. 33 (1933): 16–17.

——. "Xixing riji" [Journal of a westward journey]. *LZ* 7, nos. 9–11 (September, October, and November 1933): 47–53; 31–48; 47–62.

Shuai Yucang. "Kunming manji" [A travelogue of Kunming]. *LZ* 13, no. 8 (August 1939): 25–31.

Si Jiaozhi. "Xi'nan minjian yunshuye 'Maxiangyue'" ["Maxiangyue": The popular transportation service in the Southwest]. In *Zhonghua wenshi ziliao wenku* [Library of Chinese historical data], edited by Wenshi ziliao yanjiu weiyuanhui [Research committee of historical materials], 13:971–84. Beijing: Zhongguo wenshi chubanshe, 1996.

Sihong. "Chongqing shenghuo pianduan" [Snippets of Chongqing life]. *LZ* 14, no. 4 (April 1940): 7–10.

Simpich, Frederick. "Manchuria, Promised Land of Asia." *National Geographic Magazine* 56, no. 4 (1929): 379–428.

Smith, Graham. *Photography and Travel*. London: Reaktion Books, 2013.

Stanford, F. E. "Mokanshan." In *With Our Missionaries in China*, edited by Emma Anderson et al., 323–30. Mountain View, CA: Pacific Press, 1920.

Stewart, Gordon. "The Exploration of Central Asia." In *Reinterpreting Exploration: The West in the World*, edited by Dane Kennedy, 195–213. Oxford: Oxford University Press, 2014.

Strassberg, Richard E., ed. and trans. *Inscribed Landscapes: Travel Writing from Imperial China*. Berkeley: University of California Press, 1994.

Su, Rongyu. "The Reception of 'Archaeology' and 'Prehistory' and the Founding of Archaeology in Late Imperial China." In *Mapping Meanings: The Field of New Learning in Late Qing China*, edited by Michael Lackner and Natascha Vittinghoff, 423–50. Leiden: Brill, 2004.

Su Lan. *Kangzang suijun ji* [My journey with the army to Tibet]. Shanghai: Zhongguo lüxingshe, 1953.

Sun Fuxi. "Xi'nan shi jianguo de tianyuan" [The Southwest is the field for state building]. *LZ* 12, no. 11 (November 1938): 5–7.

Sun Peigan. "Tan zhanshi youlan shiye" [On wartime tourism]. *LZ* 15, no. 7 (July 1941): 91–94.

Sun Qing. "'Xin' youji huibian yu jindai Zhongguo 'kongjian' biaoshu zhuanbian chutan" [A preliminary exploration of the collections of "new" travelogues and the shifts in the expressions of "space" in modern China]. *Xin shixue* [New history], no. 11 (2019): 58–59.

Sun Yat-sen. *The International Development of China*. New York: Putnam, 1922.

Tai, Jeremy. "The Northwest Question: Capitalism in the Sands of Nationalist China." *Twentieth-Century China* 40, no. 3 (October 2015): 201–19.

Taiwan lüxingshe [Taiwan Travel Service], ed. *Taiwan fengjing* [Taiwanese scenery]. Taipei: Taiwan lüxingshe, 1946.

——, ed. *Taiwan lüxing zhinan* [Guidebook to Taiwan]. Taipei: Taiwan lüxingshe, 1947.

Tang Weibin. "Shengli hou zhi Zhongguo lüxingshe" [The China Travel Service after victory]. *LZ* 22, no. 4 (April 1948): 81–85.

——. "Zhonglü ershisan nian" [The twenty-three years of the China Travel Service]. *LZ* 20, no. 1 (January 1946): 91–100.

Teng, Emma Jinhua. *Taiwan's Imagined Geography: Chinese Colonial Travel Writing and Pictures, 1683–1893*. Cambridge, MA: Harvard University Asia Center, 2004.

Terry, Thomas Philip. *Terry's Japanese Empire*. Boston: Houghton Mifflin, 1928.

Theobald, Ulrich. "Southwest China: Local Conditions and Economic Trajectories." In *Southwestern China in a Regional and Global Perspective (c.1600–1911): Metals, Transport, Trade and Society*, edited by Ulrich Theobald and Jin Cao, 1–41. Leiden: Brill, 2018.

Thomas Cook Ltd. *Chinese Government Railways: Travellers' Notes and Through Booking Arrangements, Return Tourist Tickets, Circular Tours in China, Chosen and Japan*. Shanghai: Thomas. Cook & Son, 1918.

——. *Cook's Handbook for Tourists to Peking, Tientsin, Shan-Hai-Kwan, Mukden, Dalny, Port Arthur, and Seoul*. London: Thomas Cook & Son, 1910.

——. *Peking, North China, South Manchuria and Korea*. 4th ed. London: Thomas Cook & Son, 1920.

"Three Hangchow Bore Excursions to Be Held through Week-End." *China Press*, September 9, 1927.

Thurin, Susan Schoenbauer. *Victorian Travelers and the Opening of China, 1842–1907*. Athens: Ohio University Press, 1999.

"Tiedaobu xunling" [Instructions from the Ministry of Railways]. *Tiedao gongbao*, no. 48 (March 1930): 37–40.

Tighe, Justin. *Constructing Suiyuan: The Politics of Northwestern Territory and Development in Early Twentieth-Century China*. Leiden: Brill, 2005.

Tong, Dickson. "Tourism in the Northwest—a Neglected Opportunity." *China Weekly Review*, January 9, 1932.

"Tongao, di eryi-wu hao, Tongao, di eryi-liu hao" [Announcements, no. 21.5 and no. 21.6]. *Yousheng* (July 1932): 5.

Towne, Herman. "Summer Resorts in China: Six of China's Major Scenic & Health Spots." *China Press*, July 8, 1936.

Trinkler, Emil. "Explorations in the Eastern Karakoram and in the Western Kunlun." *Geographical Journal* 75, no. 6 (June 1930): 505–15.

Tsurumi, E. Patricia. *Japanese Colonial Education in Taiwan, 1895–1945*. Cambridge, MA: Harvard University Press, 1978.

Tu Zheyin. "Cong Shanghai dao Harbin" [From Shanghai to Harbin]. *LZ* 2, no. 1 (January 1928): 15–17.

"Tuanyuan xuzhi" [Notice to members]. *Yousheng* (July 1923): n.p.

Turner, Louis, and John Ash. *The Golden Hordes: International Tourism and the Pleasure Periphery*. London: Constable, 1975.

Walton, John K. "Tourism History: People in Motion and at Rest." *Mobility in History* 5 (2014): 74–85.

Wan Feng. "Cong Guilin dao Yangshuo" [From Guilin to Yangshuo]. *LZ* 16, no. 8 (August 1942): 13–14.

Wang, Chuchu. "Appendix: Distributing *Liangyou*." In *Liangyou: Kaleidoscopic Modernity and the Shanghai Global Metropolis, 1926–1945*, edited by Paul Pickowicz, Shen Kuiyi, and Zhang Yingjin, 248–58. Leiden: Brill, 2013.

Wang, Liping. "Tourism and Spatial Change in Hangzhou, 1911–1927." In *Remaking the Chinese City: Modernity and National Identity, 1900–1950*, edited by Joseph Esherick, 107–20. Honolulu: University of Hawai'i Press, 2000.

Wang Guoyuan. "Jingzhang tiedao zhi lüxing tan" [On travel via the Beijing-Kalgan Railway]. *Dongfang zazhi* 6, no. 11 (1909): 31–34.

Wang Hongdao. "Guilin xiaozhu" [Sojourn in Guilin]. *LZ* 13, no. 11 (November 1939): 20.

Wang Huanwen. "Zhongguo jiying tichang lüxing shiye" [China should promote tourism business immediately]. *LZ* 5, no. 11 (November 1931): 65–67.

Wang Huimin, ed. *Xin dongbei zhinan* [Guidebook to the new Northeast]. Chongqing: Shangwu yinshu guan, 1946.

Wang Jianshi. "Zhongyin gonglu quancheng xing" [Travel experiences along the entire Sino-India motorway]. *LZ* 19, no. 1 (January 1945): 62–65.

Wang Ke. *Dongtujue sitan duli yundong 1930 niandai zhi 1940 niandai* [The East Turkestan independence movement in the 1930s–1940s]. Hong Kong: Chinese University of Hong Kong Press, 2013.

Wang Wenru and Ling Guiqing, eds. *Xin youji huikan* [Collectanea of new travelogues]. Shanghai: Zhonghua shuju, 1921. 5th ed., 1928; 6th ed., 1932.

Wang Wenru and Yao Zhuxuan, eds. *Xin youji huikan xubian* [Collectanea of new travelogues: The sequel]. Shanghai: Zhonghua shuju, 1923. 4th ed., 1935.

Wang Xiaoting. "Meili de Zhonghua" [Beautiful China]. *Dazhong huabao*, no. 12 (October 1934): 30–31.

——. "Sanxia tianxian" [The Three Gorges as a natural barrier]. *Dazhong huabao*, no. 1 (November 1933): 14–15.

——. "Shiqu de Rehe" [The lost Jehol]. *Dazhong huabao*, no. 8 (June 1934): 20–21.

Wang Yang. *Taiwan shicha baogao shu* [Travel report of Taiwan]. Shanghai: Zhonghua shuju, 1928.

Wang Yinqiao. *Xijing youlan zhinan* [Tourist guide to Xijing]. Xi'an: Tianjin Dagongbao Xi'an fenguan, 1936.

Wang Yuting. *Dongbei yinxiang ji* [Impressions of the Northeast]. Shanghai: Shixian she, 1933.

Wang Zhaosheng. *Xibei jianshe lun* [On building the Northwest]. Beijing: Qingnian chubanshe, 1943.

Warren, Lynne, ed. *Encyclopedia of 20th Century Photography*. London: Routledge, 2006.

Wei Lihai. *Menggu renmin gongheguo lüxing ji* [Travelogue of the People's Republic of Mongolia]. Shanghai: Zhongguo lüxingshe, 1952.

——. *Zhandou zai kenteshan shang* [Battles on Mount Khentii]. Shanghai: Zhongguo lüxingshe, 1953.

"When You Go Home." Advertisement for the Travel Department of the American Express. *Chinese Students' Monthly* 16, no. 8 (1900): 5.

Whitfield, Susan. "The Dunhuang Manuscripts: From Cave to Computer." In *Books in Numbers: Seventy-Fifth Anniversary of the Harvard-Yenching Library; Conference Papers*, edited by Lucille Chia and W. L. Idema, 113–42. Cambridge, MA: Harvard-Yenching Library, Harvard University, 2007.

Will, Pierre-Étienne. "Xi'an, 1900–1940: From Isolated Backwater to Resistance Center." In *New Narratives of Urban Space in Republican Chinese Cities*, edited by Billy K. L. So and Madeleine Zelin, 223–74. Leiden: Brill, 2013.

Winichakul, Thongchai. *Siam Mapped: A History of the Geo-Body of a Nation*. Honolulu: University of Hawai'i Press, 1994.

Wu Chenyi. "Taiwan youxue" [Some travel impressions of Taiwan]. *LZ* 22, no. 9 (September 1948): 21–27.

Wu Jen-shu and Imma Di Biase. *Youdao: Ming Qing lüyou wenhua* [The Tao of travel: Travel culture in the Ming and Qing era]. Taipei: Sanmin shuju, 2010.

Wu Jingyan. *Chen Guangfu yu Shanghai yinhang* [Chen Guangfu and Shanghai Bank]. Beijing: Zhongguo wenshi chubanshe, 1991.

Wu Liande. "Wei Liangyou sheying tuan fayan" [Speech for the *Liangyou* Nationwide Photographic Tour]. *Liangyou*, no. 69 (September 1932): 28.

——, ed. *Zhonghua jingxiang: quanguo sheying zong ji* [China as she is: A comprehensive album]. Shanghai: Liangyou tushu yinshua youxian gongsi, 1934.

Wu Peisong. "Canjia Moganshan denggao jingzou ji" [On my participation in the hiking competition at Moganshan]. *Yousheng* (May 1935): n.p.

Wu Pishi. "Xianggang qingnian hui huadong youlantuan youji" [Travelogue of Hong Kong YMCA's East China Tour]. *Qingnian jinbu*, no. 140 (1931): 91–114.

Wu Silan. "Yuanye you riyuetan" [Touring the Sun-Moon Lake]. *LZ* 23, no. 4 (April 1949): 46–47.

Wu Wenzao. *Menggu bao* [Mongolian yurts]. N.p.: Pingsui tielu guanli ju, 1935.

Wu Yishou. "Lanzhou youji" [Travelogue of Lanzhou]. *Keda zazhi* 6, no. 4 (1935): 7–9.

Wuyong. "Ha'erbin yipie" [A glimpse of Harbin]. *Liangyou*, no. 39 (1929): 19–23.
Xia Ying. "Huabei lüxing guilai" [Returning from a North China tour]. *Shengli* 1, no. 1 (1936): 14–15.
Xiang Wenhui. *Hangzhou lüyou ji qi jindai mingyun* [Hangzhou tourism and its modern fate]. Hangzhou: Zhejiang daxue chubanshe, 2018.
Xiao Qian. "Pingsui xianshang (youji)" [On the Ping-Sui line (travelogue)]. *Xinsheng zhoukan* [New life weekly] 1, nos. 39–41 (1934): 804–7; 826–27; 844–46.
"Xibei kexue kaochatuan kaocha jingguo" [The scientific research expedition to northwestern China]. *Tuhua shibao*, February 3, 1929.
"Xibei kexue kaochatuan tekan" [A special issue on the scientific mission to the Northwest]. *Shijie huabao*, January 27, 1929, n.p.
"Xiezhu yousheng lüxing tuan juban diqi diba liangci huabei lüxing" [Assisting the Unison Travel Club's seventh and eighth North China tours]. *Jinghu huhangyong tielu rikan*, no. 1922 (1937): 132–33.
"Xili zhaiyao" [A summary of Western etiquette]. *LZ* 3, no. 8 (August 1928): 11–14.
"Xi'nan fengjing xian" [The scenic line in the Southwest]. *LZ* 12, no. 10 (October 1938): n.p.
"Xi'nan gesheng gonglu piaojia biao" and "Xi'nan gesheng gonglu lianyun lichen biao" [The prices and distances of through travel on the Southwest Motorways]. *LZ* 13, no. 5 (May 1939): 67–68.
"Xi'nan zhuanhao shuhou" [Postscript to the special issue on the Southwest]. *LZ* 12, no. 11 (November 1938): 109.
Xingchu [Shen Xingchu] and Juexue. "Xibei youcheng jilue" [A brief itinerary of the Northwest]. *Zhonghua*, no. 27 (1934): n.p.
"Xinjiapo fenshe kaimu jisheng" [The grand opening of the China Travel Service Singapore branch]. *LZ* 8, no. 12 (December 1934): 2–3.
Xu Bingchang [Xusheng]. "Jiangyan: Xinjiang zhi bokeda shan" [Speech: Mount Bogda in Xinjiang]. *Zhongfa jiaoyujie*, no. 25 (April 1929): 1–7.
——. "Xibei diaocha yuxian ji" [Mishaps in the scientific mission to the Northwest]. *Yanjing daxue xiaokan*, no. 22 (1929): n.p.
——. "Xibei kaocha ji" [Exploration of the Northwest]. *Jiaotong daxue rikan*, March 27, 1929.
——. *Xu Xusheng xiyou riji* [Xu Xusheng's journal of a westward journey]. Vols. 1–3. Beijing: Zhongguo xueshu tuanti xiehui xinei kexue kaochatuan lishihui, 1930.
Xu Hongtao. *Xinan dongbei* [The Southwest and the Northeast]. Hangzhou: Dafeng she, 1935.
Xu Ke. *Beidaihe zhinan* [Guidebook to Peitaiho]. Shanghai: Shangwu yinshu guan, 1922.
——. *Jigongshan zhinan* [Guide to Chi Kung Shan]. Shanghai: Shangwu yinshu guan, 1922.
——. *Lushan zhinan* [Guide to Lü Shan]. Shanghai: Shangwu yinshu guan, 1922.
——. *Moganshan zhinan* [Guide to Mokanshan]. Shanghai: Shangwu yinshu guan, 1922.
——. *Shiyong Beijing zhinan* [Practical guide to Beijing]. Shanghai: Shangwu yinshu guan, 1919.
——. *Zengding Moganshan zhinan* [Expanded guide to Mokanshan]. Shanghai: Shangwu yinshu guan, 1930.

——. *Zengding shiyong Beijing zhinan* [Expanded practical guide to Beijing]. Shanghai: Shangwu yinshu guan, 1926.

Xu Pudong. "Xihu bolanhui choubei de jingguo" [The process of preparations for the West Lake Exposition]. *Dongfang zazhi* 26, no. 10 (October 1929): 27–28.

Xu Renhan. "Taiyou guan'gan" [My impression of Taiwan]. *LZ* 23, no. 1 (January 1949): 65–69.

Xu Shizhen and Shao Yuanchong, eds. *Xibei lansheng* [China's Northwest, a pictorial survey]. Nanjing: Zhengzhong shuju, 1936.

Xu Shizhuang. "Shu Taiwan lüxing" [My travels to Taiwan]. *Xuesheng* 8, no. 12 (1921): 67–68.

Xu Yifang. "Wo canjia Dianmian gonglu xiujiangongcheng de jingguo" [My participation in the construction of the Burma Road]. In *Zhonghua wenshi ziliao wenku* [Library of Chinese historical data], edited by Wenshi ziliao yanjiu weiyuanhui [Research committee of historical materials], 13:740–46. Beijing: Zhongguo wenshi chubanshe, 1996.

Xu Yinxiang. "Dong Taiwan xingjiao" [Travel in eastern Taiwan]. *LZ* 23, no. 7 (July 1949): 53–54.

Xu Zhaofeng. "Benshe banli liuxuesheng fangyang shouxu zhi huiyi" [The recollection of our agency's assistance to study-abroad students]. *LZ* 3, no. 8 (August 1929): 42.

Xue Cishen. Preface to *Sannian lai zhi xi'nan gonglu* [The Southwest Motorways over the past three years], edited by Ding Songchang, 1. China: s.n., 1940.

Xue Guilun. *Xibei shicha riji* [The journal of an inspection tour in the Northwest]. Shanghai: Shenbao, 1934.

"Xuedoushan zhaodaisuo" [Xuedoushan guesthouse]. *LZ* 9, no. 9 (September 1935): 33–35.

"Xuzhou zhaodaisuo" [Xuzhou guesthouse]. *LZ* 9, no. 9 (September 1935): 47–51.

Yang Lingde. "Cong Suiyuan dao Bailingmiao" [From Suiyuan to Bailingmiao]. *Shenbao yuekan* 3, no. 8 (1934): n.p.

Yang Rennong. "Chuandian gonglu jixing" [Travel notes on the Chuandian highway]. *LZ* 14, no. 7 (July 1940): 15–17.

Yang Weibei. "Dalian caifeng lu" [Travel record of Dalian]. *LZ* 4, no. 12 (December 1930): 11–13.

Yang Zhongjian, *Xibei de poumian* [The cross section of the Northwest]. Beiping: Dizhi tushuguan, 1932.

Yao Songling. *Chen Guangfu de yisheng* [Chen Guangfu's life]. Taipei: Zhuanji wenxue, 1984.

Yao Yuangan. "Yousheng lüxing tuan zhi guoqu yu weilai tuanyuan zhi zeren" [The mission of the Unison Travel Club's past and future members]. *Yousheng* (December 1926): n.p.

Ye, Weili. *Seeking Modernity in China's Name: Chinese Students in the United States, 1900–1927*. Stanford, CA: Stanford University Press, 2001.

Ye Gongchuo. "Duiyu Liangyou quanguo sheying lüxingtuan de ganxiang" [Some thoughts on the *Liangyou* Nationwide Photographic Tour]. *Liangyou*, no. 69 (September 1932): 27.

Yeh, Catherine. *Shanghai Love: Courtesans, Intellectuals, and Entertainment Culture, 1850–1910*. Seattle: University of Washington Press, 2006.

Yeh, Wen-hsin. *Shanghai Splendor: Economic Sentiments and the Making of Modern China, 1843–1949*. Berkeley: University of California Press, 2007.

Yen, Hsiao-pei. "From Paleoanthropology in China to Chinese Paleoanthropology: Science, Imperialism and Nationalism in North China, 1920–1939." *History of Science* 53, no. 1 (2015): 21–56.

Yi Junzuo. "Jincheng qi ri ji" [Seven days in Chengdu], *LZ* 14, nos. 5–6 (May–June 1940): 3–14; 19–28.

Yi Shengbo. "Chongqing de fengjing qu: nan wenquan" [The scenic areas of Chongqing: The South Hot Spring]. *LZ* 17, no. 4 (April 1943): 13–16.

Ying Xing. "Yungang shiku" [The Yungang Grottoes]. *Dongfang zazhi* 27, no. 2 (1930): 91–104.

"Yiyue yitan" [Monthly words]. *LZ* 11, no. 7 (July 1937): n.p.

"Youhang zhuanche zhiqu" [The special tourism train to Hangzhou]. *LZ* 2, no. 1 (Spring 1928): 2.

Yousheng lüxing tuan [The Unison Travel Club]. *Yousheng lüxing tuan jianshi* [A brief history of the Unison Travel Club]. Shanghai: Yousheng lüxing tuan, 1947.

Yousheng lüxingtuan dishijie zhengqiu dahui tekan [Special issue for the tenth recruiting conference]. Shanghai: Yousheng lüxingtuan, 1935.

"Yousheng lüxing tuan jinyao qishi" [An emergency announcement from the Unison Travel Club]. *Shenbao*, September 25, 1931.

"Yousheng lüxing tuan Shanghai zongbu weiyuan yilan" [A list of committee members of the Unison Travel Club in Shanghai]. *Yousheng* (November 1936): 16.

"Yousheng lüxing tuan Sichuan lüxing tonggao" [Sichuan tour announcement from the Unison Travel Club]. *Shenbao*, August 14, 1935.

"Youshengtuan faqi wusheng changtu lüxing" [The Unison Travel Club initiates a tour of five provinces]. *Shenbao*, February 6, 1937.

"Youshengtuan, huabei lüxing zhi xingcheng" [The itineraries of the Unison Club's North China tour]. *Shenbao*, August 24, 1934.

Young, Arthur Nichols. *China's Nation-Building Effort, 1927–1937: The Financial and Economic Record*. Stanford, CA: Hoover Institution Press, 1971.

Young, Louise. *Japan's Total Empire: Manchuria and the Culture of Wartime Imperialism*. Berkeley: University of California Press, 1998.

Yu Songhua. "Shengping zhi jianghu quwei (san)" [The fun of traveling the country (3)]. *LZ* 8, no. 1 (January 1934): 2.

"Yu zhongguo lüxingshe heban gudu xiaohan youlantuan" [Our railway collaborates with China Travel Service to bring you the "winter holiday tour to the old capital"]. *Tielu zazhi* 2, no. 8 (1937): 116.

Yufeng. "Jiadao nongyin zhong: youshang Taibei shi" [A tour of Taipei in the shade of trees]. *LZ* 20, no. 7 (July 1946): 43–45.

Yun Yintang. "Wushan guo yan ji" [A brief view of the Wu mountain]. *LZ* 7, no. 2 (February 1933): 1–6.

Zhang Baoyu, ed. *Du Zhongyuan* [Biography of Du Zhongyuan]. Urumqi: Xinjiang daxue chubanshe, 1987.

Zhang Changgong. *Xijing shengji* [Scenic spots and historic relics in Xijing]. (Xi'an: Shaanxi sheng diyi tushuguan, 1932.

Zhang Henshui. "Chongqing lügan lu" [Impression of my sojourn in Chongqing]. *LZ* 13, no. 1 (January 1939): 49–52; *LZ* 14, no. 1 (January 1940): 31–33.

——. "Xiyou xiaoji" [Brief account of a westward journey (1–11)]. *LZ* 8, no. 9, and *LZ* 9, no. 7 (September 1934–July 1935).

Zhang Lili. *Jindai Zhongguo lüyou fazhan de jingji toushi* [An economic study of tourism development in modern China]. Tianjin: Tianjin daxue chubanshe, 1998.

Zhang Qinshi. "Lü Shu dao shang" [On the way to Sichuan]. *LZ* 3, no. 1 (January 1929): 25–31.

Zhang Qiqu. "Taiwan you cheng ji (xiapian)" [Touring Taiwan (part 2)]. *LZ* 22, no. 4 (April 1948): 37–40.

Zhang Shichao. "Dong Taiwan lücheng" [An itinerary of eastern Taiwan]. *LZ* 23, no. 3 (March 1949): 23–29.

——. "Taizhong xingjiao" [A trip to Taizhong]. *LZ* 22, no. 7 (July 1948): 56–57.

Zhang Shuiqi. "Ouzhou Riben zhaoyin youke qingkuang" [Tourism promotion in Europe and Japan]. *LZ* 5, no. 4 (April 1931): 15–18.

Zhang Ying, ed. *Benguo xin youji* [The new travelogues of our country]. Shanghai: Shangwu yinshu guan, 1915.

Zhang Yuanfeng. *Dao Xinjiang qu* [Going to Xinjiang]. Shanghai: Zhongguo lüxingshe, 1952.

——. "Xihu shi shuyu laodong renmin de" [West Lake belongs to the working people]. *LZ* 25, no. 6 (June 1951): 23–26.

Zhang Yuanhen. "Hetao zhong de mimi guo" [A secret country in the Hetao region]. *Liangyou*, no. 92 (August 1934): 6–7.

Zhang Yunfen. "Cong Shanghai dao Yunnan fu" [From Shanghai to Kunming]. *LZ* 8, no. 1 (January 1934): 59–67.

Zhang Zhihong. "Dao Xibei qu" [Head to the Northwest]. *Libailiu*, no. 584 (1934): n.p.

Zhao Junhao. "Bianjishi shinian ji" [Ten years of my editorship]. *LZ* 10, no. 1 (January 1936): 197–200.

——. "Chen Guangfu xiansheng fangwen ji" [Interview with Mr. Chen Guangfu]. *LZ* 10, no. 9 (September 1936): 83–86.

——. "Dongbei jihen ji" [Travelogue of the Northeast]. *LZ* 3, nos. 7, 9–12 (July, September–December 1929); *LZ* 4, nos. 2, 5, 6 (February, May, June 1930).

——. "Gu Shaochuan xiansheng fangwen ji" [An interview with Mr. Gu Shaochuan]. *LZ* 9, no. 7 (July 1935): 75–80.

——. *Kunming daoyou* [Guidebook to Kunming]. Shanghai: Zhongguo lüxingshe, 1939.

——. "Pan Gongzhan xiansheng fangwen ji" [An interview with Mr. Pan Gongzhan]. *LZ* 9, no. 5 (May 1935): 89–91.

——. "Wang Rutang xiansheng fangwen ji" [An interview with Mr. Wang Rutang]. *LZ* 9, no. 10 (October 1935): 65–68.

——, ed. *Xi'nan lansheng* [Scenic beauty in Southwest China]. Shanghai: Zhongguo lüxingshe, 1940.

——. "Zhongguo lüxingshe fazhan jianshi (shang)" [A brief history of the development of China Travel Service (part 1)]. *Lüguang* 2, no. 1 (January 1941): 2–4.

Zhao Renqiao and Meng Shi. "Canjia Xibei shiye kaochatuan zhi jingguo ji qi gongzuo" [Processes and work of the group exploring the industries of the Northwest]. *Qinghua zhoukan* 38, no. 5 (1932): 52–54.

Zhao Shuyong. "Pingsui yanxian daoyou zhilüe" [Guide to the sites along the Ping-Sui Railway line]. *LZ* 9, no. 9 (September 1935): 97–99.

Zhao Zheshi. "Huanan lüxing ji buzheng" [Corrections on "My travel in South China"]. *LZ* 8, no. 9 (September 1934): 27–29.

Zheng Bajia. *Fuzhou lüxing zhinan* [Travel guide to Fuzhou]. Shanghai: Shangwu yinshu guan, 1934.

Zheng Bicheng. *Sichuan daoyou* [Guide to Sichuan]. Shanghai: Zhongguo lüxingshe, 1935.

Zheng Jianfeng. "You Xi'an yan gonglu zhi Chongqing" [From Xi'an to Chongqing via motorways]. *LZ* 11, no. 6 (June 1937): 17–19.

Zheng Nuofu. "Beiyou riji" [Journal of a North China tour]. *Yousheng*, nos. 6–11 (1935–36).

Zheng Yijun. *Chuanqi rensheng: ji Cai Fujiu zouguo de lu* [Legendary life: Cai Fujiu's path]. Hong Kong: Haifeng chubanshe, 1997.

Zheng Zhenduo, Rong Geng, and Jiang Endian. *Xibei shengji* [Great sites in the Northwest]. N.p.: Pingsui tielu guanli ju, 1935.

Zhenju jushi. "Lushan zhen mianmu" [The true colors of Lushan]. *LZ* 5, no. 6 (June 1931): 3–18.

Zhichao. "Lüxing xinxun: Susheng dangbu rexin tichang dongbei lüxing" [Travel newsletter: The Nationalist Party in Jiangsu Province enthusiastically promotes travel to the Northeast]. *Yousheng lüxing yuekan* 5, no. 3 (March 1930): 53–54.

Zhiheng. "Bishu zhong zhi Moganshan" [Mokanshan during summer]. *Jiaoyu zhoubao*, no. 132 (1916): 34–35.

Zhiqin. "Wuxie shan tansheng tu" [Travelogue and photographs of our adventure at Wuxie mountains]. *Zhonghua*, no. 15 (1933): 10–12.

Zhongguo lüxing huace [Traveling in China]. Hong Kong: Jingji daobao she, 1956.

Zhongguo lüxingshe [China Travel Service], ed. *Moganshan daoyou* [Tourists' guide to Moganshan]. Shanghai: Zhongguo lüxingshe, 1932.

——, ed. *Qianyou jilüe* [Travel guide of Guizhou]. Shanghai: Zhongguo lüxingshe, 1934.

——, ed. *Shoudu daoyou* [Tourists' guide to the capital]. Shanghai: Zhongguo lüxingshe, 1931.

——, ed. *Xizihu* [The West Lake]. Shanghai: Zhongguo lüxingshe, 1929.

——, ed. *You Shanghai zhi Maijia* [From Shanghai to Mecca]. Shanghai: Zhongguo lüxingshe, 1933.

"Zhongguo lüxingshe baogao Pingshen tongche jingguo shishi" [China Travel Service report on the process of the through traffic arrangement on the Beijing-Mukden Railway]. *Zhonghang yuekan* 9, no. 3 (September 1934): 134–36.

"Zhongguo lüxingshe fashou guanchao piao" [China Travel Service sells excursion tickets for tidal bore tour]. *Shenbao*, September 25, 1928.

"Zhongguo lüxingshe jingban gedi zhaodaisuo ji shitang" [The guesthouses and dining halls run by CTS]. *LZ* 17, no. 3 (March 1943): n.p.

"Zhongguo lüxingshe Xigong fenshe kaimu" [The opening of the Saigon branch of CTS]. *LZ* 15, no. 1 (January 1941): 5.

"Zhongguo lüxingshe zuzhi youlantuan" [CTS organizes a tour club]. *Shenbao*, August 4, 1935.

Zhongri tielu zhouyou zhinan [Guide to railways in China and Japan]. Tianjin: Shangwu yinshu guan, 1916.

"Zhongyang junxiao dongbei kaochatuan di Ha'erbin chezhan sheying" [A picture of the arrival of the Central Military Academy investigation group at Harbin station]. *Zhongyang huakan*, no. 7 (September 1929): 8.

Zhou Xiuquan. "Dianmian daozhong" [On the Burma Road]. *LZ* 14, no. 1 (January 1940): 51–53.

Zhu Chengzhang. "Zhongguo lüxingshe jianshi" [A brief history of the China Travel Service]. *Haiguang* 1, no. 11 (November 1929): 3–4.

Zhu Mei. "Dong Taiwan huanyou ji" [A tour of eastern Taiwan]. *LZ* 21, no. 6 (June 1947): 21–25.

Zhuang Zhujiu and Xu Zhaofeng. "Zengbie you mei xuesheng" [Parting words for students going to America]. *LZ* 2, no. 2 (Summer 1928): 60–69.

"Zhuiji xihu youlantuan" [On tour to West Lake]. *LZ* 4, no. 2 (February 1930): 67–68.

"Zi Yaohua jun lüxing dongbei" [Mr. Zi Yaohua's visit to the Northeast]. *Haiguang* 2, no. 4 (April 1930): 5–6.

Zong Yumei. "20 shiji 30 niandai baokan meijie yu Xibei kaifa" [News media and the development of the Northwest in the 1930s]. *Shixue yuekan* [Journal of history], no. 5 (2003): 54–58.

Zuelow, Eric G. E. *A History of Modern Tourism*. London: Palgrave, 2016.

——. *Making Ireland Irish: Tourism and National Identity since the Irish Civil War*. Syracuse, NY: Syracuse University Press, 2009.

——. "Negotiating National Identity through Tourism in Colonial South Asia and Beyond." In *Cambridge History of Nationhood and Nationalism*, edited by Matthew D'Auria, Cathie Carmichael, and Aviel Roshwald. Cambridge: Cambridge University Press, forthcoming.

INDEX

CPSIA information can be obtained
at www.ICGtesting.com
Printed in the USA
LVHW090052041121
702322LV00007B/235

9 781501 761041